PERMACULTURE PIONEERS
stories from the new frontier

edited by
Kerry Dawborn & Caroline Smith

MELLIODORA PUBLISHING
www.holmgren.com.au

Published June 2011
by Melliodora Publishing
16 Fourteenth Street
Hepburn, Victoria 3461
AUSTRALIA
www.holmgren.com.au

Copyright © Kerry Dawborn and Caroline Smith

The editors assert their moral rights in this work throughout the world without waiver. All rights reserved. No part of this publication may be reproduced, stored in a retrieval system or transmitted in any form or by any means (electronic or mechanical, through reprography, digital transmission, recording or otherwise) without the prior written permission of the publisher.

Reproduction and Communication for educational purposes
The Australian Copyright Act 1968 (the Act) allows a maximum of one chapter or 10% of the pages of this work, whichever is the greater, to be reproduced and/or communicated by any educational institution for its educational purposes provided that the educational institution (or the body that administers it) has given a remuneration notice to Copyright Agency Limited (CAL) under the Act. For details of the CAL licence for educational institutions contact: info@copyright.com.au

National Library of Australia Cataloguing-in-Publication entry:
Dawborn, Kerry and Smith, Caroline Janet (editors)
Permaculture Pioneers: Stories from the New Frontier
1st ed.
ISBN: 9780975078624 (pbk.)
Includes index.
Permaculture--Australia--History.
Agriculturists--Australia.
631.580994

Cover design and images by Richard Telford
Text and typeset by Richard Telford
Proof read by Elizabeth Wade, Maureen Corbett and Su Dennett
Printed in Australia by BPA Print Group Pty Ltd

Paper: Australian made Precision Offset (PEFC)
Cover: Precision Card (PEFC)

This book has been printed on Australian offset paper certified by the Programme for the Endorsement of Forest Certification (PEFC). PEFC is committed to sustainable forest management through third party forest certification of responsibly managed forests.

PEFC/21-31-50
Promoting sustainable
forest management

Disclaimer
The views expressed in this book are the authors' own and do not necessarily reflect the views of the other authors, the editors or the publishers. Credits to the photographers were included where available. All photos were supplied by the author of the story unless otherwise credited.

Dedication

We dedicate this book to those with spirit and courage, past, present and future, who show us that a sustainable and just world is within our power. We simply have to choose it and make it happen.

The royalties from sales of this book will be donated to the Permaculture Pioneers Fund, to support permaculture-related environmental and social justice projects and initiatives around the world.

Contents

Acknowledgements — vii
Introduction *Caroline Smith* — viii

Part 1   The Future in Our Hands — 1

Chapter 1 The New Frontier: Embracing the Inner Landscape
Kerry Dawborn — 2

Part 2   Pioneering Spirits — 17

Chapter 2 The Long View
David Holmgren — 18

Chapter 3 Trend is Not Destiny
Terry White — 30

Chapter 4 Healing the Rifts: Building Intentional Community
Robyn Francis — 40

Chapter 5 Walking the Talk
Max Lindegger — 52

Chapter 6 How Permaculture Made Me Whole
Vries Gravestein — 66

Chapter 7 A Personal Revolution
Jeff Nugent — 76

Chapter 8 Thinking Big
Geoff Lawton — 86

Chapter 9 Gardening Community
Russ Grayson & Fiona Campbell — 110

Chapter 10 Building and Living in a Food Forest
Annemarie & Graham Brookman — 132

Chapter 11 Cherish the Earth
Rosemary Morrow — 146

Chapter 12 Experience is What You Get When You Don't Get What You Want
Martha Hills — 158

Chapter 13	My Seat on the Train Janet Millington	164
Chapter 14	An Edge Species Robin Clayfield	178
Chapter 15	Permaculture for the Spirit Alanna Moore	190
Chapter 16	Finding the Inner Balance Naomi Coleman	204
Chapter 17	Ethical Decision-Making for Secondary School Students Virginia Solomon	224
Chapter 18	Straw and Greywater Ross Mars	236
Chapter 19	Bringing Knowledge to Life Jill Finnane	246
Chapter 20	From Urban England to Rural Australia via Permaculture Ian Lillington	262
Chapter 21	To My Great Grandchildren with Love and Hope Jane Scott	272
Chapter 22	The Magic of Gardening Josh Byrne	284
Chapter 23	Learning from Other Cultures for Sustainable Development Tony Jansen	294
Chapter 24	Think Global: Eat Local Morag Gamble	306
Afterword	Four Key Features of Permaculture (applicable to 'everything'); and an Opportunity for the Future (also applicable to 'everything') Stuart B. Hill	324

Glossary	334
References Cited and Further Reading	336
Index	344

Acknowledgements

This book could not have been possible without the insights and support of the following:

- The contributing authors for their cheerful willingness to share their inspirational and sometimes very personal stories, and for their patience and encouragement while the book went through its long gestation period. Through their stories we can all take courage to make the world a better place.

- Bill Mollison and David Holmgren for their inspiration, courage and steadfast determination in developing and sharing the permaculture concept.

- Stuart Hill, whose invitation to the permaculture community to begin to look within as it charts its journey into the future, was the initial inspiration for the book. Stu has been unfailingly willing to help us wrestle with the interpretation of complex ideas and to work with us when the going got rocky.

- Vries Gravestein, permaculture elder extraordinaire, for believing in the importance of telling the permaculture story, and for his wise encouragement, advice and support throughout.

- Virginia Solomon and Ian Lillington for giving us wise, knowledgeable and creative advice when we needed it.

- Our families - the Dawborns, McDonalds, Lawrences and Heaths - especially Kerry's parents, Anne and John, and aunt, Anne - and the Smiths - Aidan, Graham and Alice - for their support, encouragement and patience, especially at times when the project seemed overwhelming.

- Lisa Jobson and Jane Scott who were always there to provide creative ideas, support and feedback.

- Richard Pitman and Mark Williams for timely and sensible legal advice.

Our thanks to you all,

Kerry Dawborn & Caroline Smith
February 2011

Caroline Smith

Dr. Caroline Smith is a teacher, permaculturist and organic farmer who was born in England and lived for some years in South Africa. Caroline is passionate about local food production and operates an organic box scheme with her husband Aidan in the Dandenong Ranges just outside Melbourne. Caroline has worked as an agricultural scientist, secondary teacher and teacher educator, and currently teaches at the National Centre for Sustainability, Swinburne University of Technology in Melbourne. She is also a member of the editorial committee of *EarthSong* journal. Caroline has published widely in the area of sustainability education, and her PhD thesis explored personal empowerment through learning permaculture. She has two grown-up children, a cat and a horse.

Introduction

This collection brings together for the first time the stories of 25 remarkable people, young and old, who have in myriad ways pioneered the extraordinary design system for sustainability known as permaculture. Permaculture thinking and practice is part of the global movement that ecological economist Paul Hawken has called "The Movement with No Name," fast growing, loose and broad, that is working, often unnoticed by mainstream society, to lay the very foundations for a new way of being on Earth.

Permaculture is a child of Australia, so it is in Australia that those with the longest experience are to be found. The experiences, insights, struggles and triumphs of the writers make a fascinating and important contribution to the growing body of writing on transitions to sustainability in response to the ecological and social challenges of our day. While the authors are Australian permaculturists acting both at home and abroad, their experiences will be equally relevant and accessible to those working for sustainability, whether it be in urban planning, community development, sustainability science, public policy, agriculture, education, business or international development.

The permaculture concept was conceived and developed in the small southern state of Tasmania in the early 1970s by co-originators Bill Mollison and David Holmgren. Since then it has spread around the world to more than 160 countries. A key feature of the concept is that the principles are firmly based in an ethical ecological and social justice framework:

- Care of the Earth: Provision for the wellbeing of life systems.

- Care of people: Provision for people to access those resources necessary to their wellbeing.

- Share the surplus: This is sometimes re-stated as 'return the surplus' or 'reinvest the surplus'.

Since they were first described, the principles of permaculture have changed and evolved as they have been worked through and adapted by the growing number of practitioners (see Holmgren, 2002). The principles require much contemplation, reflection and experimentation, and the writers clearly demonstrate the varying and creative ways in which these can be applied.

The genius of Mollison and Holmgren's work is that through their deep understanding of integrated systems, they were able to produce a synthesis – an understanding of natural ecosystems, traditional small-scale mixed agriculture, low impact technology, and social justice into an interconnected dynamic system of design principles for creating self-sustaining human settlements. This is an extraordinary achievement, given the dominant form of education in the West is largely through disconnected and bounded disciplines. It is not surprising that some writers describe permaculture as 'brain-scrambling'.

It was undoubtedly the larger-than-life personality of Bill Mollison who first put permaculture on the map, and it is debatable whether it would have grown into the worldwide movement it is today without Bill's dedication, courage, determination, passion and belief in a better world that both drove him and attracted many of the movement's early pioneers. Equally, permaculture could not have survived and flourished without being able to stand on its own merits as a workable system, and a holistic, accessible and practical framework for action. Complementing Mollison's work, Holmgren and others have engaged in and continued the hard and often grinding work of testing it and putting it into practice. Over time, permaculture design principles have now been variously applied to inner city projects, suburban gardens, farms, communities and larger bioregions. They have even been used to restructure the social relationships in a school (for example, see Harney, 1997).

Inspired by Mollison and Holmgren's extraordinary life-changing synthesis, combined with Bill's charismatic promotion, the early up-takers of permaculture embarked on the first of Mollison's Permaculture Design Certificate courses (PDCs). These early 'permies', as many permaculturists call themselves, were by no means sustainability experts, indeed few such creatures existed in those days. Instead, they were ordinary people from a variety of backgrounds and ages, who recognised the danger that Earth was in, and were determined to make a difference. They learned to grow food, to work in their communities, to educate themselves and others and to stay hopeful. They made many mistakes but experienced much joy as they found themselves engaging in the deep learning and transformation that comes with practice, reflection, vision and commitment. In doing so they showed how different people bring their own energy, creativity and perspectives to permaculture, transforming and renewing it as they do so.

The stories in this book provide a fascinating look through the eyes of those who have not only thought long and hard, but have embraced, often in the face of ridicule and indifference, the work of transition to a better future. These permaculture pioneers have much to teach about the difficult but critically important and ultimately exhilarating task of creating a sustainable future.

A number of the writers have spent long years pioneering permaculture on the land. Others have chosen to work within the mainstream, struggling to bring permaculture thinking into education, decision-making, planning and design. Still others have contributed to overseas development. They have learned that being humble, dealing with paradox and uncertainty, and making mistakes is all part of learning, often a difficult lesson for Westerners. Though they reflect very different personalities and approaches to permaculture, the writers are united in their belief that to have any hope of a sustainable future, there needs to be a deep shift in personal and cultural values. They all share a passion, a commitment and a sense of purpose, which has enabled them to experience the sheer joy and sense of empowerment of being actively involved in the creation of their vision and dream for a better future.

Through being part of permaculture networks such as local groups, convergences and courses, the writers show how they have been able to support and affirm each other, as well as having strong disagreements, through difficult times, and so have been able to build the movement together. The level of personal cross-referencing in the chapters is a reminder of how close-knit the permaculture community was, particularly in its early days, and largely remains so today.

Permaculture attracts people from all walks of life for many reasons (see Smith, 2000). It appeals to those interested in moving from dependency to direct participation in the most basic aspects of human existence – the provision of food, energy and shelter, and active and conscious cooperation with others, on different levels, in order to make this possible. Working with permaculture principles is as much personal, cultural and for some even spiritual, as it is technological and scientific, and many of the writers talk about a deep recognition, a sense of resonance with permaculture, of working with instead of against nature - somehow it 'feels right'.

A number of writers in this collection have paid homage to many others who have inspired them, in addition to Mollison and Holmgren. They include P.A. Yeomans, inventor of the Keyline water harvesting system; Geoff Wallace of the Wallace (aero-plough) fame; ecologist Howard Odum, who pioneered the difficult work of measuring embodied energy in ecological and human systems with his brother Eugene; Gregory Bateson and Buckminster Fuller, two of the most original thinkers of the 20th century; and Declan and Margrit Kennedy, who pioneered eco-architecture and alternative ways of thinking about money. The story of permaculture, as in most great movements, is a story of standing on the shoulders of giants, as well as producing considerable giants of its own.

This book has three intertwined strands that weave through the stories. The first stems from the 2006 Australasian Permaculture Convergence (APC8) in Melbourne. Here, social ecologist Professor Stuart Hill invited and challenged the permaculture community, having spent the last 25 years transforming the outer landscape of our homes, gardens, farms and communities, to begin to think about what a corresponding transformation of the inner landscape of hearts and minds might involve. How might the experiences of permaculturists, and the practice of permaculture itself contribute to this and what promise might such an exploration hold for permaculture and for the wider community as we strive for a better world? As a design system based in ethics, permaculture provides a process of consciously creating our world out of the best in ourselves. If the outer world that we build reflects our inner state, then exploring our personal and cultural inner landscapes to enable this transformation can only be immensely powerful.

This first strand, then, traces how individuals and communities, in their own ways, have thought about and grappled with the difficult inner personal and cultural transformation towards ecological sustainability. The writers reflect in their own particular ways on what it means to challenge the destructive paradigms by which most of us live. They provide us with glimpses of how they have engaged in the process of transforming the way they live and see the world.

The second strand is a telling of a history of the Australian permaculture movement for the first time. Together, the stories weave a very personal account of this 33-year-old social innovation movement, from its origin in early 1970s in the small Australian state of Tasmania, to where it finds itself today – an extraordinary, diverse worldwide community with representation in over 160 countries.

Lastly, the stories provide the reader with multiple perspectives on how ordinary individuals and their communities can engage in the creation of a sustainable, life-affirming future. Through the lives of these permaculturists, we can see that we all have the tools for change within and between us. Personal growth, awareness and transformation are not limited to one type of person, occupation, region or nationality. The writers' strengths, commitment, mistakes, pitfalls, struggles and triumphs provide powerful models for our own attempts to live sustainably in a world that exhorts us to do just the opposite. We hope that readers will find these stories personally empowering and gain inspiration, hope and strength from these extraordinary people to make a difference, however small, in their own lives. Ordinary folk *can* become empowered to be experimenters and doers, and we don't have to wait for experts and governments to lead; far from it.

So what can we learn from the stories of these remarkable, resilient, dedicated and insightful individuals about the transformation to a sustainable culture? They don't all live on the 5-acre dream, indeed some would argue that this is not the most sustainable way to live. Many are decidedly suburban and urban, living where most of us live, while others are part of intentional eco-communities. Their stories are highly individual, reflecting the character, personality and widely diverse backgrounds of the writers. Some accounts are quite factual, others more reflective. Many writers reminisce about a childhood where they worked in the vegetable garden with parents or grandparents, caring for chickens or foraging for wild plants. Growing up in wild places, or the inspiration gained from time spent in the natural world, has been formative for many, pointing to the importance of bringing up children in ways that are close to nature. A number of stories show the impact on a child of growing up with sensitive adults who have a reverence and a respect for nature, a conservation ethic and often, an unconventional, maverick even, view of life.

Then again, there are some writers for whom none of these things have been features of their upbringing. That they too have become permaculturists suggests a possibility of something much more, a deep connection to the wellsprings of life itself, that has somehow drawn them to permaculture. This is speculation, and we leave it to readers to draw their own conclusions.

Refreshingly, the writers have not been afraid to point out permaculture's shortcomings and blind spots as well as its remarkable insights; to kick a few sacred cows, as David Holmgren puts it. There have been times when the movement, possibly in self-defence, has been in danger of seeming like a cult, unable and unwilling to be questioned or criticised. There are even suggestions that the movement has ludicrously split into a 'Bill' camp and a 'David' camp. Implosion is a recipe for stagnation and rigidity. If it is to move forward and evolve, permaculture must be open and receptive to new influences and insights, and be capable of accepting and acting on criticism as well as praise. More than most, permaculturists should understand that.

Permaculture has traditionally been most successful in the small, the local and the personal. This is both a strength and a weakness. Thoughtful permies have long recognised that while local action is crucial to pathways to sustainability, it is far from the only sphere of influence. More systemic forms of organisation, from regional to national and global, are critical to developing sustainable futures, and Geoff Lawton's descriptions of the beginnings of large-scale permaculture projects shows that this indeed should be possible.

A number of permies are now involved in the Transition Town movement. Originated in Ireland by permaculturist Rob Hopkins, the Transition movement is an example of the way permaculture has influenced new thinking. Many 'Transitioners' have never studied permaculture or even heard of it, but the zeitgeist – the spirit of the moment – is right, with peak oil, climate change, the rise of the local food movement and Green politics becoming cemented in mainstream consciousness.

In *Part 1 – The Future in Our Hands*, Kerry Dawborn explores the relationships between the outer world that we create and the inner landscape of our thoughts, feelings, beliefs, attitudes, values and worldviews. Permaculture is about working with nature while meeting human needs, through ecological design, actions that promote cooperative stewardship, and respectful understanding of our place as part of the Earth community. Kerry argues that transformation of consumer culture to sustainable culture depends on parallel changes in our emotional and cultural inner world, and on our ability to avoid simplistic, linear responses in situations where embracing and dealing with complexity is critical to our ability to bring lasting and meaningful change. As Kerry observes, as a design system based in ethics that draws its inspiration from natural systems, permaculture is uniquely placed to help us do this. Drawing on the authors' experiences and reflections, she asks what might a 'permaculture of the inner landscape' involve, and how might it help achieve sustainability? What can we learn from the struggles and triumphs of the people who have chosen sustainability through permaculture as their way of life? We invite readers to reflect on these questions as they read *Part 2*.

The authors' chapters form *Part 2 – Pioneering Spirits*. Many of the writers are well-known high profile permaculturists, including some of those early up-takers, and a good number have been extensively published in their own right. Others are not as well-known, but have nevertheless experienced their own transformation through permaculture and are active in social change at the local level. The writers represent a range of voices – male and female, old and young, from widely different backgrounds, geographical locations and, most importantly, ways of being permaculturists. We could have compiled a different book with different authors which would have been just as inspiring, and indeed some of the permaculturists we would have liked to have included were unable to participate. As difficult as it was, we had to draw the line somewhere.

The chapters have been organised into three broad groups, which enable the personal history of the movement to be told. David Holmgren is the author of the first chapter in Part 2. As the co-originator of permaculture, David's insights and reflections provide a fascinating and provocative introduction to the movement, in particular its early days, and his perspectives on its evolution. In addition to his personal background and the events and conditions that set him on his path, David's unique position as co-originator of the permaculture concept, and his observations on its development, invite us to reflect on the role of leadership within social movements during times of change. David's writing provides the context for the other stories, and is followed by fellow Victorian Terry White, another of the early permaculture pioneers.

Terry's long career in permaculture has included being the first editor of the *Permaculture International Journal* (sadly no longer in print), as well as being instrumental in bringing permaculture perspectives into a range of mainstream environmental projects, such as catchment management and soil conservation through tree planting. Terry is followed by Robyn Francis, a permaculture elder who herself learned much from the wise old folk of Germany. Robyn's many significant achievements include the development of 'Jarlanbah', a very successful intentional eco-community in Nimbin, New South Wales, as well as her tireless work with others in the development of Accredited Permaculture Training (APT).

Next we read about Max Lindegger, father of the ecovillage movement in Australia and designer of the Crystal Waters permaculture community in Queensland. Like a number of the elders, Max's formative years in the harsh environment of post World War 2 Europe as well as learning from the old Swiss farmers, shaped his direction in life, and like Robyn, Max is in demand both nationally and internationally as a teacher.

Vries Gravestein, who has entered his 9th decade, is the oldest contributor, a true elder of the movement. Vries was strongly shaped by the forces of war and traces his interest in permaculture to his upbringing where he had to deal with the deprivations of living in Holland during World War 2. It was here that he learned to be smart about surviving on very little. An educator of long experience, Vries is also one of the few permaculturists who has worked directly with broadacre farmers, specialising in soil fertility and sustainable agricultural development.

Jeff Nugent took part in Bill Mollison's first PDC, and his chapter contains a fascinating vignette of one of Mollison's early PDCs, which he taped. Jeff is now one of the movement's foremost plant experts, and he has researched, identified and tested an extraordinary range of plants that can be incorporated into permaculture systems for their ability to provide food, fibre or building materials. Like Rosemary Morrow, Ross Mars and Josh Byrne, Jeff comes from Western Australia where he still lives and works.

Another of the early PDC participants was Englishman Geoff Lawton. Influenced by a frugal upbringing in England and with a 'can do' personality, Geoff has gone on to design and commission some very large-scale permaculture projects in some of the most difficult settings on earth.

Sydneysiders Russ Grayson and Fiona Campbell are a permaculture partnership in all senses. They have worked together and supported each other over many years as they negotiated the minefields of the tricky and frustrating 'invisible structures' of community building through community gardens, sustainability fairs and education. With Tony Jansen, they remain actively involved in overseas development, particularly in Melanesia.

Graham and Annemarie Brookman are pioneers in medium-scale permaculture development, architects of the well-known 'Food Forest' in South Australia. The Food Forest has been highly important in demonstrating how a family can work with others to carve out one of the best models around of a viable sustainable food production system. The Food Forest is a living, vibrant demonstration of thoughtful permaculture in action, and serves an important role in education in all aspects of sustainability.

Rosemary Morrow grew up in Perth and went on to study agriculture. Her story of her rejection of modern agriculture's reductionist, chemical-based, life denying message echoes many others of that time, and Rosemary went on to work in overseas development. She is a prominent writer and is much in demand as a teacher. Influenced by her Quaker upbringing, Rosemary, like many others, brings a gentle, humble and deeply thoughtful spiritual approach to her work.

Martha Hills, originally from the US but now living in suburban Melbourne, has written a short but delightful piece on how permaculture continues to inspire her community activism. Martha's work has been helpful in showing how permaculture principles can be applied to living in the suburbs, an increasingly important aspect of survival as the world once again moves towards having to deal with relocalisation and urbanisation of food production.

Janet Millington, a much-travelled child of 1960s Sydney, came to permaculture through a desire to feed her children healthy food. She has gone on to work with others in education and is a pioneer of APT training and the first Transition movement in Australia on Queensland's Sunshine Coast.

The next group of authors, a little younger than the first, continue to work alongside the elders to further develop and evolve the movement. Some, in particular Robin Clayfield, now also living at Crystal Waters, and Alanna

Moore, now based in Ireland and Australia, have found that permaculture has led them on a spiritual personal development path. There has been much heated debate about whether there is a role for more esoteric personal and spiritual considerations within permaculture. Certainly the co-originators saw permaculture as a secular, rational system for the design of sustainable human settlement, and Mollison in particular has argued forcibly against any notions of spirituality creeping in. To this day, whether a spiritual dimension has a place in permaculture remains hotly contested. People will always adapt and adopt permaculture within the framework of their own worldviews and life experiences, and there is certainly no 'one true way'. We have deliberately chosen to include Robin and Alanna's writings in this collection, not just because they are significant contributors to the permaculture story as educators and designers, but because we believe their particular interpretations will strike a chord with many readers.

Next, two more very significant educators, Victorians Naomi Coleman and Virginia Solomon, tell their story. Naomi has been a passionate and extremely committed permie for many years, working both overseas and within Australia. With her partner Rick, Naomi has facilitated many PDCs in Victoria, and continues to do so, as well as being the first to adapt the permaculture curriculum for the Vocational Education and Training (VET) system. Naomi's story is a bittersweet and very honest one as she reveals the painful paradoxes and contradictions of trying to educate about environmental sustainability while personal sustainability becomes more and more difficult to maintain.

Virginia Solomon is one of the very few permies who has managed to introduce permaculture into a mainstream education setting. Based on her work on the APT project, Virginia has achieved the difficult task of bringing permaculture into an elite Melbourne private secondary school, and her reflections on the students' reactions and learning make fascinating and informative reading.

Originally a science teacher like Rosemary Morrow and Jill Finnane, Western Australian Ross Mars has significant practical and academic achievements in the design of greywater systems. Ross continues to bring together his science and environmental training to design and develop a range of practical solutions to living sustainability.

Jill Finnane, now working at the Edmund Rice Centre in Sydney, tells a story of a fascinating amalgam of a background in human rights issues, ethics and science teaching. For Jill, permaculture embodies and allows expression of all her passions and her strong Catholic faith. Like Rosemary Morrow, Jill is an example of someone who has been able to weave her strong Christian faith with permaculture, in contrast to Robin Clayfield and Alanna Moore who have found that permaculture strengthened their sense of an earth-based spirituality.

Ian Lillington, another migrant from England, has been active in an extraordinary range of community based projects, all informed by permaculture, and is an energetic educator, author and catalyst for change. Ian has lived variously in South Australia and Victoria, and has worked closely and taught with David Holmgren for many years.

Jane Scott has chosen to write her chapter as a letter to her grandchildren, a fitting conclusion to this group of writers. Jane is a creative, talented and passionate educator who has facilitated many PDCs where she lives in the Dandenong Ranges east of Melbourne. In her chapter, she tells her grandchildren that her dream for them is that her life as a permaculturist will have played a small part in bringing about a more hopeful and saner future. Remarkably, Jane has been able to relate permaculture to her perspectives on the cancer that she is currently recovering from.

The final three chapters are contributions from younger permies. They seem to share a more natural and pragmatic attitude to permaculture than their elders as issues and understandings around sustainability have become more mainstream as they were growing up. Perhaps they haven't had to fight as hard, but what is so uplifting and inspiring in their stories is that they seem to have an innate sense and belief that sustainability is possible, they seem to just step in and do it. This gives us great cause for hope.

Josh Byrne, well-known presenter on ABC television's *Gardening Australia*, brings a youthful passion as well as a pragmatism and practicality to his interpretation of permaculture. While Mollison and Holmgren were the inspiration for a generation of thinkers on the fringes, Josh, through his boyish charm and accessible personality, has been successful in bringing permaculture to a mainstream audience.

Tony Jansen is another young, talented and very thoughtful permaculturist. Tony's strengths are in overseas development, latterly in Solomon Islands where he initiated a now very successful NGO, the Kastom Gaden Association. Tony's writing shows that he understands very well the importance of cultural understanding and working slowly and carefully with people, acknowledging and valuing the existing knowledge, insights and cultural context of the community he's working with, rather than rushing in as the 'expert with all the answers'.

The authors' collection ends with Morag Gamble's story. She too has travelled widely and has studied sustainable systems in a number of countries, working with some very well-known sustainability thinkers and activists. Morag has also researched the conditions needed for effective teaching of permaculture for her Master's thesis and now runs a very successful permaculture education business. Like Max Lindegger, Morag lives in Crystal Waters and is in much demand as a teacher. With her partner Evan Raymond, she is also a notable activist in promoting food localisation through their company SEED International. Morag brings a very fresh, heartfelt and carefully thought-through approach to her permaculture activism.

Finally, in his afterword, Stuart Hill, a long time critical friend of permaculture whose challenge to explore the inner landscape provided the impetus for this book and much food for thought as the book evolved, reflects on these remarkable stories. Drawing on his wide experience as an agricultural scientist, psychologist, social ecologist and educator, he contrasts current, dominant and unhelpful patterns of thought and response against the kind of inner landscape out of which we can build a better world. In doing so he offers the permaculture movement and the wider society carefully articulated and thought-provoking frameworks, insights and directions that can inform our evolution toward a sustainable future.

These diverse stories support the idea that with sufficient care, sensitivity and thoughtfulness, permaculture has the potential to be applied at all levels from the local to the global, from the broadacre to the urban setting. We hope that readers will find resonance in these writers' reflections and experiences and find their stories personally empowering. We hope that readers will gain inspiration, hope and strength from these extraordinary permaculture pioneers to make a difference, however small, in their own lives. Their stories offer much to the rest of us who are struggling to make sense of a world of peak oil, climate change and ecosystem disruption. The great eco-theologian, Thomas Berry, wrote that the only interpretation of recent Western history now left to us is one of irony, where blind so-called 'progress' towards an 'ever improving' human situation is bringing us to waste-world rather than wonder-world. Permaculture is an attempt to create that wonder-world, with eyes wide open.

A Note on Bill Mollison

As the co-originator of permaculture, Bill Mollison was invited to contribute a chapter to this collection. Sadly for readers, he declined to do so for reasons of his own. Bill's extraordinary vision, influence, leadership and genius as well as his big personality and idiosyncrasies have been referred to by a number of writers, so the reader is able to build a picture of the man behind the movement. Permaculture and sustainability in general owe him an enormous debt of gratitude.

Bill Mollison during a plant stock collecting trip around Tasmania in 1975. Photo by David Holmgren. David Holmgren and Bill Mollison's close and intense working relationship during 1974-1976 brought together the ideas and the practice which came to be called permaculture.

Part 1: The Future in Our Hands

Kerry Dawborn

Kerry Dawborn is a permaculturist with post-graduate qualifications in Social Science, Environmental Urban Planning and Secondary Teaching. She has worked in both sustainable and conventional agriculture and in the organic produce retail industry. Kerry is passionate about ecological economics, small-scale sustainable farming and land-use, community food security, social justice and related public policy and education issues. She has been a university tutor, researcher, permaculture teacher and political candidate. Living in the Yarra Ranges with her three dogs, her chickens, muscovies, fruit trees and vegie garden (when she has time), Kerry looks forward to a world in which humans understand and honour their place as part of a diverse, healthy and biologically rich earth community, and in which access to healthy, sustainably produced local food, and vibrant, connected human communities, are a right, and not a privilege, for people all over the planet.

Chapter 1 | The New Frontier: Embracing the Inner Landscape

> *To have any real chance of dealing successfully with climate change certainly requires the redesign of infrastructures and the rapid development of a whole new raft of technologies... Yet for any of this to happen, and to happen in time, we need to become more aware of the ways that the outer world is mediated by the inner world of people and cultures.*
>
> Richard A. Slaughter

On December 1st, 1955, Rosa Parks, an African American woman, was arrested in Alabama for refusing to give up her bus seat to a white person. Her action sparked a community wide boycott of the transit system. This crippled it financially, and became part of the wave of action leading to equal rights for African Americans in the American Civil Rights Movement. Rosa Parks had been born into and lived all her life, within a story that said African Americans were lesser citizens than whites, and must yield and accept humiliation and maltreatment, in everyday life, and in the laws of the nation. On that day, Rosa Parks chose to tell a new story about the position of her people, in American society.

In Bangladesh more than 30 years ago, economist Professor Muhammad Yunus saw that the very poor were mainly poor because the story told in mainstream society – through the banks – was that the poor could not be credit-worthy and therefore must not receive loans. Like Rosa Parks, Yunus chose to entertain a different story. On looking closer, he found that the poor he encountered, excluded from mainstream opportunities to obtain credit to finance the micro-businesses by which they survived (but barely, from day to day), were forced into a form of enslavement to loan sharks, from which they could not escape. Shocked at the ridiculously small amounts for which these people were compelled to sell their lives, he started a credit scheme with just $US27.00 and began a micro-credit movement through which the very poor worldwide, continue to transform their lives.

We create our world every day out of who we are inside. Our intentions, ethics and worldviews are the foundations on which our choices and actions are laid, and they dictate the form and nature of what we create. This is our great strength; it is our power and our hope, if we choose it. Often, as individuals and as societies, the words and actions we struggle with are those that are the most important. Yet sometimes they feel so much at odds with what we find in the world around us that we fear how they will be received, or we struggle to fully grasp what we know in our hearts is needed. This is the teetering point of creativity and empowerment. It can be a gateway to the deepest gift of love, courage and inspiration that we have within us to share; or it can be the point at which our heart begins to shrivel, and the seed of what we could do, or become, goes hungry.

Every day many of us make choices and take actions that in small and large ways change, or have the potential to change, the world. Extraordinary as the authors in this book are, they are not unique. Yes, they are all pioneers – individuals who have felt the need for something different, and had the courage, determination and passion, to take a stand. They have all engaged with a vision and begun the work of building a new way of living in ecological harmony on this Earth. Yet they are not alone. Permaculture is one wave on the tide of social and environmental change – wrought by individuals and groups the world over – that perpetually nibbles at the edges of the mainstream, reshaping its shoreline, assisting it to redefine itself, challenge its own preconceptions, and adapt to changing conditions.

Several key things make permaculture interesting however. As a design system for sustainable human settlements, beginning with clear universal ethics and environmental principles, and grounded in the importance of individual responsibility and action, permaculture empowers people worldwide with a holistic vision, with skills in observation, reflection and design, a clear path for action, and the belief that they can make a difference. It encourages people to develop and trust their intuition. Though grounded in science, permaculture equally nourishes and supports vernacular, or 'everyday', wisdom. With nature's web of interconnected ecosystems, patterns and flows as its primary inspiration, permaculture teaches people to engage and work effectively with complexity and not be afraid of it. It helps us to recognise the often unhelpful, deceptive simplicity of narrow, linear responses to complex or 'wicked' problems – 'deceptive' because such simplistic responses often just lead to more problems. While teaching people to instead embrace the complexity that characterises many of our problems, permaculture also helps prepare them to recognise and trust the profound simplicity of many of the solutions that are needed. Finally, permaculture is an inspired yet imperfect social movement made up of courageous and determined, yet 'everyday', individuals. It is a movement that, like many others, has arisen out of the very best in humans, and which, like all of us, struggles with the challenges of 'being human'. Because of permaculture's strengths and yet also because of the movements' vulnerabilities, I believe the stories shared here offer insights on individual transformation and social change that can help carry us all forward as we negotiate the difficult path to a sustainable and just future for all.

Rejecting Deceptive Simplicity

There are a number of streams and rivers in the region I call home. Just a couple of hundred metres from my front gate, a seasonal creek burbles through thick, moist scrub and over muddy bogs, to join the more permanent Cockatoo Creek. About 45 minutes away there is the Little Yarra River that joins the much larger Yarra River – the river on which the city of Melbourne was built, and which over the last 200 years has quenched the thirst of our land and absorbed the waste and rubbish we have consigned to her flow. For me, somewhere in a corner of my mind like the pages of a fairytale picture book, the notion of natural streams brings up images of clear, fresh, cool mountain water gushing in crystal tones over stones and pebbles, enticing me to plunge my hands in, bathe my face, and drink. It is a dream I hope to fulfil one day. To find such a place and feel the cool water against my skin. Meanwhile I live by streams and rivers whose waters I hesitate to touch even with my hands, so concerned am I about whether the water is safe. So much water, yet I feel the need to wash after contact. How is it that I have grown up my whole life with this bizarre contradiction, yet I only started to feel outraged about this relatively recently? *How is it possible* that we can pollute our waterways – *our life blood* – or through lack of outrage and action, simply go along with it?

Actually it isn't difficult to understand. All it takes is ignorance, or refusal to see the intimate connections between people and their activities and needs, the river, and the natural environment. All it takes is a deceptively simple, linear way of looking at the world, and individuals, organisations and governments acting without considering the complex systems we are part of. Add diverse social and individual needs operating without recognition or acknowledgment of the intimate connections between all of them, and what you get is contradictions. Yet what about the environment, and the people, plants and animals who depend on it? What about the need we all share, to live in an environment uncontaminated by poisons, to drink clean water, breathe clean air, and eat clean food? Who or what is taking care of that need?

With so many seemingly conflicting needs and such serious problems in today's world, deceptively simple, short-term, linear and often technologically focussed solutions, can seem attractive. Short of water? Build huge energy-sucking polluting plants to desalinate the sea. Worried about greenhouse and other pollution, but don't want to use less electricity? Go nuclear. Or use geo-sequestration. Never mind the waste or the unknown consequences – just bury it. Wondering how to feed the entire world's people when it seems like there isn't enough? Go monoculture, genetic modification, use poisons, big machinery, heaps of water, to grow more rice or wheat or other crops. Never mind that much of the grain grown is used either to feed animals, not people, in polluting and inhumane factory farms, or increasingly to make biofuels to quench the thirst of an inefficient, poorly thought out global energy and transport infrastructure.

Depending on deceptively simple measures to solve deep, endemic problems is like bingeing on sugar when you feel depressed. You get a quick hit, and you feel better. It seems like you are solving the problem. For a little while. For a little while, you have a burst of energy; the sadness is smothered by a dulling of the senses, a fogging of the brain. Simple, short-term responses have a place, like the emergency department in a hospital. Yet emergency measures do not resolve deeply entrenched long-term problems. Sometimes you get improvements. Few would argue that using penicillin to treat disease is a bad thing. But drugs should not replace the promotion of a healthy immune system in the individual, and systems and ways of living that ensure this for whole populations. Likewise, environmental and social techno-fixes often do address isolated symptoms. They do not however, address the problem at its source. Depending on quick fixes without matching examination and action to address the deeper individual, political and cultural causes, is like giving medicine to a sick person while leaving them in the living conditions which created the illness in the first place.

Deceptively simple fixes often wind up making the problem worse by smoothing the way for more of what caused it. Building freeways ultimately helps generate more car use. Often within very little time, freeways become clogged until it seems like the problem is that there are not enough of them. The urban environment comes to revolve around cars. It feels 'inevitable'. What are people to do? They have to get around. With more roads and cars people can travel further, so dormitory suburbs, where people sleep but do not work, become possible. Now people have to travel in their cars to work, because work is not near home. Local shops accessible on foot or by bike are replaced by huge shopping centres – islands of commerce and consumption in a car park sea. Too far to walk. Too dangerous to cycle.

So people stop walking and cycling and grow fat. They don't know their neighbours because they drive out of their garages, out the gate, and down the street without pausing as they might if they were on foot or cycling, to wave or lean over the fence for a yarn, or simply make eye contact and smile, as they move through their community to the local shops or to work. If you don't know your neighbours, or the people down the street, can you trust them? Can you feel safe walking down a street of strangers? Will your community take care of you if something goes wrong? Call the police if they see a burglar? Or will your neighbour *be* the burglar? Better to put up a high wall around your house so that no one can get in, or see in, then you don't have to look out at the wasted, empty streets used mostly by those who can't afford a car because they're socially and economically stressed, insecure and uncared for.

So you put up a high wall and for a while you feel safe, except that by putting up the wall you've made it so the neighbour or passer-by can't see the burglar lugging your hi-fi out of the window, having chosen your place to rob since the wall both says you have something you want to protect that is probably worth stealing, and provides the ideal cover for a successful heist. Furthermore, although the thief isn't fully aware of it, the wall helped him or her become a thief because it and all the other walls in the street helped make the street unsafe and un-nurturing for those with no choice but to use it – those whose lives you couldn't see from behind your nice, solid, high walls. But the street isn't safe, the world isn't safe, and you know it, and you put up your wall, and drive instead of walking, so more roads are needed, more pollution is created, more greenhouse gas, so the weather gets more extreme, so we 'need' more air-conditioners and more electricity to make them run, so we can't burn less coal, we just need to sequester the greenhouse gases in the earth, just as we sequester our confusion and sadness with more consumption, in an unending self-feeding cycle. And it's all 'inevitable'.

Embracing Confusing Complexity

The power of 'inevitability' in our lives is hardly surprising. If you can't see the intricate connections between things in this complex world; if you can only think in terms of deceptively simple linear processes, without recognising the web of actions and interactions all around us; if you are blind, or simply don't pause long enough to observe, things may appear to occur in lines, and not webs, of cause and effect. If you don't have the courage or humility to stop and look at the bigger picture without leaping into reaction, all you will see is the unsafe street and the dangerous people that you feel compelled to protect yourself from. You will see that people are hungry and want to feed them, but you won't see that the power structures, the economic structures, the social conditions, the land use patterns, are what have helped to create and perpetuate that hunger. In any case, you feel unsafe NOW, and the people are hungry and thirsty NOW, so there isn't time to go back and delve deep and figure out the structural, psychological, and social conditions that have created the problems. So most of us, most of the time, just keep reacting.

Treating linear, one-dimensional responses to the challenges we face as if they were meaningful measures for the long term, is mistaken, yet it isn't difficult to understand why we do it so often. The alternative is hard. Embracing the complexities instead of leaping into deceptively simple, more easily digested responses, requires a willingness to delve into the personal, social and economic quagmire, and patiently and lovingly untangle the knots. It demands that we go back, dig deeper, and try to better understand the underlying causes. It asks us to develop strategies and actions that face up to, rather than avoid, the web of causes and effects. It demands that we work with systems and complex relationships, rather than simply applying band-aids to weeping wounds.

As many of the authors of the stories that follow note, permaculture offered them a way to stop reacting and start building solutions. Permaculture is about looking at the whole in all its complexity, and making connections. Indeed, it is *through* the complexity of the systems we work with, that productive relationships, harnessed through careful observation and responsive, thoughtful design, are achieved. Beginning with a clear ethical framework, permaculture enables each of us to powerfully engage with the rich web of connections within which our lives are embedded. I have found it useful to characterise the process with this simple equation:

INTENTION + DESIGN + ACTION = THE FUTURE

The importance of the quality and nature of our intentions is clear. We do not think, feel or act in isolation. Our culture and experience shapes the beliefs, values and worldviews out of which our intentions are born. Out of these, through our choices and actions, we create the future every day. Permaculture's focus on ethics and key principles as the foundation for planning, design and action, provides a way to link, as Ian Lillington points out, our ethics to our everyday lives. Moreover, as Professor Stuart Hill has observed in his afterword, if as individuals and societies, we see and respond to the world through a veil of past personal and cultural hurts – fear of failure, of commitment, need to be 'right' or to be accepted, need to compete in unhelpful ways for example – we will tend to create a world that reflects those hurts, often to our detriment.

Early in its development permaculture focussed on agriculture and food systems. Yet increasingly the focus is broadening to include the 'invisible structures' – the elements of our cultural landscapes – the frameworks and worldviews – around which we organise our communities and societies to either support or hinder our efforts to live sustainably with each other and the planet. Critically, in the holographic way in which things found to be true in one area of our lives often also prove true in others, permaculture helps us to recognise and embrace, as we transform our gardens, farms and communities, the power that is in each of us to design and build a better world. We cannot continue to ignore the need for healing – of ourselves and of the planet. How much more disempowered can we become, than to believe that the challenges we face, and their outcomes, are 'inevitable'? Our authors stories show us that this belief is a diversion from a simple truth: that we own the choices we make, the beliefs, worldviews and attitudes that drive them, the actions that result, and their impacts. These are not our captors. We get to choose.

Choosing Profound Simplicity

Consider Rosa Parkes in that moment as she knew she was going to refuse to give up her seat on the bus to the white person on that Alabama day in 1955. Consider the concentrated energy as she sat firm. The focus. Such a small action. A refusal to move. Not even an action. Rather, an attitude. Built certainly out of anger. No doubt shaken to the core by fear. But also born out of a clarity of values, purpose, honesty, empowerment and love, so fundamental that they can hardly be adequately named. *Freedom!* Such a little action. Yet the impact of that moment of truth reverberated across time and space and changed everything. Forever. Such power is in each of us – it is our love, and our clarity, our wisdom and our honesty; it is our deep potential and our truth – and the courage we have to find it, stand firm in it, and live and create our world out of it.

People who are standing confidently in their integrity often have a lightness of spirit, a glint, a depth, a solid, grounded 'present-ness' about them, which others can't help but notice and be drawn to. I noticed it one evening late last year listening to interviewer Andrew Denton talking with economist Professor Muhammad Yunus about his work in bringing finance to the very poor through the Grameen bank (ABC Television, 07/12/2009). In doing so Yunus broke down traditional attitudes in banking that perpetuated the marginalisation of the poor, and empowered them to change their lives. Yunus spoke with an ease, wisdom and quiet joy, that I can only term 'fullness of heart'. He seemed to believe that almost anything was possible, *if we choose it*. That even in our most ordinary, everyday lives, we can choose to see and do things differently, and for the better. In a world focussed on profit, Yunus called for business to take off its short-term 'profit-maximising glasses' every so often, and try on a pair of 'social business glasses'. To try doing business to address a social need as, for example, the company Danone does by producing low-cost, nutrient enriched yoghurt for malnourished children in developing countries. Sure, get back the money invested, but don't try to do more than that. With that glint and that lightness, Yunus beamed that making money is exciting, yes, but changing the world is much more so. The radiance of his deeply felt words touched everyone I know who saw the interview. "But isn't it all about money? What about game shows like *Who Wants to be a Millionaire?* – one of the most popular TV shows worldwide," asked Denton. "There should be another show," responded Yunus, gleaming, "*Who Wants to Change the World?*"

I saw that glint, just for a moment, in the eyes of my eighteen-year-old nephew when I asked him if he'd seen the Yunus interview. He said he'd watched it twice. Just starting out in life, my nephew says he wants to 'make a difference'. In that instant as he spoke of the interview, his energy was concentrated, focussed, energised, and yet somehow at peace, and I felt it too. I feel it, as I feel Rosa Parks' strength, in the words of the authors as they describe how they came to permaculture, how it just seemed to fit, made sense, gave them something positive to do. The stories that follow vibrate with clarity, purpose, spontaneity, determination, a joyful embracing of life, a deep integrity and a sense of empowerment in being able to turn confusion and contradiction in the world around us into meaningful and positive action.

As the stories show, what permaculture offers goes far beyond intensive, sustainable organic agriculture or food forests; it goes beyond the household, the small business or community, beyond the city or region, the nation, and the world, to the very depths of our being. More than simply a clever way to design and work with human settlements, permaculture offers diverse ways for individuals and communities to bring head, heart and hands together; to effectively move beyond contradiction, 'inevitability' and disempowerment. It offers ways to embrace the confusing complexity of the challenges we face, and provides practical frameworks to help us think about and develop genuinely progressive, often profoundly simple solutions. It empowers ordinary people in taking action in their own lives and their communities, and in living out of a universal ethic in which the whole Earth Community – the community that comprises human and other-than-human life and the Earth itself, is recognised and embraced.

The linear responses we have become used to are often disconnected from the holistic context. Their deceptive simplicity leads us to ignore the price that is paid for the mess often left in their wake. In contrast, profoundly simple responses, such as those developed from a permaculture understanding, can seem revolutionary. If you have too many snails in the garden, then the problem is not the abundance of snails, but a lack of ducks to eat them. Moreover by adding ducks to the system you not only help restore balance and bring the snails back to their appropriate numbers, you can add meat, eggs, feathers, manure, clearing of fallen fruit, and grass, weed and pest control, to the list of products and benefits from your system. Instead of treating the snails as a problem requiring aversive, energy intensive and unsustainable measures such as poison, permaculture would treat a glut of them as an indication of a system out of balance, with a range of opportunities attached; a way to use under-used resources or a stimulant to rethink or tweak the design.

It is not about avoiding technology or industry, so much as putting both squarely back into the service of people and the planet – human values and ecological balance being at the centre. Not only have we come to value technology over more simple approaches in many areas of our lives, we have become used to deferring to experts, technology and the broader economic and political systems, for responses to our problems. Many of us have lost confidence in our ability to take care of ourselves, our families and our communities. We buy almost everything, and make almost nothing ourselves. Many communities struggle to find cohesion, meet the needs of their members or solve problems.

Fundamental skills of living which are humanity's biological and cultural inheritance – producing food, building shelter, working together, thinking creatively outside the box – are largely lost to those who have been disinherited by industrialisation, colonisation, consumerism, globalisation and free trade. This affects most people in the global north, and a huge portion of the landless, migratory poor in the global south. It is not surprising that the authors universally describe themselves and others as feeling deeply empowered in learning new skills and seeing that their actions and choices can have positive impacts; learning that they really can, in a literal sense, take care of themselves and their community. One of the most touching and hopeful examples of this comes from Naomi Coleman. Naomi's children may simply be having fun, exploring, adventuring in their family's food forest, but who can doubt the confidence and resilience that will reach far into their lives?

> They know the farm better than I know it myself, they pride themselves on being able to take people on tours, and the names of plants roll effortlessly off their tongues. They know when to pick fruits and vegetables, and how to prepare them, they look after the chooks and know how to compost. They can catch yabbies and cook them for a snack. They happily swim in our dam each summer and have picnics in their treehouses, foraging from the many productive trees, vines and bushes.

Permaculture may be based in science, but it equally facilitates the development of greater reliance on and confidence in the ability of individuals to make their own sense of what they observe, to invent, design, and respond to challenges. It doesn't always lead to action however. Sometimes it leads to 'proactive inaction'. Martha Hills gives a concise account of how permaculture thinking has helped her to reflect on her own assumptions in relation to her house and garden, and to question and rethink her judgements. Essentially she has learned, through the medium of her garden, to question deceptively simple responses to problems and attitudes ("I am a bit untidy in the garden – having not yet removed last year's plants – so should push myself to be tidier"), face up to the confusing complexities ("What will the neighbours think?"), and embrace profound simplicity – "Let it be, because it's working for me" – the flowers attract bees, there is seed to collect, and the plants are shading the soil. The problem (the plants from last year are still there) is the solution (the plants from last year are still there).

When you design, build and operate a permaculture food garden, you work with the energy, patterns and flows of nature and the qualities of each element. You see how the relationships you have set up and facilitated, between sun, earth, water, wind, plants, animals and people, help to build the fertility and productivity of the whole, and meet your needs in a way that is good for you and the planet. You live in the heart of your power to consciously create our shared world. How often, and how well, in conventional life, do we use this power? How many of us, in Geoff Lawton's words, can "feel responsible for ... and morally clear about ... the footprint" of our lives? What would our everyday lives look and feel like, if we were not held captive by our belief in 'inevitability'?

In permaculture thinking, 'pioneer species', often otherwise known as 'weeds' or 'pests', are those resilient plants and animals that move into a degraded ecosystem and help to prepare it for other species to follow. Social pioneers like Rosa Parks, Yunus, and others, including many of those involved in the early stages of the permaculture movement, by taking a stand, help prepare the way for the next generation of new thinkers and change agents. It requires courage, commitment and a willingness, to stand firm, and apart.

For many of us, however, a big shift in perspective and outlook can seem to threaten our very identity, or to require such a huge sacrifice that we are unable to countenance it, even when it blatantly calls our integrity into question. Jeff Nugent relates a tragic account of an Agriculture Department employee who on the one hand publicly extolled the virtues of the pest control chemicals heptachlor and dieldrin, yet privately and passionately encouraged Jeff to fight the use of these chemicals because they were not safe and needed to be banned urgently. According to Nugent, the man implored him not to stop fighting the use of these chemicals, urging that if he achieved nothing else in his life apart from having them banned, his life will have been a productive one. When challenged for his double standards, he responded that promoting the chemicals was his job, and he had a family to take care of. Where is the 'glint' in this man's life? Where is the fullness of heart? It may have been small, but the 'glint' was in his encouragement to Jeff to keep fighting; a drop of integrity in an ocean of 'inevitability' over which he felt no control. The tragedy is that he didn't feel able to live out of this space in the rest of his life. In his perception the needs of his family were in conflict with the need for a clean environment, and he could not see a way to bring the two together.

Why is it that we believe in the limits and obstacles that seem to block our path? What would it take for us to choose something different? Hill (personal communication, April 2010) relates how in his workshops with farmers, to help them clarify their vision for the future, he encourages them to tell a 'great big lie'. According to Hill, if you ask people to talk about the future, based on what they believe is possible, their response will be controlled – limited to what they believe is 'realistic', according to the world as they perceive it. To get them to break out of that, go beyond their own limits and find again their clarity, wisdom and sense of empowerment, Hill, paradoxically, encourages them to lie. Tell a great big story, as if it were true. For many of us, this is what it takes for us to glimpse and step into our power and creativity, so deeply are we held enthralled, by our own, or by society's beliefs about how the world works, what we 'know' and what we can achieve. Surely, as a society and as individuals, to live like this is to live in a prison of our own (and our culture's) making.

I suspect that any momentous change in worldviews, for individuals and societies, will always seem 'impossible', 'unrealistic', even 'irresponsible', in the time before sufficient numbers – a critical mass – of people have chosen to make it happen. And if we live our lives out of, and create our identities out of, our worldview and beliefs, then any challenge to these can feel like a personal assault. Even if deep down, like the Agriculture Department employee, we understand the problem, the threat of persecution, marginalisation or loss of income, can be difficult to see beyond. Perhaps in situations where for individuals and societies, the stakes and the potential price to be paid, are extremely high, it requires a kind of 'Rosa Parks moment' – intense pressure, courage, and sometimes also anger to stand one's ground in spite of imminent threat.

For most of us in our everyday lives, however, the pressure isn't there in the same immediate way, even though the ultimate price of pollution, environmental degradation or social injustice may be extreme. It's easy to just keep going from day to day, even if doing so eats us up from inside. For most of us, Rosa Parks-type pressures and choices are unlikely. Perhaps, instead, what we can embrace are opportunities to 'tell the big lie' – to refuse to be limited or seduced by what is generally accepted as workable or do-able or reasonable, or 'polite'. Many of the authors describe paradigm-shifting experiences as permaculture revolutionised their thinking. Vries Gravestein seeing a picture of one of his permaculture students' children being held upside down by the legs in play,

> ... exclaimed "That is what Bill did to me!" ... Bill Mollison had ... shaken me just like that, so that all that I knew had fallen out of my head for me to gather back up again later. Up to that time, all my knowledge had been collected in small, separate bricks. ... I realised that what Bill had done was to help me build these separate blocks into a pyramid, an interconnected, indestructible structure. This is what I understand by wholism. ... Bill sat me on top of this pyramid, so that I could see far and wide over my environment.

Thoughtful, reflective, grounded willingness to believe in the 'impossible', to engage with 'the big lie', encouraged in my belief by permaculture thinking and action, can enable many of us who struggle to go against the flow of mainstream thinking, to set ourselves free from the prison of what we 'know'. Do we really want to be held back by our beliefs about what is realistic and what is possible? What is the bigger lie – that the way we are currently living is okay, or that there might be another way?

Permaculture is about 'listening'. Reading the landscape. Listening to its topography, climate, soils, watercourses, aspect, its flora and fauna. Observing and engaging with these with an attitude of acceptance and appreciation, openness to what is happening, what is needed, and what gifts and productivity, cooperation with them might bring. Learning to listen to nature is what teaches permaculturists about dealing with complexity. What could be achieved in the wider world, if this kind of 'listening' could be applied more effectively than it is currently, in other areas?

Muhammad Yunus 'listened' to the poor and to the situation in which they found themselves. This enabled him to see and understand the trap in which they were caught. By listening with openness to what he found, and being willing to respond without prejudice, Yunus' work has enabled them to change their world for themselves. For the mainstream banks it was all too hard, too confusingly complex or simply not even on their radar as worth their attention. Like our authors, by being willing to listen and embrace complexity, Yunus has touched profound simplicity. All have embraced and helped perpetuate the 'glint moments' in which we feel whole, grounded and vibrant. In these moments 'inevitability' falls away and is replaced by empowerment, integrity, and deeply meaningful change.

We are human. Perhaps the most wondrous, painful, yet beautiful thing we share is the confusing complexity of what that means. We love. We are fearful. We are amazingly creative. It is tempting in our haste to deal with challenges to succumb, as Fiona Campbell and Russ Grayson observe, to the 'urgency addiction'. To avoid dealing with the difficult 'invisible structures' – the human side of things. In the permaculture community, and the wider society, it is easy to fall victim to the deceptive simplicity of the lie that has us believe that time spent journeying in the confusingly complex inner landscape of people and cultures, is time wasted. Yet who are we fooling?

We create our world out of who we are inside. We cannot do otherwise. This is our power and our hope. If our individual and collective inner life nourishes our outer world, then how did we get where we are, and how can we build something new? The profoundly simple answer is this: The new frontier for exploration and discovery, as we work to bring social justice and ecological sustainability, is ourselves.

Part 2: Pioneering Spirits

David Holmgren

David Holmgren is an environmental design consultant, author, teacher and public speaker with an international reputation as a leading sustainability thinker. In the late 1970s David was the youthful co-author, with Bill Mollison, of *Permaculture One* (1978), the seminal book that launched the concept in a world concerned about energy and ecology.

Over three decades since the '70s David has shown by example that a sustainable lifestyle is a realistic, attractive and powerful alternative to dependant consumerism. He has developed several sustainably designed properties, consulted widely on urban and rural projects and has been recognised as an established author and educator in Australia, New Zealand, Europe, Japan and North America and Latin America.

David and his partner of all those years, Su Dennett, live at 'Melliodora' a one-hectare family home and property in Hepburn, Central Victoria.

Melliodora is one of Australia's best documented permaculture demonstration sites and has been featured twice on ABC *Gardening Australia* as well as *Landline* and innumerable magazine and newspaper articles.

As well as constant involvement in the practical side of permaculture, David is passionate about the philosophical and conceptual foundations for sustainability, which are explored in depth in his book, *Permaculture: Principles and Pathways Beyond Sustainability*, published in 2002.

Most recently David has been recognised as a futurist for his innovative Future Scenarios website, launched in May 2008 and later published as a book. This work maps the likely scenarios that will emerge globally and locally as a result of Peak Oil and Climate Change.

With an increasing profile as a public speaker and educator, David provides leadership with his refreshing and original approach to the big issues of our time.

Chapter 2 | The Long View

Introduction

To my knowledge to date there is no written history of the permaculture movement and no substantial biographical work about Bill Mollison or myself to provide a factual context for the exploration of more personal stories. These stories of people's experience with permaculture provide a deep and broad lode for that history. This exploration of the role that permaculture has played in people's lives shows some of the power and value of permaculture beyond its collection of practical ecological strategies and techniques to a way of thinking, living and even a sense of meaning in a confusing and rapidly changing world. Of course, speaking of such things rings alarm bells about dogmatic beliefs, in group psychology and even spiritual cults. These concerns were ones that I thought about in the early stages of the movement so I think it is interesting to explore those concerns and my gradual accommodation of them to both explore the issue and provide some insight into an important aspect of my personal relationship to permaculture. In the process I need to talk about my parents, childhood and something of my brief but intense working relationship with Bill Mollison from 1974-77 that led to permaculture.

A Personal Story

Writing about my personal journey with permaculture is difficult for a number of reasons. As the co-originator with Bill Mollison of the permaculture concept, I stand in a different relationship to the concept and the movement than others do. On the other hand, many of the first generation of permaculture practitioners, designers and activists were my baby-boom peers to whom I related quite strongly. Bill Mollison by contrast was a generation older, influenced by the Great Depression and World War II. My experience of the dramatic and early popular success of permaculture in Australia, when I was still in my early twenties, was that this was all happening to me as much as me being its architect.

In those early years I focussed on full-time pursuit of building my own skills in gardening, forestry, reading landscape, earthworks and building construction, all informed by permaculture design principles. While shunning the limelight of media attention and the credibility of further academic qualifications, I never totally ignored the role of communicator and educator about permaculture.

The Permaculture Design Course process that evolved from Bill Mollison's teaching in the early 1980s created a cadre of self selected designers, activists and further teachers that, combined with Bill's not insignificant media presence, spread permaculture throughout Australia and the rest of the world.

That early success of permaculture also involved some downsides that left me as a sceptic and outsider to the movement. I now understand that this role that I took on reflected not only the deficiencies and downsides of the permaculture movement (as I saw them), but also my own history. Although I never rejected permaculture conceptually, in the 1980s I felt as much the movement dissident as its father, despite being given an honorary diploma from the Permaculture Institute at the first International Permaculture Convergence in 1984. At the invitation of several established permaculture teachers in the early 1990s I was progressively drawn into permaculture teaching and began to write about permaculture origins, education and the movement. With the publication of *Permaculture: Principles and Pathways Beyond Sustainability* in late 2002, my influence on permaculture teaching and activism worldwide has significantly increased. The nature of that influence is a subject best left to others to decide.

Family History

I was born in 1955 to parents who were experiencing the strains of challenges to their strongest beliefs. My father, born in 1916 to strict Protestant parents, had lost his religious faith and joined the communist party at the age of 18 during the turbulent years of the Great Depression. Through 'The Party' he met my mother, a young Jewish 'firebrand', also of lost religious faith. During the 1940s their shared belief in the revolution and the common good led them to (amongst other things) sell a block of land for below its market value because they regarded speculation as a social evil. Of course they were not alone in believing that Communism was the pathway to the 'workers paradise' of cooperative control of workplaces, sensible allocation of resources for the common good, an end to racism, and the recognition of the rights of women and minority groups.

In fact, a significant proportion of the intellectual elite of western societies described themselves as Marxist at that time. But the evidence of the evils of Stalinist Russia and the Australian Communist Party's following of the Moscow line led my father to leave the party before I was born, straining the marriage in the process. Like so many idealists, my mother left the party in 1956 after the Soviet invasion of Hungary. My parents never lost their commitment to social justice and human rights and were some of first Australian activists against the Vietnam war on the principle that the U.S. was waging a war of aggression against Vietnamese nationalism. I was raised with the idea that you should always think for yourself, stand up for what you believe, resisting both peer pressure and the seductions of apparent authority and expertise. I have passed those ideas on in whatever way I can, not least in raising children.

The Outsider

As the only child (in a primary school of 600) who didn't attend religious instruction and who refused to stand up for the national anthem during the nationalist fervour of our early involvement in Vietnam, I took for granted the fact that I didn't really belong and that my family was different by choice. Even my strange lunches of wholemeal bread sandwiches, carrots, celery and dried fruit were alternatively a source of wonder and derision. My Swedish name and more particularly, my Jewish heritage, gave me reason to understand difference and resultant victimisation, but I was never picked on for differences that I could do nothing about, unlike the 'wog' children of Mediterranean migrants, let alone Aboriginal kids. Our difference was like the Seventh Day Adventists in being chosen or 'self inflicted'. On the other hand my family was unlike them, in that our values were not bolstered by a strong community with shared (and unquestioned) beliefs. I took for granted that we did not belong and was more or less comfortable and proud, even arrogantly proud, of the questioning of orthodoxy of every kind.

In my teens I gained a genuinely collegiate circle of intellectual, mostly male fellows in which I saw the generation gap grow into a yawning gulf over rock music, drugs, sex and of course the war in Vietnam. Many of my closest friends despised their parents or at least their values, while I maintained the communication throughout my teenage years, despite the challenges of mind-altering drugs, something outside the experience of my radical parents. I came to see the difference between my friends and myself as that between what I called 'first and second generation alienated'.

My parents were first generation alienated. They had made a wrenching break in the '30s and '40s from their parents' traditional values; my maternal grandmother wore black when her youngest daughter married outside the Jewish faith. 'The Party' provided meaning and purpose for their youthful energy and passion. They then suffered a second loss of faith but remained committed to core values that had drawn them to Marxism. Further, they gave their children the freedom and encouragement to define themselves and, in effect, accept alienation as a normal, even natural existence in a world so full of knowledge and possibilities as well as injustice and contradiction.

As the only academic dux of John Curtin High School not recorded on the roll of honour (as punishment for my dissident attitudes), I left WA for a year hitchhiking around Australia during which I abandoned the idea of a conventional university degree course. While in Tasmania, I discovered Environmental Design (ED) at the Tasmanian College of Advanced Education (Mt Nelson campus), an academic school established by Barry McNeil. McNeil was a Hobart architect who saw that it was pointless to just train future architects, landscape architects and urban planners in the technical skills of their professions. In a world of rapid change, those skills would be redundant by the time the students became practitioners. Instead, McNeil's goal was to teach students thinking and problem solving skills that would enable them to deal with constant change. To that end Environmental Design had no fixed curriculum, self assessment up to the submission of a thesis, consultancy and workplace experience, a large role for visiting teachers and practitioners as well as student involvement in staff selection and running the school. The school attracted radical and dissident teachers and students from other design schools around Australia and ran for ten years (1970-80) as perhaps the most radical experiment in tertiary education in Australia's history, before it was emasculated and turned into a conventional design training course.

After nearly a full year of immersing myself in a wide variety of projects and activities within Environmental Design, my interests had gravitated to how landscape design and ecology could be applied to agriculture. Around this time I made a connection to a middle-aged man whose comments at an ED seminar had resonated with my own thoughts. I was immediately attracted to his ideas without having any knowledge of his position (actually Senior Tutor in the Psychology Faculty of the University of Tasmania) or his infamous reputation in Hobart academic circles. This was the beginning of an intense and productive working relationship with Bill Mollison, Hobart bohemian, ex fisherman and wildlife researcher, environmental activist and political candidate.

Over the following two years, 1975-76, I shared house with Bill, his second wife Philomena and her son on a small urban fringe property. Bill and I developed an extensive garden and arboretum of useful plants and our relationship was that of student and mentor. The permaculture concept emerged from the seed of an idea that Bill suggested as a possible subject that would fit my interest in design, ecology and agriculture. I wrote the permaculture manuscript, maintained a tenuous relationship to Environmental Design, adequate enough to be awarded my degree but not enough to ever have a sense that I belonged there in that gathering of radical designers and activists, or to stay for postgraduate studies. I also found myself the outsider amongst the radicals. I put my passion into more practical work, building, hunting and gardening and handed the permaculture manuscript to Bill for edits and additions that eventually became *Permaculture One* published in 1978, to substantial fanfare and even acclaim. But my days of working with Bill Mollison were limited to those three years that my self-effacing youth allowed me to ignore his difficult personality. I had already concluded that an evolution into a truly collaborative relationship was not possible, so I quietly disengaged.

For Bill, who was tiring of the university and looking for a larger, more dramatic stage, the book was the stepping stone to launching the permaculture concept as a world-changing movement. While pleased with the public reaction to the book and the concept, I had plenty of doubts. Some of those were about my own lack of practical experience to back up the conceptual nature of permaculture. Some were about Bill's tendency to talk up our achievements and those of others as evidence supporting permaculture. My focus turned to building my own skills in gardening, hunting, farming, forestry building, reading landscape, event organisation and farm design as a kind of continuous self-training program that included helping initiate Permaculture New Zealand, and developing a bush property and building a passive solar house in southern NSW. While I was busy 'doing permaculture', I watched while Bill's energy and charisma tapped the feelings of those years to build the foundations of the movement.

Those foundations were: The Permaculture Association and its quarterly publication, the *Permaculture International Journal* (1978), brilliantly edited and produced by Terry White at Maryborough, Central Victoria; an intentional community and publishing business in Stanley Tasmania (Tagari) in 1979; a second book (*Permaculture Two* in 1979); and most importantly, the three week residential course that laid the foundations for the Permaculture Design Course in 1981.

For many first generation alienated young people of my age, permaculture provided a framework of positive direct action to create the ecologically integrated and anti-authoritarian world they imagined. Mollison was the wise father figure warning of the catastrophes ahead and the solutions to be found in a reinvigorated Garden of Eden. As well as being excited and engaged by meeting such fellow travellers, I also heard lots of warning bells. Although Bill denied the mantle of Permaculture guru, I knew too well that his dynamic combination of polymath, 'jack of all trades', charismatic yarn spinner, and huge ego were a dangerous mix to be at the head of such a movement.

Several colleagues who were aware of Bill's failings tried to raise my sense of responsibility to 'save the permaculture movement'. I was adamant at the time, that while I was the co-originator of the concept and more than willing to give my views on any and all aspects of permaculture, the permaculture movement itself was not my responsibility. While I recognised Bill Mollison's faults, I also saw acolytes willing to believe anything Bill said and at his behest launching themselves hopelessly unprepared on the real world as design consultants and activists trying to create the permaculture world. I found my own ideas being turned into ideology that I attempted to correct privately or publicly. My habit of kicking the sacred cows of middle class Australian society, the design professions and the environment movement reflected my upbringing as a sceptic. It was very ironical that at age 25, I was adding permaculture sacred cows to the list to kick.

I always tried to do this in ways that were diplomatic and respected people's good intentions and, in some cases, substantial achievements. But it was as plain as daylight to me that many of the problems of ideological faith in Marxism that had failed for my parents' generation were emerging, albeit in small and more benign forms, in the environment movement for my generation. While this may seem a ridiculous comparison now, I felt at the time, like many others, that the environmental/energy crisis would break the world we knew within a decade, and movements like permaculture could be catapulted into that chaos to provide leadership on a larger scale. That was certainly Mollison's aim and expectation.

The reasons why we (and so many others) were wrong about the timing of the fundamental environmental crisis is a complex and interesting issue, but my point here is that I saw the seeds of dogma and rigid ideology developing in the permaculture movement from the early days.

In 1983 I founded Holmgren Design Services as a permaculture consultancy based on nearly a decade of intensive but organic and self-directed training. I actually seriously considered not using the term 'permaculture' because of my disgust with its over-promotion. In the end though, the positives outweighed the negatives. In 1984 I attended the first International Convergence of Permaculture 'Designers'. That event was a landmark for my relationship to the permaculture movement. I took two prepared and documented presentations, one a case study design about the NSW bush property, later published as *Permaculture in the Bush*, 1993; the other about skills in reading the landscape

David at Jackeys Marsh 1978. Photo by Bruce Hedge.

that I saw as an essential prerequisite to effective permaculture design in rural areas. My intention with those presentations was to set an example of real, if modest, practical projects, documentation and skill development. It didn't surprise me that there was not as much substance behind Mollison's international cadre of permaculture designers as might have been imagined by the public front Mollison had built. On the other hand I met an extraordinary number of like-minded people, people who thought about the world as I did. I actually felt at home, almost at one with my tribe! And of course I was an honoured participant. This sense of belonging which many people gain from their extended family, workplace, profession, recreational clubs and nationality, is mostly taken for granted as with fish swimming in the sea. For me as a second generation alienated individual, this was a new experience.

Despite my open criticisms of the permaculture education system that Mollison had set up (with its fixed curriculum in ironic contrast to Environmental Design that had given birth to permaculture), I was presented with an Honorary Diploma of Permaculture Design which I graciously accepted (without asking why Bill didn't need an honorary diploma to be part of the tribe). I heard the discontent with aspects of Bill's leadership from some of the more experienced activists and sensed their interest in me as some sort of alternative leader. I can even remember the feeling that I had the power to push the guru off his pedestal but I made a very clear decision that I was never going to do that because it would be destroying something without replacing it. As a primary school child I had learnt the lesson that bringing down a group leader without offering a viable alternative leadership was destructive for the whole group.

At that moment I learnt to accept the powerful value that the permaculture movement provided in focussing the lives of capable and sensitive people towards holistic goals and that Bill's leadership, despite his faults, was clearly essential to it, at least at that point in history. My anarchist idealism about a leaderless world of self-realised individuals was set to one side as one that required a lot more work.

In the following years Mollison was the subject of two successful TV programs, received the Right Livelihood award, published his classic work *Permaculture: A Designers' Manual* and continued to spread permaculture around the world. Despite the sense of connection through the 1984 convergence, I continued on my independent path but now with less angst about the faults of permaculture education and the movement. In 1990 I accepted an invitation to co-teach on a Permaculture Design Course and over the following three years I co-taught with some of the more experienced of Mollison's students. After sitting in on all sessions of three PDCs presented by six different graduates of Mollison's early courses I felt I had as good a grasp of the whole process as anyone. I then embarked on my own evolution of the PDC with a group of colleagues in Hepburn, Victoria, during the 1990s to better reflect my own understandings and perspectives developed in the years since *Permaculture One*. Ian Lillington, a permaculture activist from the UK who was trained by leading British permaculture teacher Graham Bell played a strong role in initiating that process with two courses in 1993.

Those years included the first decade of my relationship with Su Dennett, the homebirth of our son Oliver in 1986 and the establishment of our home 'Melliodora'. The timeline of the development of the property and the deepening of our relationships in the local community are marked by Oliver's years. The process by which we became locals, connected to community and the soil are interwoven with our increasingly strong connections to permaculture networks and the wider movement. On the one hand the personal and local relationships represented a counterbalance to the somewhat disconnected nature of conceptual thinking and global networks of permaculture, while also providing authenticity and strength to our wider connections. Similarly the cross-pollination of ideas and culture through the networks kept us from becoming too parochial and added a hybrid vigour to our local activities and projects.

While the warrior types seem to survive if not thrive, in the lonely world of public and global action, I have found that without the grounding of family and home, I would have quickly been burnt by the flames of action in the wider world. At one level this is an admission of weakness, at another it is the assertion that groundedness is one of the foundations necessary to the ecological world we seek to create. How can we reach out with true strength if our centre is not balanced and strong?

There are many other personal landmarks in my growing sense that I can feel at home and exercise a collegiate leadership role in permaculture without the paranoia about becoming a guru defending permaculture dogma. Some of that is simply the process of growing older, being more prepared to accept the leadership that I had the wisdom to refuse in my 20s. Some of that leadership comes not only from having been there at the start but also from the credibility that accrues to the 'tortoises' of this world long after the 'hares' have done their dash and given up on the long road of a social revolution based on ecological principles. Certainly permaculture history has included more that its share of flashy projects that didn't last, movers and shakers looking for bigger rewards and generous souls who burnt the candle of their own evangelism at both ends. Some of my sense of belonging and leadership in permaculture comes from being an observer of the wider environmental sustainability scene over two decades as it has gone through various cycles and fashions without producing many coherent frameworks that have given rise to the organic diversity and strength that is permaculture.

Perhaps my next great test was the decision in 2000 to use permaculture as the title and identity unifying my major work *Permaculture: Principles and Pathways Beyond Sustainability*. Again there are enormous conceptual strengths in this choice but there are also downsides. Because permaculture has such a high profile as a 'brand' of environmentalism in Australia, everyone thinks they know what it is, generally a cool form of organic gardening or self-reliant living. This acts as an impediment to it being taken seriously as a conceptual framework for redesigning our world to adapt to the energy descent future. These days I accept the baggage of simplistic interpretations, bad experiences and hostility to permaculture as minor issues in the overwhelming positive results that continue to flow from permaculture activism worldwide. Most recently on the Peak Oil and Permaculture public speaking tour of Australia with Richard Heinberg we experienced some of those problems as well as powerful examples of permaculture activism from the grass roots through to significant, if subtle, influences on public policy. I now see these problems as simply some of the 'collateral' damage from the trails cut by Bill Mollison through Australian popular and intellectual culture. I walk along those trails marvelling at the energy of the warrior and his foot soldiers as I help re-clear overgrown parts and contribute to repairing the damage.

Learning to accept my place in permaculture has also strengthened and grounded my internationalist idealism that I inherited from my parents. Six month long study and teaching trips overseas in 1994 and 2005 have given me a deep sense of permaculture as a global movement no longer even centred in Australia. But this experience and the complementary one of being a host to many WWOOFers and other visitors from overseas has also allowed me to acknowledge another aspect of belonging that I rejected in my youth; that of being an Australian.

By the age of 12, I had come to the conclusion that Australians were mostly small-minded racists and bigots unable to think past beer and football. I thought that as a nation, Australia was ripe for the descent into fascism, should times get tough. I have not dropped my 'black arm band' view of Australian history and there are many dark clouds on the horizon that suggest my childhood nightmare is getting closer. Ironically the fact that many more Australians now have the same feelings about their country gives me some hope for its future. I now accept this darkness as part of the heritage along with some very positive aspects of Australian history and culture, some of which are well-known and celebrated, other less so. The necessity to explain to foreigners (and Australians) my interpretation of our history as a context for the emergence of permaculture in Australia has allowed me to gradually come to terms with the obvious and undeniable fact that I am an Australian. Further still, I have come to accept that I am just another Australian trying to work out how we move beyond camping in this land to being rooted to it, part of a diverse and multicultural process that draws on many sources, ancient and recent, global and local. Permaculture has been my long sustained search to find home.

Terry White

During the 1980s Terry White founded and edited the *International Permaculture Journal*, coordinated Project Branchout, which was at that time Australia's largest community based revegetation project and initiated the Saltwatch, Frogwatch and Ribbons of Blue community environmental monitoring programs.

During the 1990s Terry managed Vicwatch, the Victorian State Government Community Environmental Monitoring support program, co-authored *Listening to the Land*, Australia's first Directory of Community Environmental Monitoring Groups. He also coordinated national research on Community Environmental Monitoring for the Commonwealth State of the Environment Unit, co-authored *Environmental Indicators for Community and Local Uses* and was actively involved in promoting roof top gardens for the Cities of Port Phillip and Melbourne.

Since 2000 Terry has developed a triple bottom line audit manual for small communities, coordinated local native vegetation preservation and emission reduction programs and has worked to establish regional Greenhouse Alliances in Central Victoria (CVGA) and Melbourne's Western Suburbs (WAGA).

Chapter 3 | Trend is Not Destiny

I come from a working class family, one of seven children living in a housing commission estate in the suburb of Williamstown in Melbourne. As children, my brother and I would go to the rifle-range, a no-go wilderness area because of the shooting, but we'd crawl under the fence and spend a lot of our time making homemade rafts and kayaks and paddling around among the mangroves on the coast there. For me it was a haven of quietness but also a place where lots and lots of things were happening. I liked that world. It was the world I grew up in.

Mum was orphaned at a very early age and hadn't been educated beyond 4th grade. But she was absolutely affirming of anything we wanted to do. If I had an idea she thought was a good one she'd say, "Ring the Prime Minister," or, "Go and see him!" She believed everything was possible for her kids and we came to believe that too. We were never reticent about talking to people, whoever they might be; we never had the mental barriers that other kids had about who you could talk to and what you were allowed to do.

Being the second youngest in the family, my brothers and sisters gave me a whole range of options about how to be in the world. My eldest brother rode around India on a water-cooled motorbike just after he finished his printing apprenticeship, and I just loved reading his airmail letters sent back to the family. My eldest sister was the first of our family to go to university, so I naturally thought it was possible for me to go to university too. All my brothers and sisters did interesting things, and when they were away I had access to their libraries. So I guess my primary influences and inspiration came from their lives and their books. I still love books dearly to this day.

As a teenager I got involved with a fundamentalist church through an open air Sunday school in our housing commission area. I got heavily into the Bible, and as I got older I became interested in the philosophical arguments for and against that sort of worldview. I'd left school early but later went to night school to get the qualifications necessary to go to university as a mature-age student on a teaching bond. At university I studied philosophy and history – because I was interested, not because I wanted to spend a lifetime teaching or reading. I taught to pay off my bond, telling myself right from the start that I would give up teaching after three years and do something else. I've always been attracted to doing what I think is important, rather than what someone else thinks is important. If I get paid for it, well and good, if I don't, I'm still getting what I want out of life.

I have always been interested in universal themes and fundamental principles, so thinking about whole systems really attracted me. From about age 26 to 36 I became interested in the thinking of people like Eric Fromm, Buckminster Fuller, Howard Odum and Gregory Bateson. I became an avid reader of publications such as the *Whole Earth Catalogue* and *Co-evolution Quarterly*, as well as superbly practical local publications like *Earth Garden*, *Grass Roots* and *Owner Builder* magazines that were around at the time. For me they resonated with the biblical approach of James who said, "show me your faith by your works," and Jesus, who said, "by their fruits you shall know them." In other words, it's not what you say that matters, it's what you do.

I first encountered permaculture when I was working as a community liaison officer at the Maryborough Christian Community College in central Victoria. My job involved identifying the social and educational needs of our region and trying to do something about them. I first heard Bill Mollison on an ABC radio interview with broadcaster Terry Lane in about 1976 or 1977. It was an amazing interview and lots of people were electrified by it because Bill joined the dots. We had all been pottering around with different bits of the system, like alternate technology and organic agriculture, but here was somebody who put all the pieces into one parcel and showed how they could fit together. What attracted me most about permaculture though, wasn't the usual, 'permanent agriculture' angle, as much as the idea of a 'permanent culture'. That's what got me going. Here was a word that summarised what I was searching for.

I was also very impressed that David Holmgren and Bill Mollison were big thinkers who not only built word castles in the air, but rolled up their sleeves and worked hard to put real foundations under them. And what's more, they encouraged others to do the same. That was the zing I got from those two guys.

Some time after that, a colleague of mine had managed to get Bill to come to Ballarat to speak, so I said, "If he's coming to you in Ballarat, let's see if we can get him to come on to Maryborough." And he did. We asked for a gold coin donation at the door and at the end of the meeting we proclaimed everybody to be fully paid-up members of the Maryborough Permaculture Association and decided when we were next going to meet! That night one of the first Permaculture Associations in Australia was formed.

I'd mentioned to Bill that I'd put student newspapers together, had experience in editorial and art work and was happy to produce the *Permaculture International Journal*. He immediately gave me $500 and a member list and said, "Go for it!" I edited the *Permaculture International Journal* for 10 years, unpaid for most of the time, but nevertheless a richly rewarding experience.

This was a really exciting time. The work enabled me to subscription-swap the permaculture magazine for an eclectic collection of similar magazines from all around the world. I'd just about run to the mailbox to see what was coming in. Species lists, designs, domes, vaults, local currencies, aquaculture, biotecture, vertical gardens, roof gardens, living fences, owner-producer co-ops and more. I regarded it as an extension of my education. Today, you can google a key word or phrase and immediately find somebody who's thinking what you're thinking, and is maybe on the road to a book. Back then, it was in journals that you found the freshest ideas.

Those journals were written by 'doers' from all around the world. A lot were in the third world, a lot came out of the United Nations and, strangely enough, they included many great applied research publications on subjects like plant usage, renewable energy and aquaculture, completely free, from the US government. In fact I found those North Americans who responded to permaculture to be pretty exciting thinkers. There is a zeitgeist in America that is very positive (sometimes to the extent of being rather sickening), but it's also an extremely good antidote to the apathy, stoicism and cynicism that's prevalent in much of Australian culture. I also found the Germans to be hugely exciting. You begin to realise that what we've been doing for so long is creating consumer cities rather than producer cities. But way before Havana was forced to re-invent urban agriculture after the collapse of its access to cheap Soviet oil, urban designers like Margrit and Declan Kennedy were doing great work on rooftop gardens, local currencies and vertical greening in the heart of Berlin. They came out to Australia to talk with Bill, and visited Maryborough. They're still very active today.

Those were indeed very heady times. Every issue of the permaculture magazine developed a theme depending on what was exciting me most.

Bill was very enthusiastic about P.A. Yeomans' book *Water for Every Farm*. We had 'P.A.' here in Maryborough doing keyline work around our sewage settling ponds. Like Bill, he was a very independent and creative thinker. Both Bill and P.A. could be abrasive, but they were in the business of putting sand in oysters, so were great catalysts for thinking. I did about three of Bill's early courses. They were ten days long, but I never got bored. In many ways this was familiar territory for me. Growing up in a fundamentalist church I was used to listening to missionaries coming to us from challenging circumstances in Nepal or India or Africa and bring us fervent messages full of challenge and commitment. What united these two worlds was earth care and people care – two very strong ways of saying people matter and the world matters.

I was impressed by the real generosity of permaculture people; their readiness to share and to help one another. I was awe-inspired and sometimes taken aback by the amount of trust and permission-giving that Bill gave – basically fire at will! Later he went back on that a bit, with his insistence on certification, but fundamentally I think he was on the side of generosity. He was saying we've got important work to do and we can't be bothered with the niceties. You'll learn on the job, so go and do it. So I really love the ethic of permaculture, the generosity, the altruism, the openness to ideas, the commitment to work in hard places, the risk of being embarrassed, the risk being wrong. I love all that about permaculture. In other circles you often find the opposite, a culture of stinginess – "These are my concepts, this is my intellectual property." So when I compare and contrast, there's a heck of a lot still going for permaculture, permaculture still has a life.

But there have also been casualties as people tried to apply 'one size fits all' solutions and came to grief. I think a lot of us went through that phase. The basic principles of zonal and sector thinking provided useful handles for analysis. But particular strategies don't always work in every situation. For example, when we applied newspaper and carpets at home in central Victoria as a sheet mulch veneer on top of our clay and quartz, we bred a heck of a lot of earwigs which got to feast on the new plantings while the mulch blew around the backyard! Understandably, a lot of long-time gardeners in our town were sceptical right from the start. Our next-door neighbour was an excellent gardener who always had a meal on the table out of his garden. Looking over the fence I don't know what he made of us! I was botanically illiterate and couldn't tell couch from kikuyu, even while I was editing the permaculture magazine. Fortunately, with a crowbar, compost and double digging we managed to get a good garden going. And my wife Fay is a fantastic gardener.

Permaculture is just one pair of glasses for looking at the world and there are some blind spots. For example, the too-casual attitude toward the treatment of exotics has got up my nose on many occasions. Right from the start I've never liked that 'green guerrilla' approach of popping a seed in here and a seed in there, as if just any seed is OK. I think David Holmgren would suggest that nature will sort it all out, but my view is that what we are seeing happening as a result of the spread of rampant species is a tragic loss of diversity and richness and a rapid simplification of ecologies worldwide. Australian paperbarks are taking off in the American Everglades, and marine ecologies are being disrupted as species are dispersed around the world via the ballast water of boats. It's happening at an alarming rate and its being spread by globalisation. I have great respect for David but on this we disagree.

I also believe there's another real blind spot in terms of governance. Bill and David tend to be very pessimistic and disparaging about the role of state and national governments. They are all for grassroots action, while I believe there are some issues like the state of the oceans and the atmosphere that just can't be dealt with by grassroots action alone. To stabilise the climate and protect the oceans and rivers we desperately need strong inter-governmental agreements and protocols based on the common good. The longer we leave those blind spots vacant, the more they will be occupied by multinationals wanting to turn the commons into commodities. Of course we need the grassroots but we need good governments as well. It's got to be both-and, not either-or. Top-down as well as bottom-up. Not to mention sideways as well.

Permaculture and whole systems thinking has been a continuing influence on my work. While I was in Maryborough editing the *Permaculture International Journal* I was also establishing a community based revegetation program called 'Project Branchout' to combat dryland salinity. By the end of the '80s, Project Branchout was operating across 33 municipalities and was the largest tree planting project in Australia. We made successful applications for around $850,000 of State and Federal funding and planted thousands and thousands of trees. The Project Branchout team raised both native and exotic trees for Central Victoria in our nurseries, and we employed David Holmgren to write *Trees on the Treeless Plains* at that time.

On the strength of my involvement with Project Branchout, I got a job as a statewide educator with the Victorian Government Salinity Control Program. It was in this role that I started almost a decade of work on community environmental monitoring. Unfortunately, the observational skills that give Bill and David so much potency as teachers are not taught in schools. We teach literacy and numeracy but we don't teach kids how to read the land.

I started with a water quality monitoring program called Salt Watch, where students were given salinity meters to monitor the salt content of their dams, creeks, irrigation channels and rivers. My later work in designing the guidelines for the national Water Watch program came out of that. These programs gave whole communities a hands-on understanding of the quality of aquatic ecologies and what could be done to improve them.

Later, I was involved in compiling *Listening to the Land* – a directory of community environment monitoring groups across Australia who were observing everything from birds to bats, ants to earthworms. I did this with Jason Alexander of the Australian Conservation Foundation (ACF). It was great working with Jason, and we went to a lot of the early permaculture courses together. From that I moved on to consultancy work for the Commonwealth investigating the possibility of community groups being able to collect data for the national *State of the Environment* report, and we had pilot programs operating in all seven states.

I've never liked the idea of teaching people about stuff I know nothing about. While I'd been to Bill's courses and helped edit *Permaculture Two*, I didn't know the material personally. So I was quite happy to teach the water component in lots of Melbourne permaculture courses. This was great fun because I was able to work on my water module and refine it over time, so that every time I was invited to teach I could go through my overheads and find a way of incorporating a new concept or a better way of illustrating a design principle. I bought a lot of books and maintained contact with some of the hydrologists that I knew from the Department. I just loved that, I love teaching and love putting together material in ways that I think comes across as interesting and engaging.

When I was asked by Permaculture Melbourne to teach a whole 12 day course at La Trobe University with Vries Gravestein, it was a real challenge but I enjoyed it, Vries and I were able to teach to our strengths and two of our students went on from there to Havana, Cuba. But I've not been tempted to do full courses since. I think the model Jane Scott used in Kallista (Melbourne), was excellent. She facilitated the course and taught some core modules but also invited a range of guest experts to add depth and breadth.

If the '80s were largely consumed by fostering land literacy, the '90s were a time for roof gardens. An early black-and-white issue of the *Permaculture International Journal* had featured a drawing of a block of flats in Berlin with vegetation growing up the walls and shepherds and sheep on the roof. All pretty fanciful, but compelling nonetheless. Sufficiently compelling for me to suggest to the Cities of Melbourne and Port Philip that they should look at the viability of vegetating the roofs of their CBDs. At that stage, all the literature on rooftop gardens was in German and French, but by enlisting the help of tertiary students from Melbourne, Monash and Deakin Universities we were able to organise a 'Gardens in the Sky' exhibition. This made it clear that we were 20 years behind Europe and showed just what could be done by inviting nature back into the city. Better air quality, greater storm water control, enhanced biodiversity, improved liveability. At the same time, things began to happen in Canada and in Chicago, and now there is a substantial English literature in this field. Hopefully we will begin to see extensive re-vegetation of our cities.

From green roofs I became interested in climate change, and in 2000 initiated the Central Victorian Greenhouse Alliance (CVGA), building it up in a similar way to Project Branchout. What began with four people quickly expanded to 14 councils, and is now the largest regional grouping of cities for climate protection in Australia, covering 22% of Victoria. As well as councils, the CVGA membership includes the Bendigo Bank, Origin Energy, Bendigo Access Employment, Ballarat University, La Trobe University, and St. Luke's Anglicare. This consortium takes climate change seriously and has a target of reducing regional emissions to zero net by 2020. That work continues.

After six years I resigned as coordinator of the CVGA and in 2006, I took up my present work as coordinator of the Western Alliance for Greenhouse Action. My work explores the way in which the western suburbs can reinvent themselves and move from 1000 megawatts of coal-fired electricity to 1000 megawatts of green energy, and how every sector might transform itself within a ten year time frame from now to 2020. It's about how we re-design the way we live. Actually I believe if we took the time frame seriously we'd be better off looking at 2015 rather than 2020, but that stretches the friendship a little with conservative councils. My business plan indicates that if we take the science seriously, we should be carbon-neutral by 2020.

And talking of the future, when I think about what might take place, the rational half of my mind says that we have already passed a number of physical tipping points that are irretrievable. We will see, as James Lovelock predicts, vast transmigrations of people and huge waves of conflict as people fight for resources. We'll see the incumbent wealthy protecting their patches, and we'll see gated communities on a global scale. The other half of my mind says that trend is not necessarily destiny. What we have learned about the planet just since 2000 is just awe-inspiring. In 2000 scientists didn't understand that the summer Arctic sea ice could melt so quickly. They gave it 100 years, but what we're seeing now is that it could all be gone by 2012. The speed at which the scientific community has been getting to grips and modelling the way the world works as Lovelock predicted, is breathtaking, really exciting and extremely positive.

But for permaculture to be a formative influence in all this it needs to do a lot of things it's not doing now to fast-track the necessary transformations. I am seeing the start of it with the permablitzes that are happening in Melbourne – weekend makeovers where backyards are being transformed into edible gardens in a day by people-power. I see communications as central to that. In my view there's got to be a Wikipedia approach to permaculture, where concepts are launched and local and regional transition strategies can be contributed to by people from a thousand different quarters.

So this other part of my mind says we can change more quickly than we ever thought possible. We can enjoy life just as much as we did yesterday without being weighed down by so much stuff. Lots of people are already proving that in their own lives now, leading very fulfilling lives with very little stuff. We've all got to get to that place.

If we have the time to do it, the curriculum of universities should be totally different from what they are now. I'm talking to Victoria University (VU) about their whole student body being tuned to address these issues. VU is dedicated to community connections, so we're talking about carbon coaches in schools and in industry, and we've already got students working on renewable energy futures for the western suburbs.

We're not going to get 'permanent culture' unless we get permanent human settlements. The tendency has too often been reductionist, to separate out. As always in permaculture, we're trying to find the 'best of' around the world, putting it together and see what synergies arise from co-locating them, and trying to do it all at once. Yes, I can be critical of permaculture, but there are very strong positives in there. And they were never more needed.

Terry and Fay White with their wicking gardens.

Robyn Francis

From a background in organics, sustainability and community activism, since 1983 Robyn Francis has worked throughout Australia and internationally as a permaculture teacher, designer, facilitator and presenter. She has taught hundreds of courses including PDCs, specialist training and Accredited Permaculture Training (APT). She edited the *Permaculture International Journal* for five years, was a founding director of Permaculture International Ltd (1987), and contributed to the design and successful accreditation of APT. Since 1993 she has developed one of Australia's leading purpose-designed education and training centres, Djanbung Gardens, now home base for the bioregional campus of Permaculture College Australia Inc. Robyn has trained and mentored numerous permaculturists, many of whom are doing significant work internationally and in their local communities and bioregions. Robyn is a passionate communicator, cook, gardener, poet, singer/songwriter and composer who loves gourd crafting, building with bamboo and generally walking her talk.

Robyn's property in Northern NSW, Djanbung Gardens, in 2005. It was a barren cow pasture in 1993.

Chapter 4 | Healing the Rifts: Building Intentional Community

Getting involved with permaculture was for me a natural progression in my life. I grew up in Inverell, on the western slopes of northwest NSW, between Glen Innes and Moree. I was a product of the 1960s; the Vietnam moratoriums, the peace movement, 'back to the land' and self-reliance movements that were happening during my teenage years had a really big impact on me.

My parents also had a very important influence on me, on two levels. One was that they were very active in the Salvation Army and in social work, and they instilled in me that the most meaningful thing to do in life is to be involved in something that not only takes care of yourself but also helps other people, as well as creating a better world – thinking in terms of more than just oneself. The second thing that I really appreciated about my parents was that they were, and still are, such incredibly resourceful people. They grew up on farms through the Depression and even during the fifties and sixties there wasn't much disposable income. I grew up in a home that had an incredibly productive quarter-acre backyard with lots of vegetables, fruit trees, chickens, ducks, bees, and a milking goat that we tethered out to mow the neighbour's lawns.

Few people had as much in their gardens as my parents had. Most of the neighbours had a little tiny vegie plot and a great big lawn, so there was heaps of grazing for the goat and we were glad to mow their lawns. Before we were connected to town water we had to survive on a 2,000 gallon tank, so I grew up being really conservative with water – a half a cup of water to brush your teeth. Everything was rationed so that we'd get through the summer, because in those days you couldn't just order in a truckload of water if your tank ran dry. We couldn't afford to buy new clothes, we would just buy the cloth and sew our own on an old treadle machine. I remember the first thing I ever made was a 'moo moo' – a simple straight dress with a ruffle at the bottom, popular in the early '60s. We used to bottle fruit. We'd go out to the local orchards to harvest boxes of plums, peaches and nectarines and come home and bottle them up in vacola jars. We were always making big pots of jam and pickles. There was so much stuff like that I grew up with. I didn't see it as anything special, it was just normal. As a result, I now tremendously value these things. So it was a bit of a shock for me when, in the late '60s, I finished high school and went out into a world that was so different from the one I'd grown up in.

Later, in the early seventies, I left home and lived in Sydney for several years. There, connections I made with people who were involved in the alternative movement, the so-called counter culture of the time, also had an influence on me. But it was my travels overseas from 1972-79 that had a major impact. I travelled widely in Asia and Europe, and lived for three and a half years in Bavaria. It was a wonderful time to be there because the old generation farmers were still farming traditionally. The only thing that had changed was that instead of horses they were using tractors, and the only thing they imported onto the farms was diesel for the tractor. They weren't bringing in fertilisers and they weren't using agricultural chemicals. They were still using the natural system of crop rotation, with storage of the sugar beet and cow beet for fodder. Some of the old farms still had their old apple juice presses to brew their own cider and make their own sauerkraut. There was a growers' market every Thursday in our local town that had been going for 900 years!

I managed to learn the local dialect and spent a lot of time with the farmers' wives. I even got married there, and my Bavarian husband and mother-in-law introduced me to many of the local wild foods – mushrooms and herbs. At one stage I was collecting over two hundred medicinal herbs from the wild. I even got to meet an old traditional herb woman, a village crone, who shared a lot of her information with me and who was so excited to meet a young person who was actually interested. I kept coming across incredible old people in their seventies and eighties who had an amazing wealth of knowledge and skills. They would burst into tears when they found somebody young who was interested in what they were doing, thinking that all their knowledge and skills would die with them because they hadn't found anybody young to pass them on to. There was a beautiful old fellow who used to make birch brooms and grew the biggest leeks and cabbages I've ever seen in my life. He'd talk to me for hours about the art of making birch brooms and the fact that he'd never had anybody to teach his craft to. All the young people were going to university and were just not interested.

I collected wild herbs and identified them using my Collin's *Field Guide to the Wild Flowers of Northern Europe* (Fitter et al, 1983). Then I'd ask the farmers' wives their local names. I was constantly coming across plants that they hadn't seen since they were children, mainly because I was getting into little hidden out-of-the-way places where these plants had survived, next to the creek or the few square metres that had been missed by the mowing or farm crops. I began to be really concerned that there was a mass extinction happening in the herbaceous layer. And there was also the 1970s oil crisis while I was living there. I began to feel a real sense of urgency that things needed to change, so that's why I believe that permaculture was just a natural next step for me.

My first encounter with permaculture took place when I returned with my husband to Australia in 1977. We were coming back to get some land because I wanted to create a farm that was a botanical garden of useful plants and organisms planted in a way that recreated their natural environment. I'd also learned a lot about traditional farming systems through my travels in the Middle East, India and the tropics. We returned to Australia to pursue our vision of integrating my observations in nature with the wisdom of traditional systems. My husband and I purchased some land up on the mid north NSW coast near Wauchope, and set up a herb farm. We did all the usual 'back to the land' stuff, putting in dams and orchards and gardens, keeping chickens, doing some market gardening, building a place to live, and raising a family.

Not long after we returned to Australia there was an organic festival just out of Sydney at Upper Colo. I went along to reconnect with what was happening in Australia. The Henry Doubleday people were there with their seed bank, and also Esther Deans, of no-dig garden fame. And then there was this bloke from Tasmania, Bill Mollison, talking about his new system – permaculture – and promoting *Permaculture One*, which was to be published the following year. My Inverell High School education had omitted to teach me words like 'ecology', 'ecosystem' and 'symbiosis', so I didn't have names for the things I'd been observing. Bill really blew me out because he was talking about all the stuff that I was thinking about and what's more, it had a name and a methodology. I couldn't wait for the book to come out. Then I found out about the *Permaculture International Journal* and subscribed to that too.

I was at Wauchope for six years. A couple of my neighbours had done one of the early permaculture courses with Bill, and it was great seeing how they turned their places around using permaculture design. So in 1983 I also did the PDC. I'd also been motivated because I found myself surrounded by the 'back to the land' movement of the '70s, with people moving out of the city having wonderful ideals but absolutely no idea, no knowledge or skills to actually make it work. I wanted to be able to teach and work with people designing their properties and empower them with information, getting them off to a good start so that they could actually realise their vision, so their dream wouldn't turn into a nightmare.

The PDC I did was actually the first women's course, up at Tyalgum on Lea Harrison's property, taught by three women – Lea Harrison, Susie Edwards from the Southern Highlands, and Judith Thurley from Canberra. I found it somewhat bizarre that there were several women who knitted their way through the course and didn't write down one word. Maybe they go back to a particular scarf and come across a certain colour, and say, "Oh yes, that's when we were talking about...!" Otherwise it was just a standard PDC except that there were no men other than a couple of partners supporting the women and providing for children. This was excellent – one of the biggest impediments for women doing PDCs was a lack of access to childcare. They had tried to really keep the costs down, so we were all camping or accommodated on camp beds in the cow bales, with a large tent for the classroom. It poured with rain for most of the two weeks and nearly everybody got sick. I was in a camper van so I was a bit separate and was one of the few who managed to get through without getting sick.

A few years later I taught on a women's PDC in the States that was much the same except that they had one session on women in permaculture, facilitated by a local woman. It ended up being a one and a half hour bitch session about men causing all the problems of the planet, and how if women were running the planet everything would be much better. I said, "Well yes, like Margaret Thatcher," and I got some very black looks! I actually found it really insulting to the male friends and colleagues that have supported me in my journey. I would not be here if I had not been encouraged by my male peers. They gave me a lot more support in those early years than any of my female friends, which is interesting because often people think it's the other way around. So I spoke up and said, "Look, I've got to defend my brothers here."

From my experience as a permaculture teacher and designer, I think people are attracted to permaculture for all sorts of reasons. A lot are motivated because they just want to get their own life in order. Probably fifty percent of people would be coming into courses because they want to apply it in their own lives in some shape or form, live more sustainably and become more self-reliant. They want to reduce their footprint. Whether they're urban or rural, many are looking to buy, or have just bought land. The other fifty percent are really there because they want to pursue permaculture as a possible career path or as a way of augmenting an existing skill set or profession, in a more sustainable way. These are the two key motivators, but I've had people come for all sorts of other reasons as well. There was a psychiatrist from Brisbane who did a PDC with me because his wife was pregnant. He found that working with depressed people had given him a very black outlook on the world, and he wanted a more positive view on the world if he was going to nurture a new human life.

Permaculture also attracts activists, especially people who are 'burnt out' from protest – Greenpeace burnouts who just really want to do something constructive. They've done their time as warriors. I think it's important that we do have the warriors, I don't devalue that at all, but you can't fight forever. You've got to take breaks, you've got to nurture yourself, and you've got to build as well as fight against. Permaculture is very attractive in that respect.

I'm passionate about being engaged with my community, I see myself as a part of a greater community, a bioregion. For me that's a really critical thing. I'm concerned about my community's wellbeing, its future, so I've always been involved in community building activities like organising local markets at Rolands Plains and community forums here in Nimbin. I really feel blessed and privileged to live in the community that I am in, because it's very proactive. I've given presentations about my 'eco-neighbourhood' over in the States. They were completely blown out. Even the Transition Town people say, "Actually you're light years ahead of us." There isn't much documented here under the banner of 'Transition Towns' but there have been a lot of initiatives over the fifteen years I've been living here that would come under that general banner. I've facilitated a number of creative participatory planning processes for the community. More recently I've been involved in a working group looking at Nimbin's food security.

Other projects are to get Nimbin plastic bag free and the solar purchase group. We had fifty households who came together and got a special deal from Rainbow Power Company to install solar systems on their roofs to obtain the government rebate. Back in the nineties the community got together and in 18 months raised $140,000 to purchase the old school as a community centre. I could talk for days about the wealth of community initiatives that have happened here that I've been privileged to be a part of, as well as the things that preceded my involvement here, and plans for the future.

So there's the wider community but there's also the intentional community movement. I've been working in that since the '70s. Back in about 1978 the multiple occupancy issue emerged in NSW and I was involved up till the mid '80s on this. Then, the early intentional communities were all illegal till the state government realised it needed legislation to enable and regulate this form of rural settlement. I was involved in seminars and conferences leading to the state planning policy for rural land-sharing communities, as well as lobbying state government to create Community Title as a preferred form of land tenure (like rural strata title), which came through in 1989. I have been involved with many communities, not only as a designer but workshopping with them on governance issues and the physical environmental management and maintenance of their land. The design of 'Jarlanbah', NSW's first rural Community Title, has been one of my major achievements. I broke a lot of ground with this, which has influenced other developments, council planning ordinances and State Planning guidelines for rural settlements. A central part of this arises from the Eco-Social Matrix (ESM) tools I developed for human settlement and bioregional planning. These process tools enable people to identify patterns of human settlement and their relationships to watersheds, land, networks and regional demographics.

Permaculture has certainly given me a whole new framework to filter my view of the world and information through. Something that comes to mind is that though I spent 17 years as a vegetarian, it was through permaculture and thinking ecologically and holistically, and especially working with indigenous communities, that I started to question my dogmas around diet. I'm now a 'selective omnivore'. That was a major change for me at that particular time. Permaculture thinking is subtle in so many ways and I'm constantly coming up with new connections and new realisations. The magic of permaculture is that as your mind thinks more laterally, and as you become a more acute observer, you start to see connections, relationships and interrelationships, causes and effects, that you were oblivious to before. That is a constant, ongoing process and refinement.

Another area of my life influenced by permaculture was stimulated by teaching about patterns in nature and patterns in communication in traditional societies – how traditional and indigenous societies used the patterns of symbol, song, and dance to communicate vast areas of knowledge about the environment and survival, all deeply rooted in the seasonal cycles and patterns in nature. When I was teaching this subject I used examples from all sorts of different cultures such as North American Indians, Solomon Islanders and Australian Aboriginals. Then one day I thought, "Well, what about my own roots? Surely if I go back far enough, I must have a heritage like this too?" That motivated me to research my Celtic roots. I was pretty amazed with what I found and how earth/nature-based spirituality pervaded the practical and philosophical realms.

In my teens I'd checked out all the different 'isms' but never felt inclined to join any particular one. Beyond the dogmas, I found they all had a basic common ground embodying a sense of connection with the universe and something that is greater than oneself, along with principles or values for right living and relationships with others. My previous searches had not yielded any indigenous earth-based spiritualities, sometimes referred to as animism. I was struck by the incredible richness of this heritage and the fact that it's been so denied us. I realised that what the European settlers did to Aboriginal culture, the Roman political and religious empire had done to my Celtic ancestors and their culture. Celtic culture is very rich in philosophy, culture, poetry and music. What I really love is the eight-fold year, the eight stages of the year and the seasonal festivals and celebrations. This is really common in so many indigenous cultures. There's such a strong connection to nature's rhythms and cycles. But in our society we've completely divorced ourselves from nature, so it's a tremendously healing and empowering thing to discover, and it stimulated me to create my Celtic eight-fold year seasonal calendar and those wisdoms, and transpose them on to our southern hemisphere to live within the Australian landscape.

This project opened up communication with some of the local Aboriginal elders and the Aboriginal community. Working with Aboriginal communities also had a huge impact on me; very moving on so many levels. As one of my Aboriginal friends, Burri Jerome, said, "The problem with most whitefellas is that they have got no idea of their own dreaming. They've got this big cultural vacuum inside, and they think that they're just going to suck up our culture to fill that void. They'll never understand it unless they know their own dreaming first." He said, "Robyn, you've got an appreciation for your own dreaming, so you can understand ours." They've shared a lot of beautiful stuff with me, and I feel incredibly enriched by this. But I don't think I would have ever gone down that path if it hadn't been for teaching permaculture and patterns in nature and human culture, and seeing a lot of the wonderful things in the pattern chapter in the *Designers' Manual*. This gave me a really beautiful framework to bring together practicality and spirituality. It was that earth-based spirituality; being one of a whole universe, the whole planet and a whole system; seeing and recognising the life in everything. Mountains grow, they just have a different time frame and sentience to us.

I also really appreciate the times I had with Bill Mollison in the mid '80s when I ran the permaculture centre in Enmore, Sydney. Whenever Bill went overseas to teach PDCs he'd stay with us for a few days to get through his jet-lag and print his photos. He was writing the *Designers' Manual* at the time, and we had many long conversations about patterns and other things until three or four in the morning. When I open up the different chapters in the *Manual* I remember, "Oh, that was that wonderful rave that we had for six hours after we ate quail in the Turkish restaurant down in King St." Bill used to love that Turkish restaurant and their quails!

Up until about 18 months ago, I was very optimistic about the future. I still try to be, and believe I have an obligation to be, but I am very, very concerned about the tardiness of our society to take what's happening seriously and do what needs to be done. The impact of climate change – we've only got one chance, we don't get a second chance at getting this right. I think back to 1987-88 when I was writing articles about climate change and global warming in *Permaculture International Journal (PIJ)*. Now it's going so incredibly fast. I've followed the signs, I'm a bit of a news/current affairs junkie in terms of knowing and following what's happening with the world, the science of our planet. I think it's really important to know what's going on out there. I come across many people who are ostriches. They know things are bad but they don't really get the science – maybe they find it too depressing or maybe it confronts their comfort zones because they're not quite doing enough. But it's just so important to be informed. As the druids said, "There is responsibility that comes with information." You have to respond to it and you have to share it. So when I feel overwhelmed, all I can do is do what I can do as an individual. At least I can look my children in the eyes and say, "I did what I could." Although sometimes I give myself a hard time that maybe I should be doing more. I've only got so many hours in a day, so maybe I need to find ways to use them more effectively – I don't know.

My students keep me going as I see what they're doing, the positive impacts that they are having on the world. I am honoured to have been part of that pivotal process that set them on their path and to have their thanks to me for that. And I honour their work and commitment through my own ongoing work and commitment. I feel if I fail, I'm letting them down, so even if I do feel depressed, I can't succumb to it because I've got a responsibility to more than myself. I find it always happens that when I feel down, I'll get an email or a little message from somebody saying, "Look, thank you so much – you've inspired me or empowered me." And then they say what they've been doing and I just get this huge warm rush, and think, "How silly of me to have felt like that."

There's another side of me that uses music and nature to replenish myself, because I have burnt out a few times. I've found during those burn-out times that music is a very important part of my healing process. It's one of the ways I am able to create some degree of balance in my life. I probably need more time for music to balance myself a little better because I do have a tendency to 'workaholism'. My music is an important part of processing my inner stuff, and just being in nature and being surrounded by such incredible natural beauty is as well.

It's a pity that 'permies' don't tend to be asked to comment in the media on what's going on regarding sustainability. What's mainly managed to get into the media is the garden side of things rather than the big picture. It's very difficult to sell such a complex concept such as permaculture, and it's always been something that we've talked about – how do we actually market permaculture to do it justice? In 2005, the UN announced the Decade of Education for Sustainable Development. I was reading an article about the launch and one of the things that had them stumped is the same dilemma – that sustainability is so huge and encompasses so much, that it's very difficult to put it into a nutshell that can be easily marketed. It's a very complex system and people like really simple things.

We've got to learn to think in such a totally different way. Our whole society is conditioned to a certain mindset. We must, on a very large level, rethink the priorities that we make and what informs our decision-making. We must move beyond the selfish gene, from 'me' to 'we' thinking, that's a very powerful thing. I really like what eco-psychiatrist Joanna Macy says about the dichotomy and the dynamics between the inner and the outer world. There's got to be a healing of that rift, because people are so disconnected from the natural world. One of the things I find with the PDC is that people get unexpected bonuses out of it, things they weren't actually looking for. They may have their particular motivation for doing the course, but then they find that other things happen to them in terms of their way of thinking, their fulfilment, the healing of the rift of their separation from nature.

I've been incorporating a lot of that into my work and it resonates with all the Celtic stuff too, it's really interesting. I wish I could take three years off from my work to write. There's just so much that I've been processing and accumulating through my journey and it's hard not having time to put it all into formats that can be shared, apart from verbally in my courses. But even the courses don't give me long enough for a lot of those things.

I'd like the future to be a cross between Cuba and Djanbung Gardens, or between Cuba and Nimbin. Cuba was really amazing. I have been there twice and found a lot of challenging food for thought. It's a powerful reminder of how mindlessly affluent people have become, even conscious people in western society. Affluence now is just so much greater than it was in the 1970s and '80s; so much has changed; people's expectations, the amount of stuff, the short life cycle of things, the amount that's thrown away, the mindless consumption which is still accelerating. Cuba provided a stark contrast. It's amazing to be in a place with no real estate agents and no billboards. It's a relief to not be bombarded with advertising everywhere. The Cubans live so simply and frugally. If we really want to look at living within our planetary footprint, we need to live more like Cubans. And that means cutting back on an awful lot of stuff and living incredibly simply.

Cuba is still very vulnerable and has a lot of dry areas and a lot of poor soil. It's not all lush and wet and green. It's not perfect and I think there are many ways that the quality of life in Cuba could be improved without increasing their footprint. That's why I say my idea of a perfect world that's actually sustainable would be a cross between Cuba and Nimbin.

One of the things that is really missing in Cuba is a local economy with small enterprise and cottage industries – apart from the hidden black economy. That's a layer that I would just love to see there, it would make a huge difference. Here in Australia we need to work on food security a lot more. Where I live we're in an incredibly fertile area where we could be growing so much more of our own food, but everybody's still caught in the economic debt cycle, and it's very difficult to break out of that treadmill because you can't live without money. That seems to be a big obstacle for a lot of people, being able to earn a living income from meaningful or productive work.

Cubans can't really go into debt very much and they don't have credit cards. They have a phenomenal 85% home ownership and speculation on housing is illegal. If people buy a new house, it's from the government and their repayments are no more than ten percent of their income. They don't sell but swap houses. There's a whole little plaza in Havana where people come on the weekends. If, say, a couple is divorcing, they're looking for a couple that have just married who are looking for something to live in together. They might have their two single units on the market that the divorcing couple might be interested in. All they do is just talk to each other and check the houses out if they're interested. Then they just shake hands on a deal and go to the registration office to register the change of ownership.

Whether that could ever happen in Australia depends on how far things crash. When you look at what happened in the States with the sub-prime mortgage crisis, wouldn't it make sense for the government to simply buy out those properties? Then the government would own them and people can continue to live in their own homes and still pay them off, this time buying them from the government, but at a reasonable rental.

The biggest problem is greed. The greed factor is *Homo sapiens*' undoing. Once we start talking about sustainability we've got to start talking about zero growth economies, where we're just making enough. We cover our costs, and then we just make enough to survive. We've got to look at survival wages. We've got to look at paying real prices for our food. We've got to go back to a far more simple, less consumptive, lifestyle. And that involves a huge change in thinking. It was really interesting in Cuba that so many things there provoked flash-backs to my childhood in the fifties and early sixties, in terms of the simplicity of life, and the lack of stuff and superfluous appliances in the houses. And especially the sense of community that people have with their neighbours, the conviviality.

This would require a huge change in mindset, but changes can happen very quickly. A lot of it really comes back to what people are sold by different vested interests who are selling their mindset to people, so they'll buy and consume more and more. People are being actively marketed to, about how they should think. Some people find that aspect of permaculture really confronting, and it can be very difficult for some students in a course because they just don't like their worldview being challenged. Some try to challenge back, or rationalise or debate, but that in itself is often a healthy precursor to change.

One of the strengths of permaculture is that it provides an environment where people, through their gardening or through their work in community using a permaculture approach, can begin to explore, in a safe way, changes in worldview and changes in the way they understand things. What a lot of people really appreciate about permaculture is that it's more than just a garden system, it's something you can apply to your personal life, to your professional life, to your community interactions. It's not just dogmas and techniques.

Our world is such an incredibly special and beautiful planet. I simply want to do absolutely everything I can to ensure that it continues. For me, part of that is working with other people. I'm motivated by concern, but that comes from a very, very deep love, the love of life, the love of nature, the love of the incredible diversity of plant and animal and fungal life, and all the rest of it that we share this place with. It's not only a huge responsibility, it's an incredibly rewarding experience to make that paradigm shift to become a steward, a custodian and a nurturer. We can all be afraid of what the future scenarios might be out there, but I'm more motivated by love than by fear – that's really important to me on a deep personal level.

One of the things I find really special about permaculture is that it heals the chasm between head and heart and hands. Head, heart and hands can be working in tandem. We live in a very schizophrenic society which separates all these things out, so it's beautiful to find something where you can bring them all together in a work of love, and something that rewards you emotionally and spiritually and intellectually and also physically. I've been living it for the past twenty-five years and that's what keeps me there. It is a whole of life, a whole of culture concept. It's something that pervades absolutely every aspect of living and working. What more could you want?

Max Lindegger

Max Lindegger is an internationally respected and sought-after teacher and designer of ecological communities and sustainable systems. His reputation is born of 20 years of hands-on experience and leadership in the design and implementation of practical, workable solutions to the challenges of sustainability.

As the creator and director of the Oceania/Asia Secretariat of the Global Eco-village Network, Max participates in and contributes to the international flow of current thinking and best practice in the fields of sustainable systems design and education.

A key element of Max's design philosophy is the recognition that sustainable systems must first and foremost fit and tread lightly on the environment in which they are placed. He gives primacy to the importance of sensitivity to local custom and culture, utilisation of local resources, ongoing benefit to the community, and positive impact on the local environment. Max was recently awarded the Australian Prime Minister's Centenary Medal for distinguished achievement in the field of developing sustainable communities.

Chapter 5 | Walking the Talk

Growing Up

I grew up in Switzerland in a farming community, a rural environment surrounded by small farms. I have always had a love of the land and the environment, it was not something that just started one day, it was something I grew up with. But my parents were not farmers though. My father worked in a factory. My mother was of Italian background and they definitely were not well off. We ate simple but exceptionally good food because my mother was an exceptionally good cook. So we all had enough to eat, but while we weren't poor, we had to be frugal. It was a very safe environment too, and my mother never had to worry about where I was. When I came home she knew where I'd been by the way I smelled. I grew up with people who were mentors, people who had high ethical values and who genuinely loved children.

We children didn't get pocket money. If we wanted to go to a fun event, we had to earn the dollar. We chopped wood, collected kindling, milked cows, worked the garden, looked after chickens, collected wild herbs and flowers and took them to the market, that's how I became economically viable. I didn't have to do a course, I was lucky enough to learn it as part of growing up, just like you learn to use a knife and fork.

From about the age of five or six, my favourite farmer 'adopted' me. His wife had committed suicide and he had no children, so he was very lonely. I probably didn't realise this at the time, but we were good company for each other. To me he seemed like an old man but he would only have been about 50. He showed me how to use a sickle and how to milk a cow, when to pick apples and how to make cider. Although I did quite well at school I didn't want to continue. Instead, I wanted to become a farmer. Even though there were no farmers in my family, farming seemed to be in my blood. My mother was the youngest of 16, my father was in the middle of 13, but none of them as far as I know ever had anything to do with farming. But they all had practical skills, which I also picked up from all those uncles and aunties and substitute uncles and aunties.

My father said to me, "Look, our family owns no land or money, so you always will have to be working for somebody else." Getting another base was good advice, so I studied mechanical engineering design for four years.

When I came to Australia in the late '60s, I couldn't speak any English, had no friends and virtually no money. For the first few weeks I lived off cheese on toast because that was the only thing I could actually ask for! Then I realised that spaghetti was the same in English as it was in German, even though it was a different type of spaghetti. I very quickly realised that if you wanted to eat good food, you had to grow it yourself. Back then it was a pretty selfish attraction, it wasn't a matter of saving the world.

My First Contact with Permaculture

It was about 35 years ago when I first heard about permaculture. I was reading a one page article by Bill Mollison – maybe it was called 'Permaculture!' – in the *Organic Gardening* magazine in Tasmania. A friend of mine showed it to me and when I read it I just felt that this was amazing, it was a sort of a 'Eureka!' moment. I was already a member of the campaign against nuclear power and things like that, but those were against something, whereas permaculture was actually *for* something. So I contacted Bill (in those days he would still respond to letters), and organised a speaking tour for him in South East Queensland. We had workshops in Nambour and a talk at the Botanical Gardens auditorium in Brisbane. It was amazing. This was the time of Bjelke Petersen (the State Premier at the time) and the electrician's strike. There had been frequent brown-outs and we were told there'd be one at the time Bill would be talking. We contacted the radio stations and asked them to get the message out that Bill Mollison's talk was still on, but people should bring candles or kerosene lamps. So here was Bill, standing at the front between two kerosene lamps with people holding candles. It was absolutely amazing.

Then there was a bloke on ABC radio, one of those talkback hosts whose name I can't remember. I took Bill there and sat outside waiting. It turned out to be the longest ever one-to-one interview and people could also ring in. The bloke got the sack over it because he overstepped the mark. But the impact was incredible. The people who were on brown-outs couldn't receive the ABC so were ringing in with the same questions. And because the radio would suddenly cut out, people actually believed that the government was censoring the interview! We all were on an incredible high because of the response we got. There are moments in our lives when things like that happen and we need to grasp the opportunity and run with it. We definitely did that. Some of the statements Bill Mollison made I knew were actually not quite factual, but it didn't really matter. I think people needed to hear that we needed to do something different, that we couldn't continue as we were. It's pretty much the same message I hear again now about peak oil and about climate change. It's nothing new. Over the years we've had a few different events which we thought were potentially critical. Thirty years ago it was acid rain. Who talks about acid rain now?

Teaching Permaculture

In 1979, Bill invited me to do the first permaculture course, and I did a second one in Buchan (Victoria) in 1980. After that I chucked in my engineering job and began co-teaching with Lea Harrison, Gahan Gillfedder and Bill Peak at the Nambour Police Citizen's Youth Club. It was sort of the beginning. I've now taught permaculture courses in around 26 different countries. Quite a number were a first for those countries – the first permaculture courses in England, Demark, Norway, Sweden, Macau and the Philippines. I've now taught permaculture courses throughout Europe, Australia, New Zealand, South America, North America and Asia. Over a period of eight years, Lea Harrison and I also taught the first teaching methodology courses for permaculture teachers in Australia, New Zealand, the USA and Europe. I've done design implementation, contract tree planting and fruit tree maintenance.

As well, I developed a curriculum for advanced permaculture courses which focussed more on design. With my engineering background, the technical and design aspects were my forté. I remember the reaction of one student in particular. She was 24 and all her life she had followed an ecological and environmental direction. It was the last evening of the course and she didn't look happy. When I asked her what was wrong she said, "I've just realised that just about all my school life, I've been cheated all the time. This is the first time the teacher has lived what he's talking about. For my other teachers it was just something from a book." I realised then that we have to be aware that we need not to be afraid to have dirt under our fingernails, that we should teach from experience, and how important it is to 'walk your talk'.

Observation and Good Design: the Keys to Successful Permaculture

I've been around long enough to be able to see some successful examples of good design that show it really works. It may take time, but you really begin to see results, sustainable results in the truest sense. What defines a successful permaculture venture is keeping it simple. Often people make designs which are much too complex, and these days I go for under-designing rather than over-designing. If we're designing a fifty-acre property, we don't have to design all fifty acres, we have to design the key elements on that property. Look for small steps rather than big impact. That's more important than using large machines. I'm not against using large machines as such, we have used them at Crystal Waters with great success. But it's important to take the time to understand the energy flows on the land, not just in the context of sun and wind sectors, but also on a spiritual level; to understand, to learn to love and respect the land.

To me, the key part of design is observation. The farmers I grew up with were exceptionally good observers. Today I still do exercises in observation in every permaculture design course, wherever I am. I usually start with some slides that I took in Sweden. They're nothing out of the ordinary until you have a closer look. I ask the students to look at them for a minute, then turn the show off. As well as asking what type of tree is in the slide, I ask them in which month of the year did I take this photo? I think what permaculture has taught and still teaches me now, is to look for the unexpected, for things which are not 'in your face'. When I go to a property, I might say, "Look at that slope over there – has it been burned?" I can see by the bracken fern and blades of grass growing that fire has occurred. And the type of grass and weeds will also show you what type of farming practices have been used there. Looking in the 'now' allows you to tell both the history and the future of the land. I show a slide I took in Norway of land I consulted on, which I felt had been treated very badly. I went back two years later and I took another photo. What I felt was would happen actually had occurred in a matter of only two years.

If you start by observing it allows you not to make big mistakes, or to prevent other people from making mistakes which could have quite disastrous effects. I also work in commercial situations, where the clients don't give you the luxury of spending days wandering around. But after a while you become quite efficient at reading the landscape and understanding what is going on. I think pointing out the importance of observation has been one of Bill Mollison's real strengths.

The Ecovillage Movement

The other thing in permaculture that was important to me was the potential of doing it in a village. In the small village I grew up in we all knew each other, and there was support whenever we needed it. It wasn't the ideal picturesque village but it had a social aspect to it that a lot of people are really missing now. After we'd been in Australia for a while, three years in Sydney having a child and watching her grow up, I realised that as a kid I had been really

Max with the co-designers of Crystal Waters at the 20th Anniversary. (left to right) Rob Tap, Geoff Young, Max Lindegger, Barry Goodman and Bob Sample, the previous owner of the property before it because Crystal Waters Permaculture Village.

lucky. The way cities and suburbia are going, being able to grow up in an environment as good and as safe as I had is not something we can guarantee for children any more.

Having a big backyard and farms around us is probably what attracted me to the ecovillage movement, and I became involved in the design and implementation of Crystal Waters. I think Crystal Waters is probably one of permaculture's real successes. We now produce more food on our lot than we need, and Crystal Waters is also a fantastic place for children to grow up in. Living in an ecovillage is for me very important; that aspect of sharing with other people without living in each others' homes. I've now worked in the ecovillage aspect of permaculture in over 50 different countries. It has taken me around the world and I have been lucky in that regard. My footprint would be very modest if it wasn't for my travelling. That is something I struggle with all the time.

Reflections on Permaculture

Permaculture has a very positive message, and there are not many positive messages out there. Most of them are about disaster. Permaculture deals with very basic human needs, so the take-up rate is probably higher in poor countries than in rich ones. In countries like India, Bangladesh or Pakistan and other places where I've been, you find that people pick it up quickly and before you know it, it has spread. They don't take it as a full package, they just take what is really needed for them at that point. Probably the best implemented design I've seen was in Bangladesh. I show this in my courses because the picture we have of Bangladesh is of starving people and children with big bellies. And there is definitely plenty of that in Bangladesh. But by grasping the concept of permaculture and having a local person to guide them through it, they have achieved amazing stuff. They are not starving. They are not rich. Their cash income is still only a couple of dollars a day, but they're wealthy by being healthy, and having good, clean housing. They deal with their own waste and have clean water. Permaculture really does touch people at the core.

We have to allow people to take what they want at that moment from permaculture, and not feel that they have to take the whole package or go away. Maybe it's the gardening part, or maybe more the energy or the social aspect, because that is where people are at that moment. I remember the first permaculture course I was co-teaching where we had a session called 'Keyline adapted to permaculture'. We got a phone call from P. A. Yeomans, the initiator of the keyline system, who said, "No-one should be adapting anything from keyline. Take it or leave it." And so people left keyline. You don't see much keyline implemented anywhere.

It's never been our aim to do everything and I like to be reasonably normal. I used to feel that the people would come to our place here and think, "Oh this pioneering permaculture teacher, he's got it all together." People come here and ask if we have our own fibre. We used to have some sheep, but they got killed by dingos. We've got some wool and have done some spinning, but it makes bloody hot underpants! It's the same with ecovillages. I really believe that if we want to make a contribution, we need to be ready for compromise. We need to be able to mainstream our ideas. Our designs have to be both productive and beautiful. They also have to be affordable, not just financially but time-wise too. How many times do you hear people say 'the designer becomes the recliner'? Bill Mollison made people believe that permaculture was a no-work way to grow food, and a lot of people found that really attractive. But it's not like that at all. All the work is in the planning, then the initial implementation and then steady, a little bit every day, to keep things ticking over.

We've had seven very, very dry years where I lost a large percentage of my citrus orchard. Robin Clayfield said she'd lost just about all of hers too. That's the way we can talk as mature permaculturists. It's not about trying to apologise if don't grow fruit or grain. We try to do what we can do, but it's not a life sentence. It's not something we *have* to do, it's a choice we make.

Empowering the Young

With young people, being able to give them a purpose for learning is what I'm looking for rather than 'teaching'. I've been working a lot with young people trying to experience what Lea Harrison called 'The High'. That's when you see something which you've seen all your life, but you've actually never seen it. That is 'The High'. Or when you teach and hear yourself saying something you didn't know you knew. Sometimes I just wander around with the students and tell them what I see. You don't have to prepare for it, you just go out in the landscape and tell people what you see with your own eyes. I realised very early on that for a lot of people, this is quite amazing. They can't believe their eyes, they look but they don't actually see. But I think it's something you can learn. Once you start looking at the landscape with permaculture eyes, your eyes really open up, it's a real hook. You see different things, you see the interconnectedness of it all. You see the problems but you see the solutions as well. I would like to see my young students experience that for themselves.

We need to slow down. We need to pull out the MP3 earphones, we need to put our mobile phones away and create a space where we can have experience which is probably not possible through just about any other means. These days that's more important than the technical aspects. My grandson helps me strip down and paint my beehives. I love working with bees with my grandson. He's eleven, so of course he talks the whole time, mainly when you do something as exciting such as re-queening. We also pick kaffir lime leaves with big thorns, which is a mongrel job. I reward him for doing that, even financially. There are certain needs he has. I maybe don't agree with them, but I don't want him not to have an MP3 player. I want him to see for himself if he wants it or not, and I also want him to see what the value of work is. It's not just someone giving him money for nothing, he has to earn it. That is the value that I would like to instil.

It's important for young people to realise how much they actually know, and to trust themselves a bit more. With all the people I've employed over the years and with the people I work with, I explain about the combination of authority and responsibility. If I give people responsibility, at the same time I give them authority. This authority includes being able to make a mistake. I use permaculture and ecovillage courses to introduce the responsibility/authority aspect to people, and I'm hoping it rubs off. We get a lot of Korean and Japanese students who have little confidence and limited ability to make decisions because they haven't grown up in an environment where they could learn from small successes but also from small mistakes.

These students, aged from about 18 to 23, usually come to Crystal Waters for around 20 days. We call their course 'Introduction to Permaculture and Cultural Exchange'. On the second day we start with a poorly maintained, overgrown lawn area and build a mulched garden. The students go and collect the mulch and get involved in the recycling of organic waste by making compost. Then on the second day we make a garden using fast-growing seedlings like pak choi, tatsoi and lettuce. They come dressed up like they're going on safari! They wear hats and white gloves – they are protected against lions and tigers and fire and floods! I don't know what they expect when they go into the garden, but they carry the compost and try to plant seedlings with their gloves on. I don't tell them to take them off. Every time they put a seedling in and withdraw their fingers, the seedling gets stuck on the glove and they pull it out again. Then I ask them, "Have you tried it without gloves?" They look at me in disbelief. And I say, "I don't wear gloves, it's okay, it's safe." Because I tell them it's safe they believe me, they put the seedling in, and then they see this thing growing.

The students have to look after the garden, water and fertilise it with a liquid manure tea, make sure the pests are kept away and the mulch is kept up. Seventeen days later, the evening before they go home, they harvest a meal. It just blows them away. They didn't know how food is grown, they didn't know that they could do it themselves. And they didn't know that it could only take 17 days. For the first time in their lives, they eat food out of a garden that they've had a part in growing. Being able to plant is important. It's not work, it's meditation. Eighty percent of the food in Japan and Korea is imported, while more than 60% of the farmland is abandoned. Nobody wants to farm any more.

Seeing the students' faces when I hold the ladder while they pick oranges for the first time and turn it into juice, I don't think they could be any happier or more joyful if I gave them the latest MP3 player. I take them on night walks where they have to walk in silence and turn off their torch. It's pushing their limits because they've never been in the dark in their lives before. They drink rainwater and use composting toilets. It's all experiential learning. Permaculture fosters that, but I don't think it's specific to permaculture. In our next course we have nine students, they will be here for four months. They're from Russia, Korea, South Africa, Turkey, Portugal, Australia, UK and the USA. Nine people, eight nationalities. It will be amazing to work with these young people, not so much to 'teach', more to help them learn from each other and by being here.

But we also have some students in permaculture who are not willing to put in the hard yards, to actually do the physical, repetitive, sweaty, dirty work. I think quite a lot of permaculturists are basically lazy. They are not willing to go out and do things day after day, to produce food for twelve months of the year, and learn from that process. A lot of people in permaculture still feel that the more books you read, the more you will know. But books are only part of the equation. I actually also believe that some of the people who are teaching permaculture should not be allowed to teach. They can do damage and I have seen that damage, especially if they go into the countries of the global South, where you haven't got the luxury of making errors. We need to make the errors at home, and learn from our mistakes and successes here, so we won't make them when we get to Haiti or Bangladesh.

Fifteen years ago I pretty much removed myself from the permaculture scene because I could see there was considerable in-fighting and arguments on issues which were not important to me. So rather than trying to argue back, I stepped out of the permaculture organisational aspect altogether.

Working with the Mainstream

In rich countries like Australia there are not enough good permaculture examples and I feel there's a lot more talk than actual doing. We need examples which are productive and look interesting, but also appearing to be under control. I once went to a workshop at the property of a well-known Australian permaculturist who I admire greatly. On the way back in the train people were saying, "Look, if this is what it's going to look like, I don't want to be part of it. We can see the abundance, but it's so messy." So I think we need to compromise a bit. Even if there isn't really, people need to feel like there's some order. And there are definitely some permaculturists who do go out there for shock value. But we need to understand where society is at, and how we can help move to the next step. For a while I had WWOOFers from another permaculture farm. They'd say that at this farm they do this and they're self-sufficient in that. I thought "Gee, this must be an absolutely incredible place, I'll have to go see it sometime." But when I asked what the people did for a living, I found that they're unemployed. I told them I have to earn a living and I haven't got the time to do just work on the farm. I had a mortgage and kids growing up and needing things.

I also find quite a number of the young WWOOFers who come to me are not able to work a full day because they have never had that work discipline. When these same people talk about disasters that are going to happen, for them playing an active part in saving the world will mean a full day's work.

Saving the world seems to be the big issue for many people, but when I read about things like the two Sea Shepherd crew who were held on a whaling boat, what they do makes me feel totally frustrated as far as public relations are concerned. It is close to piracy, so people will say, "No, I'm not going to go there." So while people might be against whaling, and I agree with them on many levels, when it starts to get extreme, people step back. I find that very much in Asia. I've been working at government level in the Philippines. To do that I've got my hair short and have even shaved my beard off, because in Asia a beard is not really acceptable. Because I'm willing to do that I now have entry at a government level. It took me a long time to get the sort of acceptance

I've got now in such places, and when you can work with these people you really start to see the possibilities. For example, in Sri Lanka I was invited to work on a government policy paper post-tsunami. They took our suggestions word for word and put them into government policy. We have now worked on other post-tsunami projects and I will be going back again. A village has been built which might not be perfect but 53 families are living there.

It's very easy to question, and in permaculture people are very quick to ask "Why do you do this? Is this permaculture?" I say, "Well, I don't really know, you tell me." But in that village, there were people who were wiped out, who lost friends, relatives, fathers, mothers, children. They had nowhere to live. Within a short time, they were living in small, modest houses that they are comfortable with. They have small photovoltaic panels for a couple of light globes, a radio and a black and white television. They have clean latrines, clean, safe drinking water, and rainwater collection off the roofs. There are food gardens in many of the houses, they're using fruit trees as street trees or for shade, and there's some small-scale commercial enterprises. They have a community centre and a spiritual purpose. In a way, all their wants are met. To me it's quite amazing that it happened so quickly. It's also something which is replicable as well, unlike most permaculture systems which need you or me to maintain them. If you walk away, they usually collapse.

In Australia it's more complex and takes a lot longer than in places like Sri Lanka. I gave a talk a couple of years back in Perth, and there were two architects in the audience. I mentioned something Clive Blazey of Digger's Gardens had said, that for a family of three you could produce most of your leaf vegetables and more all year round on 42 square metres. We've done trials here as well as in Bangladesh and he's right. I showed photos of my garden. The architects said, "Well, we design units and mass housing in the suburbs of Perth, and from now on we'll include a five square-metre garden in each unit as part of the structure." That will make a huge difference.

Even with this amazing drought in Australia, only about 10% of houses have rainwater tanks. I don't necessarily think legislation is the quickest way to fix this, but governments are learning from the carrot and stick approach. Giving people the opportunity to adjust to the stick is important. The sort of policy I would be looking for would be saying, "Look, we require any house, existing or new, to install a 5,000 litre tank. We will pay for it, and you will pay it off through your rates over the next twenty years." We would hardly notice, it actually doesn't cost you anything and of course you will use less water. Then over the next five years we will increase the cost of water, so you have an even keel again, and as a nation we will be better off. The next step after having water tanks is changing the light bulbs, this is already happening too.

The same goes if you want an air-conditioner on your roof because the house has been so poorly designed. Then you need to show that you are making other carbon sacrifices, or that you are producing your own energy to equate with the extra you're using, which I see as a luxury. We are swimming in money at the moment in this country. Look at solar hot water systems, which are still not installed on every house. I'm not saying there should be one per household, that's probably not cost or energy effective, but the equation is easily worked out. And again, government should absorb the cost and you pay it off through tax over the next twenty years. It would probably save us from having to build more coal-fired and nuclear power stations. So I do think there are easy things we can do.

If you look at the piece of paper you're writing on, it's probably recycled or you're using the back of something else. That's what we all do now, but twenty years ago we didn't. It hasn't affected your or my standard of living or quality of life at all. Changing a light bulb hasn't hurt us at all. The next step is not that we should all have composting toilets, but that we are given the choice, initially the carrot, and then the stick. As a society we don't have a choice any more. We all eventually will have to have energy efficient washing machines, fridges, freezers and hot water systems and houses.

Also with plastic bags. Soon plastic bags for shopping will be disallowed. Then people will say, "How will I line my garbage bins? How will I live without plastic bags?" Well, I grew up in a place where we didn't have plastic bags. We would take our paper bags to the shop and you'd put a few eggs on a piece of paper and roll it. They don't break. We didn't have plastic bottles and the glass bottles were all recycled or refilled. I don't think it's actually going backward, it's going forward to something much smarter than we have at the moment. We will all look back on our thinking now and say, "How could we ever have done that, that's ridiculous."

Figures show that we need to reduce emissions by 80%. In his book *Heat*, George Monbiot shows us that we have alternatives, and points out how we can cut back. But what he can't show us is what we do if people are unwilling to change their attitudes without an emergency. That's my concern and that's why I feel a carrot and stick approach would work well.

Permaculture and Spirituality

Bill Mollison might come across as a person who separates permaculture from spirituality. I've heard stories where he has reprimanded people for introducing meditation into courses. I don't mean spirituality in terms of an organised form of religion, I mean it in terms of finding the inner self, the meaning of life, the purpose. But in one of his books Bill himself said that permaculture is a way of life, and I think he is actually a very spiritual person. He's just frightened of it, maybe less so now, but he definitely was when he was in his 60s.

I find that the act of 'permaculturing' on a property is actually a meditation exercise. I've just been teaching at an ashram in Thailand. I've been there before and have participated in their meditation practice but I didn't like it at all. This time I said, "Look, I'll only be here for five days, and I would like to do a couple of the early morning meditations." What I did was not sitting in a circle meditating, but taking the students out into the Ashram area and observing. I explained my meditation as being in the garden harvesting, planting seedlings, fertilising, or watering. Maybe my peak experience is working with bees. When I'm working alone with the bees or other animals, with my cattle, I find I become one with them. Permaculture does allow for these kinds of extreme calming experiences and I have let go of a lot of the more technical aspects.

When *Permaculture One* came out, it was the most amazing book, the most incredible of all the permaculture books, and I've still got it here signed by Bill. But for someone who is just new to it, I think there are other books which can show the path better than the permaculture books can. *One Straw Revolution* is an example of that, or *Places of the Soul*. There are a number of books which have nothing to do directly with permaculture, but which maybe make you become aware of what you need to know, and then you will want to find out more.

And the Future?

There will be many different aspects to the future. People living in the global South will have no say in what's going to happen to them. They will be really squeezed, are being squeezed right now. For some it will turn into disaster, and I think there will be more hunger in the world. For us in Australia, we still have the choice of a fairly soft landing. We have gone overboard and have so much, so cutting back doesn't sound very attractive. But by learning to be comfortable with less, we will find that uncomplicating our lives is important.

When I go to Tokyo, the first thing on my itinerary is that students take me to the Hundred Yen store. It's a huge supermarket where everything costs about 100 yen. They're always surprised that I don't buy anything. I tell them, "I can see lots of interesting things here, but there's nothing I need." It makes them think. It's very difficult when you're in your early 20s to realise that you actually don't need very much. I want young people to see energy as more than something which comes out of a power point or an oil can. All around us is energy moving and being stored, so while we have photovoltaic panels on the roof and solar hot water systems which are important examples, it's equally important to see soils as energy stores. Having around 10% organic matter in my soil is probably a bigger contribution to reducing carbon dioxide in the air than the photovoltaic panels on the roof. Also the forest we've planted, and that we've made our own wooden furniture. Now I'm not one who thinks we should leave all the forests the way they are. I'm actually for the use of timber, but I'm against the abuse of timber. I think we need to use it, and store it in furniture and buildings. It's a really important carbon store.

We do an exercise in our courses where we ask people to write down what are the essentials in their lives, what is really important to them. Young people don't put down the latest CD or mobile phone, they put down things like family, religion or love. I think if you can guide that a bit, and show by example, that maybe it will lead to a positive outcome. But people may soon be forced into a situation where they can't afford things anymore because they are becoming too expensive. Maybe that will lead to an understanding that they are actually not worse off, rather they might be better off. The road to get there shows that we are not losing anything which we will miss at all. I see signs that people will be able to adjust willingly. For some it will mean having smaller cars and more public transport or a more efficient fridge rather than no fridge. It will mean more local food rather than not enough food, and finding enjoyment in more local things. I think it will lead to a more sensible lifestyle, that's what I hope for. It is really not a technical issue, it's a matter of choice and attitude. We won't feel like we're missing out.

So what keeps me going? In the end there's probably a selfish aspect to it, but hopefully it's beyond that. We're not just about producing food for ourselves, it is being part of a solution together. What keeps me going are my grandchildren and young people, those people in their 20s to mid–30s who will inherit the mess of my generation. For the next five years I would like to pass on to young people what I've learned. They don't have 25 years to learn it anymore. I think our generation has powerful skills to be an example, which is not down and nearly out, but instead happy and joyful. We can show a viable alternative rather than a hard yakka alternative. Jill Jordan, one of the Maleny pioneers, always said we've already got the solutions, we just need to learn to work together. We have to show how to work together and show that it can be fun, that's important too. I think it keeps me young. I like the company of these young students. They're great fun, and given encouragement and a bit of time, they are the future, not us. Hopefully they will carry things in a different direction. I'm sure that one day one of them will come back and tell me what permaculture really is.

Vries Gravestein

Permaculture elder Vries Gravestein was born in Holland and learned his conserver lifestyle by living through the deprivations of World War 2. After emigrating to Australia and working as a jackeroo, Vries became an educator. He first taught agricultural science and later, after completing his PDC with Bill Mollison, became a permaculture educator, specialising in improving soil fertility particularly in relation to broadacre agriculture. With his son Hugh, Vries ran very successful PDCs over a number of years from his small farm in Chiltern, Victoria. Now retired to the south coast of New South Wales, Vries maintains an active interest in permaculture, especially working with farmers on soil fertility and nutrition.

McGaffin's Keyline designed farm at Yackandandah, visited during APC4 at Albury NSW (April 1990), on a field trip that was organised by Vries Gravestein. The cows grazing on lush grass in an otherwise dry landscape made a big impression on many of the younger permaculture designers present, whose only knowledge of Keyline was from their PDC. Photo by David Holmgren. Photo of Vries courtesy of Angi High.

Chapter 6

How Permaculture Made Me Whole

European Beginnings

I was born in Amsterdam before the Second World War under Aquarius stars. I did not realise how important these aspects of my birth would be till much later after I had lived in Australia for many years. Looking back, it is clear that my journey on the road to permaculture started in the Netherlands. With 66% of the landmass below sea level, the Dutch are a very inventive and independent people. They are bordered by nations who speak different languages, and equality and self-sufficiency are of great importance. My grandfather was an especially strong influence in my life. During the war he grew a lot of vegetables in his backyard, cultivated his own tobacco and kept beehives for honey. My grandparents were never short of food.

These are traits I unconsciously absorbed. I had experienced the war as a youngster without being seriously affected, and was still a teenager when it was over. But the experience of being in a country at war meant people were short of food and largely cut off from all sources of energy. We lived in a three-story house that was freezing cold in winter. My mother had to cook on a pot-belly stove in the lounge-room, and as the eldest child, my job was to supply the wood for the pot-belly. After school I'd go scavenging and learned to find firewood in all sorts of places. To this day I still have a great interest in the timber lying at the side of the road. Looking out of my car window I always work out how many tons of firewood I could harvest, and our family homes have always had wood stoves for heating and cooking.

My father used his dentistry skills to provide food for us. This is the time when I learnt to skin a hare and how to make flour out of whole wheat by grinding the grain in a small coffee mill balanced precariously between my knees. My wartime experiences left me with a real interest in the production of food and fibre that has remained with me all my life.

And So to Australia – Love and Teaching

At the age of 20 I emigrated to Australia. While I was waiting for my migration papers, I jackarooed on an intensive mixed farm where they grew 23 types of crops, as well as dairy cows and a mob of Texel ewes. This provided a wonderful experience of polyculture farming in a cool climate. Once I got to Australia I continued work as a jackaroo, this time on a Merino sheep stud, another time in charge of a milking herd of cows for a month. I also shared the management of a small mixed farm near Ballarat for four years. All these practical learning experiences left me with a real love for the land and a growing interest in holistic and organic farming. I went on to study agriculture and graduated from Sydney University.

By the time I was 28 I had met June, the love of my life, and we naturally wanted to get married. But I had no funds – jackarooing and university had seen to that. So I had to find a steady job with a regular salary. Fortunately I was offered a position as an agriculture teacher at a boys' independent school in New South Wales, where I remained for the next 25 years. These years were the centre of my own lifelong learning. The students were mainly from rural backgrounds, farmers' sons who were boarding at the school and many saw themselves as budding farmers. My family and my Dutch background had prepared me for a holistic and organic outlook, but they had certainly not prepared me to teach these big, challenging boys! I suffered some serious stomach cramps from stress during my first year of teaching. Fortunately I had confidence in my farming knowledge and this proved enough to keep my end up as a teacher for the first few months. But the students were very determined and not a day went by without a challenge. I realised I needed to provide down-to-earth, practical examples of what I was teaching, to prove what I felt so strongly about, was correct. This, along with my thirst for knowledge about natural farming, meant that I was always searching for new information. In this I have been immensely fortunate.

During this period I travelled extensively in my local area, the Riverina (the Murray River area). My job as an agriculture teacher had evolved to the stage where a special one-year Rural Training Course was added to the syllabus. The school had bought a 100-acre farm and I designed a farm management course to run from it. English, Maths, Bookkeeping, Mechanics and other subjects were adapted to this course, which I had the pleasure of directing. It provided an agricultural education for the boys based around a 17-cow herd, a flock of 300 Corriedale ewes and five sows which were all raised for prime meat production, and marketed on a value-added basis. Through the course I came into contact with a number of farmers as well as the Rutherglen

Agricultural Research Station, which further broadened my knowledge base. My students and I were frequents visitors to the Station.

However, back at the school, major changes were taking place in management as well as in the education system in general. The syllabus for agriculture in NSW had been well thought out and was comprehensive. In the Independent school system I had the freedom to interpret it as I saw fit, and over the years I was able to attract large number of students to this voluntary subject and achieved excellent exam results. But now, a new syllabus was in place based on a 'unit' system, which, as far as I was concerned, destroyed the holism of nature-based subjects like biology and agriculture. These changes were not to my liking. My students started to remark that my smile was missing. I took heed of that, and at the age of 56, I resigned from the school.

Meeting the Thinkers-and-Doers

My passion for learning and the natural way of doing things brought me in contact with the thinkers and doers of what was then labelled 'alternative'. Practitioners of organics, biodynamics, mineral fertilisers, and homoeopathy began to cross my path. As if from nowhere, the 'right' teachers arrived for Vries, the teacher, just when they were needed. One of these was Bill Mollison. P.A. Yeomans, of keyline fame, was the first of a long line of wonderful teachers and mentors that I encountered who had a great impact on my thinking. Yeomans beckoned me to come and observe how his keyline system worked with nature's laws and was ecologically sustainable. Keyline practice has proved to be essential to broadacre permaculture systems. Through keyline, water can be made available for nearly every farm on a sustainable basis. The keyline principles of planning on the contour and treating the soil as an organic system made eminent commonsense. And commonsense (which these days is really 'uncommon sense', according to Mollison) is a key to permaculture. Looking back, I realise that the two periods when organics was on the rise in Australian agriculture did not coincide with my times of intense learning – in the '40s and again in the '90s. So while I often felt lonely with my unconventional views on agriculture, I decided that if that meant that I had to be out of phase with the rest of society's attitudes, then so be it. I'm not sure how I 'stuck to my guns' during this time but in hindsight, my Dutch heritage, my parents, my grandfather and the war, all contributed to creating my personal attitude. I also believe my Aquarius star sign made sure that I was forward-looking.

Another of those remarkable thinkers-and-doers who had such a significant influence on me and the way I taught, was Geoff Wallace, the designer of the aeration plough. Geoff was an old boy of the school where I was teaching, and the organiser of the Kiewa Field Days, held each Easter on his property under the magnificent river red gums beside the Kiewa River. I attended these Field Days religiously every year. Their appeal lay in being able to talk freely to knowledgeable, like-minded people about natural processes, with Nature herself all around us. Talking and learning in this kind of environment does not require the props of a conference centre. The red gums provided the ideal setting. These remarkable Field Days were very educational, and hosted many well-known speakers, always on organic-related topics. It was here that I met Bill Mollison and for the first time heard the word 'permaculture'.

Shaken up by Permaculture

And as soon as I had resigned from the school, my son Hugh proposed that we embark on a Permaculture Design Course (PDC) in Tasmania with Bill Mollison, held on the property where Bill had grown up. So it was that in 1986, Hugh and I took Bill's PDC, a time not quite in the middle of the two organic periods, but nevertheless very significant for me as an individual. That course turned out to be a major milestone in my life's path and permanently changed the way I saw things.

Years later I had a revelation that clearly explained what had happened to me during that PDC. By this time I had taught a number of PDCs myself and met many interesting students, who I visited often, car travel being not quite as expensive as it is now. The revelation happened at the house of one of my students where I was staying. He had a young family and a new baby. The children's mother had created a photo collection for their eldest daughter to make her feel special, and she proudly showed it to me. One of the photos was of her father holding her up in the air by her feet. This stuck a real chord. I promptly exclaimed, "That is what Bill did to me!" The daughter was highly intrigued and with big eyes asked me what I meant. I explained that during the PDC, Bill Mollison had (metaphorically) shaken me just like that, so that all that I knew had fallen out of my head for me to gather back up again later. Up to that time, all my knowledge had been collected in small, separate bricks. I had knowledge about micro-organisms, disease, contours and water in the landscape, irrigation, plants, vegetation, microclimates and so on. But they were all separate pieces of knowledge. I realised that what Bill had done was to help me build these separate blocks into a pyramid, an interconnected, indestructible structure. This is what I understand by *wholism*, with a W (but these days written as *holism*). Bill sat me on top of this pyramid, so that I could see far and wide over my environment. And Bill said to me, "Go and teach."

The PDC Years

And so I did. By this time I already had plenty of experience in the kind of teaching the practice of permaculture required; I believed I was good at it and had the confidence to tackle it on a whole farm design basis. When I returned home after the course I could see more clearly what was right and what was not right, according to Nature's laws. When I looked around our 60-acre property 'Willuna', near Beechworth in Victoria, I saw it with very different eyes. I realised how my wife June had intuitively placed her garden in the right place – Zone 1. The old house was nothing special, but we renovated it with a kitchen in the part that faced north-east and the morning sun made that kitchen special. The winter sun reached as far as the slow-combustion stove. Over the years we continued building our little farmlet into a self-sufficient unit that began to produce prime quality food. By ensuring permaculture principles were clearly visible on our property, Willuna became an interesting and effective teaching site, where the ideas of permaculture could be seen as being integral to a sustainable lifestyle.

Our first PDC was held in the garden and in the house, and was wholly residential. After completion of that first course in 1987, Hugh and I, sometimes with other guest teachers, held two to three courses a year at Willuna for a period of ten years. The students were from all different backgrounds and ranged from teenagers to people in their 60s and even older. Some were broadacre farmers who were concerned about the level of fertility on their farms. All were concerned about what we humans were doing to the environment and wanted to live their lives differently. We tried to remain faithful to the original PDC curriculum, and delivered most of the course material largely in a lecture style, with the farm and the garden both practical proof and relief from this intensive input. You may wonder how people managed to stay the course! But they did, and end of course surveys showed one thing above others; that the participants felt enormously empowered by what they had experienced.

It's not easy to convey the complexity of educating people from all different backgrounds to understand the holistic nature of permaculture. The way I developed my 8 hour lecture on my favourite topic of soil fertility over the years illustrates how I attempted to do this. I broke up the topic into 8 sub-headings: Water; Air; Temperature; Structure; pH; Minerals; Microbes; and Organic Matter, all having to be put across on a blackboard and in one day. By being down-to-earth and extremely practical, I was not only able to keep all the students awake, but I was able to make understandable the complex chemical reactions that occur in living tissues. Nobody walked out, and in an extensive survey of several PDCs, this lecture was chosen by many students as the most interesting. The deeper the understanding, the bigger the joy.

The emotional energy created in these courses was enormous. My interpretation of this is that the combination of new understanding, the down-to-earth-ness, the contributions of everybody's special interests, and the property and its plants and animals created this communal energy. All was in harmony and every course ended in tearful goodbyes. In 1994, one of our PDCs was intensively researched for a PhD thesis (see Smith, 2000). The research showed that our own authenticity as teachers and practitioners as well as the setting itself was very important to the students. And though there was a lot of information, our course also provided opportunities for bonding and participation of like-minded people both during the classes and informally over meals and in the evening. This is what made what we were doing special for people.

At the same time I was also involved in seminars and field days to disseminate holistic farming knowledge to others. With great enthusiasm two of us organised 'Permafest' in Beechworth in 1991. This was great fun; we invited as many alternative displays and stands as we could find. This was one of those times when the upsurge in interest in organics was in full swing, and I became heavily involved in that movement. I remember with great pleasure a visit by a group of organic farmers from the USA, organised by ACRES USA, a well-known publisher of literature on organic and 'alternative' agriculture. The farmers accepted me as their guide for the two days, and, some years later, I was rewarded by an invitation to visit America to learn more about the energy fields Nature uses to communicate. It was there that I learned about insect antennae. I knew a fair bit about insects through my father's interest, but in the USA I learned an amazing amount about how plants and insects communicate through energy fields. I learned how they co-exist as pests and predators, and how they defend themselves. Learning how living things work together taught me how to include Integrated Pest Management (IPM) on my farm and in my teaching. In Nature there are always checks and balances. For example, in apple trees there is a red mite, which seriously damages the trees and their ability to produce. Researchers have found that there are other mites that eat these red mites. We have learned to breed these predator mites and to use them as natural control. In other words, one can integrate these natural methods of control as a non-polluting system.

But times change, people are fickle and society changes its attitude. There was a general collapse in interest in our courses the '90s, and I'm still not sure why. In 2000, June and I finally retired, sold up at Willuna and moved to the Bega Valley in southern New South Wales to be near Hugh and my grandchildren. As I reflect on my long association with permaculture I realise how much I have learned. I am certain I can survive and be self-sufficient on much less than this present society thinks necessary. I am now an 'Elder', the stage where I can communicate, advise and help people at all levels how to best survive in our degraded Australian landscape. I want to continue to be useful to the society around me as an adviser and mentor in the practicalities of food production, whole-farm management, integrated pest management, and organic practices in general, as well as waste management and energy efficient housing.

The Soil Man

It is my understanding of soils that I feel most proud of. Looking back on agriculture over the last half century, in the years between the '40s and the '90s, technology ruled, production costs went up and most significantly, soil fertility declined. Soil acidity, salinity and compaction became the topics of discussion. My association with the independent soil testing laboratory SWEP for over 45 years has given me status as an 'expert' on soil fertility and deep insight into the profound fertility problems of Australian soils.

Getting established in a new area is always difficult, and when we first moved to Bega I began to believe that in this society the Elders are no longer considered of value. Trying to work independently rather than being attached to an institution can be a real handicap if you still want to contribute your expertise to society. As a result I asked myself many times, "What am I doing here?" Well! Something happened quite recently, very unobtrusively, and I think my question is going to be answered. Around the middle of 2005, via a contact from one of my PDC students, I was approached by the Veterinary Officer of the Bombala Lands Protection Board to start a soil testing survey of the area. So far I have taken 22 soil samples on 19 properties and the SWEP soil analysis results are on my desk. Taking the soil samples meant travelling all over the shire, and I was able to meet landowners and listen to their concerns about their operations. The soil analyses were beginning to paint a detailed picture of the mineral status of the soils in this shire. My ability to 'read the land' gave me a good overview of the environmental status of the land as I travelled though it. As in most parts of Australia, farming practices have caused major degradation.

Over the past few years I have collected over 60 soil test results which have allowed me to make some acute comparisons and deductions about the poor fertility of our local soils. In particular, I made two rather devastating observations. Firstly, there is very little or no phosphorus available in these soils for plant growth. There is plenty of phosphorus in the soil, the result of long years of fertilising with acid superphosphate ('super'), but it is locked up, unavailable to plants. Phosphorus is essential for healthy plant growth. It forms part of the structure of DNA and ATP, the molecule that stores energy for all living things. Secondly, there is virtually no air in these soils to provide soil microbes with oxygen, because of severe and uniform compaction. The reasons for the situation in this valley are to be found in its history. For a long time the Bega Valley land has been used for cattle grazing, so ploughing and crop production are not that common. The traditional fertiliser used on grazing paddocks is superphosphate. Phosphorus in 'super' is in a highly soluble form and goes into solution completely with as little as 25mm of rain. And it then gets locked up, bound by minerals in the soils, never to be released again. On average there is 20 times more phosphorus 'locked-up' than there is available for plants. The major conclusion that can be drawn from the 22 soil tests is that the available phosphorus is not sufficient for economic pasture production. The solution is to use reverted super or rock-phosphate.

Heavy animals with cloven hooves are excellent soil compactors, which explains why this land over time has become oxygen deficient. Evidence for this lack of air can be readily seen in the light green colour of pastures, contrasting with the dark green 'roughs', the areas where the animals leave their droppings but don't graze, and which are microbe-rich. The solution to this is the aeration plough developed by my old mentor Geoff Wallace. This plough cuts channels in the soil that allows air in without damaging the delicate structure of the soil. This enables the micro-organisms to breathe and do their job of breaking down organic matter properly.

As well as phosphorus and oxygen depletion, the water status in the Shire's soils has reached a serious stage. The effects of drought and climate change are going to be more devastating than in the past. This is a relatively cold area of Australia, and the microclimate at the soil surface is very harsh. These three factors are mainly responsible for the economic downturn in the animal industries that form a major part of the local economy.

During my first six years of living in the Bega Valley I added this information to my store of practical knowledge. So, where to now? I have now been accepted as the 'Soil Man' in the Bega Valley and still attend my beloved field days (this time in Candelo). Here I operate as soil advisor and organise soil tests for the local people. I recently organised a round-table discussion with all the landholders concerned to discuss these issues and introduced myself as the mentor to the concerned farmers. The first issue for discussion was the phosphorus situation. After several questions and a general exchange of experiences, it became clear that there was considerable concern about a number of aspects of farming other than soil fertility, and that these farmers possessed considerable knowledge of their situation.

The farmers decided that it would be a good idea to form a group to continue to share and benefit from each other's expertise. I suggested that the ideal outcome of this round-table discussion would be to form an 'educational' communication system, such as a set of walkabout field days. So the seeds for learning and further action have been sown. The farmers decided to form a club, and a secretary was promptly chosen and appointed. One farmer volunteered to organise the next gathering. My fervent wish is that this group of concerned, traditional farmers may grow into a permaculture association with ecological sustainability firmly in its sights. For me, bringing conventional farmers to a permaculture understanding would be a fitting way to conclude this story of my progress within permaculture as a practitioner and a teacher.

Looking back, I see that my life's work has been to educate people from all backgrounds and ages to live in a world where local self-sufficiency is practiced through gardening and small-scale food production, such as in backyards, community gardens and 'community supported agriculture'. This is the world I want for my children and grandchildren. The only way I can see this being achieved is by moving away from industrial farming systems that are based purely on making money through bulk exports. Backyard gardening for food production has always been more successful than commercial production, and it's what we've always relied on in times of hardship and war. It has a significant advantage in reducing pollution levels as well, as pesticides are only needed for large-scale monoculture farming. If my life's journey has taught me one thing, it is that improving soil fertility though proper organic management is the key to food security in Australia, and that an awareness of the holistic nature of systems, made possible through permaculture, forms the basis for this understanding and how to achieve it.

Jeff Nugent

Jeff Nugent lives on a small property in Nannup in the south west of Western Australia. In 1983 he attended one of Bill Mollison's Permaculture Design courses, which he recorded and has made available to others.

Jeff has over twenty-five years experience in teaching Permaculture Design and a longer history as a Permaculture Designer. He is co-author of the book *Permaculture Plants: A Selection*, and author of *Permaculture Plants: Agaves and Cacti*. He is currently working on other permaculture plants books and a futuristic novel. Jeff has worked on four continents and has taught courses in Australia, North America and Africa. He enjoys working with sound, and apart from the several Bill Mollison audio sets, he has also produced two music albums.

Jeff Nugent teaching a course in Kenya.

Chapter 7 | A Personal Revolution

What sparks a personal revolution? For me it was the Vietnam war. I never fought in it but I was next in line to be conscripted. I knew I wouldn't fight but risked becoming imprisoned or a fugitive. Luckily Gough Whitlam took office in 1972 and ended Australia's military involvement in Vietnam, just months before my time. The experience of opposing the war gave me an inherent distrust of governments and a need to become self-reliant. By the age of 21 I had organised myself on to a parcel of land in Nannup in the south west of Western Australia (WA). I built a modest house and planted an orchard and a vegetable garden.

There were quite a lot of like-minded people in the area, all recently moved and many from the city. These were a diverse set of people, loosely referred to by the locals as 'hippies'. The 'back to the land movement' was catching on, as increasing numbers of pilgrims found their way out of the city. In those days our local store did not cater for our exotic tastes. Sesame oil and tahini were foreign to the local diet and one only ate nuts at Christmas. As a way of supplying our basics as well as linking together and showing some vague form of solidarity we established the Nannup New Settlers Co-op. (As a testament to private enterprise, it ultimately became the privately owned Good Food Shop.)

By 1977 Jim Cairns (recently sacked from his position as Deputy Prime Minister of Australia) became the political spearhead of the Down to Earth Association in Australia. None of that filtered into my own life however because there were no information networks at the time and I certainly wasn't reading newspapers, listening to the radio or watching television to get information. Even the *Earth Garden* and *Grassroots* magazines that I was thumbing through by then were a few years old so the festival that we were about to host took me by storm.

Dennis McCarthy was working at Murdoch University, in Perth, as librarian and had made a serious collection of alternative literature. Dennis, with others, had formed the Down to Earth Association of WA. As the east coast had been having festivals it only seemed fitting that we in WA should do the same. So Dennis hitch-hiked down to Nannup one weekend looking for a likely site for the first WA 'Down to Earth Confest'.

We took him out to Cambray – an old railway siding along the St John's Brook. All of the houses had been removed but most of the gardens were still intact so it offered an ideal festival site. Much groundwork later in January 1978, we became the hosts of a 1500 strong assembly of 'alternative' people.

Our involvement had stopped at the groundwork, so I had no idea who was going to turn up but had heard whispers about 'two wise men from the east'. It sounded quite biblical and mysterious. The first wise man was Jim Cairns and he spoke about getting organised as a political lobby group, something that didn't really appeal to me. I knew Jim's politics and had no problem with them but my political energies had been spent on Australia's Vietnam War involvement. It was time for action. If a revolution was to come it would be in the form of grassroots movement. Wasn't that why I was in the bush?

Then Bill Mollison, the other wise man, came on stage. If I had followed my ego I would have dropped him before I had listened. In the first minute he had made a cutting remark about vegetarians. If you've ever been a vegetarian you will probably know how sensitive I was, as a vegetarian, to such cracks.

Bill always had the ability to be confrontational without being wrong. A kind of irritant factor that kept the listener engaged and alert. But despite all of the personal challenges that he presented to his listeners, what oozed from Bill was wisdom. Not knowledge *per se*, although he is probably the most knowledgeable person I have ever known. This was the logical application of a wealth of knowledge – wisdom.

Bill was talking about putting things in their right place. I had already at my own place, in a very vague way, attempted to locate things in logical places. I had even read up on companion planting, but that was kindergarten and Bill was way beyond university. He spoke of the soon to be published *Permaculture One*. I couldn't get my hands on it quickly enough. I read it, then read it again. Thirsty for knowledge, I read *Tree Crops* by J. Russell Smith, then I read *Permaculture One* again, then *Forest Farming* by J. Sholto Douglas and Robert J. Hart, then *Permaculture One* again. And then finally, when it was published, *Permaculture Two*.

Whenever Bill came to Western Australia I tried to be there, usually with a tape recorder. In September of 1983 I took professional recording equipment with me and attended one of Bill's Permaculture Design Certificate courses in Stanley, Tasmania, the place of Bill's birth. I had reasoned that I was privileged to attend such a course and that by recording it, I could make it available to many people. It was a great course. Bill had been writing some of the *Designers' Manual* and the ease with which he peeled off the lectures without any notes was incredible. As I went through my tapes every evening I became astounded at what I had. Here was a blueprint for saving the planet.

Toward the end of the course Bill said, "If you guys do it right, you'll never need to do a design for anyone because everyone will be a designer." I was a fairly shy person and standing in front of a group was 'not my thing', so I promptly forgot that he had ever said it. On the homeward journey I recall thinking how amazing it was that with all that valuable information on board, the tapes still weighed the same as on the way there. For the next couple of months I spent every moment that I could, going back through the tapes. I had come back with forty, ninety-minute tapes and I wanted to become familiar with them before attempting to edit them. It was great to be able to go through the course time and again. To have the luxury of hearing it all over again – and again and again. How much we miss the first time around!

Finally I did hear Bill saying, "If you guys do it right, you'll never need to do a design for anyone because everyone will be a designer." He was right and so about five months after I had attended his course, having worked through my fears and five re-runs of my recordings of Bill's course, I ran my first Introduction to Permaculture Course. Nothing too taxing the first time around. The local high school ran night classes for adults. I worked out a program to teach the course over eight weeks and taught three hours once every week. This worked well as a first course because I had a week to prepare each three hour segment. The course was well attended and well received.

The following term, the headmaster asked if I would be prepared to run the course again. With my armoury of notes from the previous course I reasoned that this would be a cruise. Imagine the shock-horror when I casually walked into the classroom on the first night, my notes under my arm, to see all the old faces and a few new ones. Thinking on my feet, I got consensus from the group to change the format. I quickly revised the whole course in the first night. Each week we assembled in the late afternoon at a different student's property. We allowed a couple of hours of daylight in which to walk the property and discuss the 'client's needs'. We then went indoors and shared a meal. After the meal was cleared away we bounced ideas and drew up rough outlines of a design for that property. It was a great process and everyone gained a lot of knowledge along the way. It was also very rewarding for me personally, seeing what each student had done on their own properties since doing the first course. Some very good systems had gone in.

In the midst of all of that we took on the Agriculture Department who were spraying our town with Heptachlor to combat Argentine Ants. This was not the sphere I wanted to be operating in, but it became a simple matter of survival. One night, after a public meeting I sat down over a coffee with the Senior Entomologist for the Agriculture Department. After a little small talk he turned to me, "Keep doing what you're doing! Don't stop. If, in your lifetime, all you achieve is to become the patron saint of fighting heptachlor and dieldrin, then you will have lived a very productive life. That stuff is not safe and needs to be banned urgently!"

I was stunned. The same man had been standing in front of a group an hour earlier espousing these chemicals and telling us all how safe and wonderful they were. When I challenged him about his seemingly double standards he pointed out that it was his job and that he had a family to raise. I wondered how he slept at nights. I was very much a babe in the woods but the lessons were coming. Over the next few weeks I witnessed many (so called) public servants lie about the health and safety aspects of pesticides. We did, after a year or so, manage to have the chlorinated hydrocarbons banned in Western Australia. But they were replaced with what? It is the approach that is wrong. For that reason I have always tried to work from that front rather than becoming reactive to individual situations.

It was finally time to edit the tapes that I had recorded in Stanley. I was not interested in trying to do anything too radical. Keep the theme going with as few distractions as possible along the way. Now that's a tall order for any permaculture course and this one was no exception. At the end of the day all I can say is that I did the best job that I could at the time. I have recently had the joy of converting the entire set to mp3 format and making it available on one DVD. That is certainly more compact than the thirty, ninety-minute tapes that I had edited my originals on to. The written notes are all on the DVD too, so happily they are both compact and complete, but most importantly, backed up.

In the process of editing I took it upon myself to check as many of the facts and figures presented by Bill as I could. As anyone who has listened to him will testify, he does have the knack of presenting the listener with unbelievable facts. Remember that this was a time before internet searches; every bit of research took real time, especially if one was living away from a city. About the only fact that I am sure Bill got wrong, despite all my research, was in his coverage of acid rain. He stated that the two acids are sulphuric and hydrochloric. All of my research was saying not hydrochloric but nitric. Several years later, when Bill was in Perth, I approached him with the same tape deck and microphone. I explained that he had said hydrochloric instead of nitric. He laughed and said that was something he was always doing, but agreed to say nitric acid into the microphone so that I could set my recording straight. If you listen to the set today you can hear a somewhat disinterested Bill humouring one of his students by saying nitric acid into a microphone!

That was December of 1985 and Bill had come across to Perth to help us out with Western Australia's first Permaculture Design Certificate course. We had given the first week of lectures and then Bill was to arrive and take the second. On his first day Bill was given three topics. He chose to launch into aquaculture. Luckily I had my recording equipment set up. I remember only an hour into the topic thinking, "I really must make an aquaculture audio set around this talk." It was a tour of aquaculture unlike any I had heard before, or for that matter have heard to this day. As I write this (February 2007), the Aquaculture set is one my most recent achievements. Finally!

The main reason why we wanted to get Bill to come and help with that course was Earthbank. Bill had coined the word to describe a whole range of strategies relating to legal structures, ethical investment, community financing, land access and city-farm links. He spent about four days working through them and I happily made an audio set (nine ninety-minute tapes) out of his talks. This will be my next conversion to digital when my recording equipment returns from servicing.

One of the things that had impressed me with *Permaculture One* was the species list at the back. Thirsty for such knowledge I began writing down all of the information that I could accumulate about useful species with potential in our own bioregion. I had adopted the philosophy that "the pen may be mightier than the sword, but for nuclear bombs you need word processors." By about 1986 I was compiling all of the information I could find on useful plants, onto a computer. Clients that I did designs for were grateful for it and students on our courses encouraged us to publish it in a book form. A colleague, Julia Boniface and I finally committed at the end of one of our Permaculture Certificate courses to have the book ready to be launched at the International Permaculture Convergence to be held in Perth in September 1996.

Although we were already well down the track with the research, there was still so much to be done in less than six months. Deciding when enough is enough was the hardest issue. Finally it was dictated by a deadline. We spent a lot of time proof-reading – something that seems to be lacking in many publications since desktop-publishing made self-publishing possible. Also, we had some neighbours spend days going through all of the botanic names, comparing them to the spelling in other texts. It really was a thorough process so that in the second printing only two very minor changes needed to be made.

With a size of book in mind we could then go to the next step: either find a publisher, or some capital to self publish. Having invested so much time and energy into the process, we were not keen to compromise it by putting it in the hands of a publisher. We sent a mail-out to all of our students calling for pre-sales. Twenty-five dollars would buy the book which was to have a retail price of thirty five dollars. Not only would the subscriber get a book at a cheaper price but it would be mailed to them, hot off the press and signed by the authors. They would also have the knowledge that they had helped to pay the manufacturing costs. The letter also called for people interested in loaning us some money as a short-term ethical investment.

Presales made the printing possible as the money poured in. One generous loan from friends made the colour pages possible. The book was ready for the convergence (with several hours to spare!). Having invested all of my own funds in the book I was in the position of not being able to afford to attend the convergence. Suddenly I had an epiphany. We had just printed our own currency. I approached Pat Dare the Convergence coordinator and put it to her that I could supply the Permaculture Association of Western Australia (the host group) with books at wholesale price, to cover the cost of the convergence fees. They could then market the books to recover their fees and make a tidy profit. Too easy.

I must confess to having gone to the convergence with a sense of holiday. My work was finished (ha!) and I was on a kind of vacation. It had been a hard push to get the book finished and at times I was working about nineteen hours a day on it. The remainder of the time I was dreaming about it.

At the convergence a common theme kept occurring amongst many of the speakers from the third world. They all required seed. It was days before it occurred to me what was actually happening. I tested my theory with some of the participants who had been calling for seed. Sure enough, they were all trying to grow temperate vegetables in the tropics. It was a legacy of colonial days. If you weren't eating European vegetables then you must be a heathen. Most temperate vegetables require changes in photoperiods to trigger flowering. The tropics experience a constant photoperiod all year. They could grow the vegetables but they couldn't produce seed.

Although I had no direct experience at that stage I did know that there was an abundance of very good vegetables from the tropics. I had tried pushing the limits with many of them over the years by attempting to grow them in my own temperate garden. Very few had proven possible but I did know there were a lot of tropical vegetables with good nutritional qualities. I decided to start researching the plants of the tropics that can be used as vegetables. This led directly to my publishing *Permaculture Plants: Agaves and*

Cacti and the almost completed *Permaculture Plants: Palms and Ferns*. There is still a large body of material to be sifted into other volumes but I won't get too far ahead of myself.

In my quest for new and interesting plants I discovered the Quito palm in a rather remarkable book called *The Lost Crops of the Incas*. I tried to locate seed but to no avail. One day the phone rang and it was Geoff Lawton, a fellow student at Bill's course in 1983. Geoff was somewhat amazed that I was still at the same address some 12 years later (I still am 24 years later). In conversation he mentioned that he was working in Ecuador and he agreed to send me some Quito palm seed. A photograph of the first germinating seed can be seen in our first book.

The Quito palm (*Parajubaea cocoides*), also known as the mountain coconut, has an edible nut, somewhat similar to coconut but much smaller. What fascinated me was that it is unknown in the wild. Other *Parajubaea* species are barely surviving in remote valleys of Bolivia and are considered endangered, but nobody even knows where the Quito palm originated from. It is only known as a street tree in old Inca cities of Ecuador and Colombia. Some argue that it was most likely selectively bred from wild species by the Incas.

In researching for the palms and ferns book I joined the International Palm Society and frequented their forum for several years. Nobody knew of anyone on the planet who was trying to maintain sustainable breeding populations of these species. Most palm growers seemed to consider that in some obscure way, they were contributing to the preservation of palm species. However a sustainable breeding population requires enough plants with a broad genetic base so that future generations do not become inbred. Few growers, however, kept more than a couple of specimens of any one species. To worsen the situation, collectors keep many related species in the same garden so that hybridisation is common. These hybrids are favoured by collectors because of their unique nature, but contribute nothing to the preservation of the species.

As far as I am able to determine I am the only person on the planet attempting to preserve the *Parajubaes* outside of their native habitat. Operating on the principle that 'nobody can do everything, but everyone can do something', I have adopted this obscure genus of plants and am doing what I can to establish a second habitat for them on the planet. It is a very small act in terms of the big picture but a huge undertaking for an individual.

I have never been a great traveller. Mostly I would just as soon stay home as go away. But some opportunities have come forward and I have rallied to the challenge. When my interest in the tropics was first kindled I had Michael Nickels come to visit from Canada. Like many Canadians he had somewhere else to be in winter. He had set up tree nurseries across Kenya and was very passionate about tree planting. He went to a seminar in Nairobi about trees and caught a talk there about permaculture. He had planned to check it out on his way home. He had already booked a flight to Perth with the intention of visiting the forests of the south-west. Armed with a WWOOF (Willing Workers on Organic Farms) farm-stay directory, he arrived at our place.

We became friends instantly and I went across to Canada to teach a course on his property. A couple of years later I worked with him on a project in Kenya. Over a five week period we put 5 km of swales across 35 acres of burned out wasteland. The transformation was beyond any local expectations. The whole area was victim to heavy overgrazing and firewood collecting. Grasses and other fodder plants rarely had an opportunity to seed. Although the area is classed as arid, they do receive significant rainfall. The problem is that one rain event is usually about 50mm of rain and it all dumps in a few minutes. The compacted soil erodes as the water runs off to the stream, carrying any loose soil and accumulated seed with it. This situation really lends itself well to swales. Once built, the swales held all of the rainfall, allowing it to percolate into the soil. As we had fenced the area, all of the trees that we planted were safe from stock. They quickly grew, as did the grasses which were now able to seed. The site changed from bare earth to full foliage cover in just one year.

The site now gets visitors weekly from all over Kenya and the process is being adopted more widely. Two years later I returned with a group of students from Japan. On that course we transformed the adjacent Ngare Ndare school, where we were staying, into a food forest. Eighteen months later it was dripping with fruit. This is in a region where students would walk 20km to get to and from a remote school because that school had a free lunch program.

During that course I was looking at the village and trying to get a grasp of their energy budget so that we could deal with energy as a topic. I sat one evening with Kariuki – a student but also the manager of our project. He related to me his own power budget. Being the youngest born to a Kikuyu family he was living with his parents, so it was an extended family situation. He would spend 31 shillings a week on paraffin (kerosene) to fuel the kerosene lamps. Tropical nights last more than 11 hours so having light makes other chores indoors possible. It is also a time for reading.

All of his electricity needs were in the form of 12 volts. He had a heavy-duty, deep-cycle battery that every three weeks he would carry on his shoulder for several kilometres to put onto a bus. This was down a very dangerous track and he made the trip in darkness to catch the bus before it left. The battery would then have to endure being bounced down a dirt road for about 26 kilometres to its destination where it would get put on charge and then finally returned. He then had to carry it back up the hill to home. The luxury of 12 volt cost him 50 shillings every three weeks.

The battery would drive a small 12 volt television and recharge his mobile phone. I had witnessed the difference that mobile phones had made to communication, since my earlier visit. People saved themselves several days walking just by sending a text message. I was less convinced about the television though. I asked Kariuki whether they watched American movies. He looked shocked at my question, "No, this is Kenyan television. We watch programs on being better gardeners and better farmers, about soil conservation and health. It is all educational." Point taken. I included the television in the basics category.

I immediately saw a direction that could be taken. I suggested that the village set up a solar club. The club could sell raffle tickets to the value of 50 shillings. A 14 watt solar panel was valued at 4,000 shillings so as soon as 80 tickets had been sold they could afford to purchase a panel. The panel would end up on some lucky person's roof and they would no longer have to breathe paraffin fumes or carry dangerous lead-acid batteries. Those batteries would last for years longer because they would not have to be constantly charged and discharged and not bumped along dirt roads. But the panel would belong to the club and a rental fee of 50 shillings a week would apply. The next raffle would be subsidised by the rental and so on. Kariuki saw the potential immediately and within a week the Ngare Ndare Solar Self Help Group had met and the first solar panel was on its way. This was a practical solution to a local problem.

I have always promoted permaculture as the only package that has potential to solve all of our environmental problems. As these problems become more obvious to the masses there are more people taking up the challenge. Tim Flannery with his book *The Weather Makers* has been a major contributor to this process. My only criticism of Flannery's book is that it is extremely light on solutions. That's where the *Permaculture Designers' Manual* steps in. The future of Permaculture, even if under a different name, is absolutely tied to the future of the planet. Let's hope for and work towards a bright future.

Geoff Lawton

Geoff Lawton is an experienced permaculture consultant, designer and teacher. He holds Diplomas of Permaculture Design in education, design, implementation, system establishment, administration and community development.

Since 1985, Geoff has undertaken thousands of permaculture-related jobs and projects consulting, designing, teaching and implementing in 17 countries around the world. Clients have included private individuals, groups, communities, governments, aid organizations, non-government organisations and multinational companies.

In 1996 Geoff was presented with the Permaculture Community Services Award by the permaculture movement for services in Australia and overseas.

In 1997, upon his retirement, Bill Mollison asked Geoff to establish and direct a new Permaculture Research Institute on the 147 acre Tagari Farm which Bill had previously developed. Geoff developed the site over three years and established the Permaculture Research Institute as a registered charity and global networking centre for permaculture projects. He is now the Managing Director of the Permaculture Research Institute.

Photo by Craig Mackintosh.

Chapter 8 | Thinking Big

Early Influences

I was born in England in Stoke-on-Trent, Staffordshire, but I grew up on the Hampshire-Dorset border on the south coast. My mother was of gypsy descent, and used to talk about all these unusual things and loved unusual foods, especially wild harvest, going out cockelling, blackberrying, and looking for nuts and wild mushrooms. She loved things that were a little bit risqué for most people in those days. Then, everybody was into modernism and science and thought technology would solve all their problems, whereas my mother was almost contentiously organic in her attitudes to health and remedies and foods – gutting her own chickens and cooking her own rabbits. It encouraged me to wild-harvest too, and I used to go rabbitting with ferrets, and shooting pigeons. I was a fisherman from when I was five years old. I loved fishing, and I still do.

I grew up in the hippy era when everybody was dreaming about self-sufficiency, peace, love and harmony and all that sort of thing. That was the great dream in the 1960s but there wasn't much room for it in England with its population and land prices. As a working class person it was not easy to move on to land. My mum often dreamt of buying a farm and got quite close a few times. She instilled that dream in me too, excited me about being on a farm with animals, though it never actually came about.

My father was a truck driver, an ex-army man from the Second World War who was a lot sterner and stricter than my mother. But Dad loved driving his truck through the countryside for which he had a shy and quiet love. My mum was a business woman running her own business, so I had to stay with my father. During school holidays I'd often be stuck in the truck and would sit there looking at the countryside.

My auntie, mum's sister, was an avid mountain walker. She was a very old fashioned, stern lady and I often used to get lumped with her too. She'd take me walking over mountains and river valleys. It was old-fashioned hiking, that sort of very Christian, old-fashioned holiday style of 'good healthy fresh air', getting out in the mountains and taking a deep breath. I stayed with my auntie quite often in summer holidays. There were lots of lakes and farms in the area, and I spent a lot of time around farms and fishing in the lakes. I actually got into quite a lot of poaching when I was younger, because I could not afford to fish in the royalty fisheries and expensive salmon fisheries. Up until I was sixteen it wasn't a criminal offence, just a clip around the ear and a march home by the local policeman.

All of these things influenced me and my love for natural systems and the wild, and having to be quiet when you're sneaking around, trying to fish or hunt. You've got to be really quiet so inevitably you see a lot of wilderness.

I eventually started to travel, and for a teenager that was a big hook into other cultures. I travelled through Europe and in the early '70s I went to Morocco for three months on the hippy searching trail. I lived very cheaply in a Volkswagen Kombi and saw extreme variations of culture for the first time. That was my first connection with Islam as well, and the start of my respect for the Islamic faith and the Islamic peoples. That's become a very long-term situation where I've had a growing understanding and since 2003 I have become a devout follower of the Islamic faith. My second wife Nadia is a Jordanian Muslim of Bedouin decent. Her father is from the Negev in Palestine, and is a tribal judge, herbalist and organic farmer. Her mother's side of the family is from the Dead Sea Valley. Nadia is the strongest, most sincere, faithful and wisest person I have ever met. In Bill Mollison's words, I have married my dictionary.

A Permaculture Life

I emigrated to Australia in 1979 with $50 in my pocket. I'm a mechanical engineer by trade but wasn't getting much work so I started a service industry business doing commercial cleaning. I worked day and night, and with my staff, ended up working day and night shifts up and down the coast. I was doing pretty well. Eighteen months later I'd paid off my first house, and two years later I bought my first farm – 34 acres between Noosa and Cooroy. So started my farming life. Typical immigrant story.

I first saw the word 'permaculture' around the Sunshine Coast (Queensland) in 1979-1980, when it was used to advertise various events. Permaculture Nambour was the largest group in the world at that time. The word was kind of interesting, I wondered what it meant – it was an enquiring word. I went to a weekend course with Max Lindegger at Nambour TAFE and then to Permaculture Nambour meetings. Then in 1983, I saw a PDC course advertised in Stanley, Tasmania; Bill Mollison was the teacher, and I thought it would be worth attending.

I sat at the back of the room and never said a word. Half way through the first day of Bill's lectures, I realised there was so much evidence that things seemed compellingly bad – it's nearly all come true now. I thought, "I can't take notes." So I threw my notepad over my shoulder and just folded my arms. Every time Bill lit up a ciggy so did I, to the 'tut tut' of all the hippies in the class. I just sat there and sponged it in.

When I got back to the Sunshine Coast after the PDC I asked Max Lindegger if I could work with him as an assistant consultant and see what he did. He only had a little motor bike at the time, and I had a nice Landrover so I said I'd drive him to his jobs if he'd let me work with him. I walked with him and listened to what he was doing. On each job, I inevitably started to consult with the wife, while Max consulted with the husband. After a while, Max offered me some payment, and after about two years of casually consulting, I started my own consultancy.

I'd already started doing earthworks on my property, and was lucky to come across a good earth mover in the local area, who I ended having a very long working relationship with. Being a mechanical engineer I wasn't afraid of directing machinery. I understood earthworks and how civil engineering worked, and consulted on and installed a lot of earthworks like dams and swales.

Around 1990 there was no more Permaculture Nambour. Crystal Waters ecovillage had sucked the energy away from Nambour and the permaculture group had collapsed. It became a magazine, which was purchased by PAWA (Permaculture Association of Western Australia). Jeff Nugent had recorded Bill's PDC and I listened to it again. Bill said, "Look, if there isn't a permaculture group in your area, someone's got to get off their arse and start one." I remember thinking, "He's talking about me." My biggest fear was the invisible structures, building community. That was something I didn't want to know about. I didn't want to start any communities. I didn't want to do any of that stuff. I hated the idea of public speaking but I had always been good at small business and I could make money. I liked the idea of starting a charity and alternative finance, and have since been involved in initiating eight charities. So about eight years after my course, after I'd developed my farm, made a lot of mistakes and learned a lot, I decided to bite the bullet. I began teaching and started a permaculture group.

I started to get a result with my students. Bill moved up to the Tagari farm at Tyalgum, northern NSW and I used to go down and talk with him. He offered me a job in Ecuador, teaching. That was the start of my international teaching, and I just kept getting asked back to teach. That's the thing with permaculture. If you get a result, they're going to ask you back again. And they're going to ask you to more difficult situations each time. If you keep getting a result they'll keep asking you.

So I kept being requested to teach overseas. I also continued to teach in Australia, and I kept getting bigger consultancies. Eventually I got asked to do some emergency aid work. Then Bill asked me to manage the farm at Tagari when he decided to retire from international teaching in 1991. One of the conditions was that I set up a charity called the Permaculture Research Institute (PRI). That's what Bill said it must be called, which was a clever strategy. Now we've recently set up the new PRI USA and PRI Jordan. It's on our website with all our new projects, our whole master plan. It feels like I'm crouched behind a very large floodgate trying to undo the bolts and we're going to drown. We're going to try and swim as quick as we can and I might need a surfboard but it's going to be a lot of fun. I think there's nothing else worth doing actually, there's no other game in town.

People say, "What's your plan?" Plan? We just work on request and with the increasing enquiries that keep coming in our direction, the best plan I can think of is our master plan to set up sustainable aid projects as permaculture education centres, and demonstration sites that network their information locally, nationally and internationally. I'm having a job keeping up with the flow. I'm trying to hang onto a bus that's taken off. Sometimes people argue about which way the bus should be going or who should be steering or who should be allowed on or what colour the bus should be painted, irrelevant things like that. I find it's just better to carry on. Just keep going. Don't even engage in that conversation. It's not relevant. You're just trying to hold the damned thing straight, as it goes off like a dragster down the street.

Permaculture in the Industrial World

So much of our lives is a contradiction if we live according to the dominant paradigm; with our ideals, with what we know is right. That takes away our dignity. It takes away our humility. It means we can't develop our own potential, our own personality, our own fingerprint of existence, our own explosion of expressive energy as a living thing. We think and feel meaningless. One of the main things we feel in the everyday of industrialised existence is that we should be cheating time, and therefore we should be gaining time within our life. But the opposite happens. You look back on an industrialised life

– the standard, modern human being living in what we call 'civilised living scenarios', and it's gone in a split second. It seems like there's no record of the very little time that we've had in our life. But when we're actually going through that experience, it's boring and it's going very slowly. We look at the clock at five to five on a Friday afternoon – it's glued at five to five. That last five minutes is an eternity of boredom. Then we finish work and the weekend goes in a split second as we go out and often abuse ourselves in some way to get over what we've experienced for the past forty hour week.

You're not even a pawn in the game, in modern life. You're an insignificant little part of a massive system that you have no respect or honour for. People in our cities don't have clean air or clean water. They have a huge job to find clean food, especially locally produced. There are barely any sensible houses at all, they're almost all made of toxic materials, they don't heat themselves or cool themselves without use of fuel, and most still don't catch their own water. The materials they're made from are not replaceable from within the bioregion. In cities, community is very limited. If you scream in the middle of the night in a Melbourne suburb someone's going to complain, not come and see if you need help. It's an outrage that we can't drink the water from our rivers. In our society we take this for granted, it's like we've gone to sleep. We accept pollution in our environment, the extreme poverty around the world and the poverty of spirit, the sadness and the destitution in our own communities, the alienation. We accept these things, we're not outraged by them. I think there's room for some righteous indignation and outrage.

But as practicing permaculturists we don't accept these things. We have the solution. It's time to step up to the plate and put on the badge of honour and take a position where you could end up being an honourable warrior who can mentor other people into that position. It's time to stop being afraid. Let's get on with it. That's how I feel. I try to help people into that position. I'm sure there are many, many people who are going to outshine all the action we've done. I see some incredible young people coming through. What is really interesting is when you *engage* in a meaningful life, and you're *really involved in it*, every single second is engaging, is involving, is something you're absolutely passionately interested in. You don't own the future because it hasn't happened yet. The experience now, what's happening *right now, this* is the experience. You don't own it because it's happening, millisecond by millisecond. What you do own is the past, and what you have done.

When you actually make a move, it's at the point when you get fed up with living with one foot inside in an inappropriate world and one foot over the line. It might be on weekends that you're living the appropriate life that you want to live. But you get so bored with having one foot on one side of the line and one foot on the other. That's where most people are when they take their permaculture design course, or by the time they finish it. They just get fed up. When you're living in the boring world, and want to live in the engaging world, you're dying every day. Many times. Because you know the difference.

The day that you decide to put both feet over the line is when you say, "Oh shit, I'm fed up with being scared, it's just bloody boring, I've had enough of it, I'm going to take on the challenge. With courage I'm going to step over the line." Now the moment you step over that line, the next step comes very quickly – and the next one after that comes even faster. Before you know what's happening, you're actually charging towards the inappropriate system to make change. You are now starting to function and to do it better and better. When you actually get into that stride, you have no fear because you're in the fury of the charge, going forward. It doesn't mean you have to be angry; you may be driven by anger but you don't have to be angry. But you have meaning and purpose driving you forward.

Now at that point something very interesting happens. Your time expands because you're starting to function very clearly – and you also lose all your fear. You have no fear. You're not dying inside every day. You have to face your one death that we all have to face, but that's it – just that one. It doesn't matter if you die in the charge, because it never matters if you die in the charge – that's what a charge is about. It's taking on the mission full bore. That's the experience. It's very hard to describe to people but it is a real experience.

I've talked to other people as they embrace permaculture and I can see it in their eyes, "You're terminal, I've hooked you, or you're hooked by somebody." You think, "Here's another warrior, I can count on this person." They're with you all the way. We're building a global peace army of people like that, terminally infecting them. It doesn't matter what culture or what religion or what race they are. As long as they're a human being, they can be hooked. They are moving towards change. It may be an evolution of human thinking. It's about pride, pride of action. It's okay; what's needed is just to make the leap. You can give it your all.

Then there's a point where you stop owning a project, it stops being your project and becomes the property of everyone else who's involved. You might try and guide and direct it but it has its own life. And that's great. Permaculture has a persona that is larger than life. That's one of the attractive things about it. We don't own any of it and we have no power, no control over it. We never will have. I haven't been to a Permaculture Convergence yet where any of the decisions that are made actually happen. People think that they've got some kind of power to make things happen but things happen anyway whether they like it or not. Whether it's the way they expect it to or not, it doesn't really make any difference.

Maybe decision-making at convergences is just part of the process. Maybe those decisions won't result in that thing being carried out, but it's the process of getting together and making the decisions that leads to the next thing. I think it should be a lot more light-hearted, just everyone getting together and sharing a few experiences and having a bit of a laugh about it really. A laugh and a cry if necessary.

Young People and Permaculture

Of course some young people get so passionate that they become a little bit too expressive, a bit too maverick, a bit too gung-ho and aggressive. I was absolutely guilty of this too. I could be violently aggressive, but I've learned to control and refine my tongue. I'm quite a verbal person and it's probably one of the things that helps me teach, but I've had to refine and sharpen the edge of my tongue so that the razor is very sharp when I need it. But I don't want to go slashing my way with my words, giving the impression of just emotion and anger. Maybe it's something that comes with experience and age. There's humility that comes as you get older, a recognition that it's important sometimes to meet people where they are. As a teacher, a lot of people take you on as a mentor, and you have to stay humble about that. I like to make sure I am giving people the right mentorship, a refined quality of action.

My first wife was Australian and I have two grown-up children born here. They're both terminal permaculturists. Often kids don't necessarily follow the same path as their parents. But I see the opposite in permaculture. I'm always surprised by that and have thought about it as a father. Everybody said, "They'll probably become bankers and real estate agents," but no. I'd never been to university until I was asked to lecture, I didn't even know how they worked. But my son and daughter went through university. My son has a Master's Degree in Environmental Science and Environmental Management. He wasn't academic at school but it was his hook into permaculture that helped him academically to get into university. Not having finished year 12, he did that through the credit he got for his permaculture diploma. From the bottom of his class he went to the top of his class during his university studies and became more and more academic because of his love of the subject. I've noticed the children of many people round the world have continued their parents' involvement and taken it even further.

Permaculture tends to foster that. It is probably one of the most meaningful ways you can approach the world. It gives a framework of knowledge that is common sense to most children and young people as they develop.

It fosters meaningful approaches and the ability to rationalise one's existence, and the creation systems that we have to live through and amongst. If your parents are leading with that example, this gives you even more security that you're doing something worthwhile that your parents also are passionate about. It's like an insurance, knowing you have a meaningful existence which was also the main direction your parents followed. It's just common sense.

Permaculture changed the way I think and see and act in the world. The framework of permaculture provides an opportunity for this kind of thing to take place, whereas our modern life generally doesn't make room for it. It gave me a gauge for my responsibilities; what I should be responsible for. I need to feel comfortable about my energy use and the waste products that my life creates. And I now have a very accurate gauge of what I can feel responsible for and I can feel morally clear about my actions. In my life before permaculture I didn't have that clarity. I couldn't really consider the footprint of my life. Now I can clearly see the footprint of my life. I can't walk out of a room, even in someone else's house, and leave a light on. Now I live on solar power and a battery bank. It's pretty obvious I can't waste energy. When I look at recycling, when I look at putting garbage in a bin, I can't just chuck garbage in one bin if there are recycle bins there.

I have a pride in the way I can live. I'm very proud of the level of health that I can achieve. I'm very proud of the quality of food that I can produce for myself and supply for my own needs. I have an endless supply of irrigation water on the farm and an endless supply of rainwater – we can bathe in rainwater. It's a luxury, but the bath water that comes out of my shower and sinks goes through a reed-bed greywater system and ends up as soaking trenches under the lower beds of the kitchen gardens. I'm very proud of that.

I mineralise all of the animals on the farm so that their manures are of a very high quality which all goes through the pastures, through the animal pens. They also get composted and turned into highly oxygenated compost tea that mineralises the landscape. We do it by spreading minerals throughout the whole farm so that the herbs we give them as forage – pigeon pea, acacia, comfrey, mugwort, arrowroot, acacia and bamboo leaves – whole mixtures of forage that we cut for the animals, are already mineralised. They have a very high quality health, they have very vital energy.

And when we butcher those animals we do it in the most humane way. I feel very comfortable about the way I can process the life systems. The living animals are dealt with very well, not only the way they live but the way we choose for them to die. I feel very comfortable about all of that. Ninety percent of our animals we knife-kill with an extremely sharp knife. I never sharpen the knife in front of the animal; I take the animal away from the other animals of its type. When it's in a very comfortable position, it feels comfortable, with the knife at hand, I will pull the animal down on its side, and very quickly cut its throat, right through into the spinal cord. I'll keep the animal's eyes covered and I will be holding it with my hand on its heart. If it's a large animal like a calf, I'll be sitting on its chest with my hand on its heart all the way until it is completely dead, and it is a piece of meat. I will not leave that animal. I am with it all the way. It's a matter of honour. I am responsible for that death or whoever does the butchering, they have to do it the same way.

When I take students through that process they are impressed at the very small amount of fear or reaction there is in the animal. It's a very passive death. It may be quite bloody at times if it's a large animal, even a bit noisy and gurgly at times because of the volume of blood. But the reaction of the body is very slight. After maybe 30 seconds you get a nervous twitch and a kick, but it's very little if you do the job right. It is a specific skill.

That's probably the epitome of pride of action, being very responsible for what you do. And feeling very proud of that, it's honourable. On a large animal it's very rarely something a woman does. It's something men need to take responsibility for, and it's very sadly missing in young men. This has certain psychological repercussions, I am sure. It does not allow them to grow up. If I take a young staff member who's been with me for a while through different sized animal kills till we do a large animal kill, they'll normally take a deep breath and a gulp of the throat afterwards and say, "Whoa, I've grown up a bit today." And that's missing if you're eating an animal that's out of a feedlot, that's gone through an abattoir. It's like a rite of passage in a way, which is really important. Your ancestors did it, it's inherently in your veins, and you need to take responsibility for honouring your ancestors in all these things, and in everything we do. All of what we do is about pride and honour in being part of existence. That changes your view; it gives you a certain amount of bravery and courage. And we need quite a lot of courage, we're in a pretty desperate damned situation. These are the kinds of changes I've observed in people through their involvement in permaculture.

Nearly everybody finds their energy level comes up, their ability to give it their all. They can see no reason why they can't feel, "We can go with this, there's absolutely nothing wrong with it, we can give it all our life's energy." People vitalise around that. And they take action. No matter whether you like it or want it to happen or not, you do get a certain amount of courage. The thing about permaculture is that it just makes sense. It's one of the few things people encounter that just fit. After his course my son said to me, "This makes the whole world make sense." Once you've done this so many things suddenly make sense. Another permaculture teacher from the Sunshine Coast told me about her work with Transition Towns and Power Down movements, and all these new kitchy words that we're using. She said that she now goes into meetings with authorities and organisations of influence, and purposely doesn't use the word 'permaculture'. And the more she doesn't use the word permaculture, the more they bring it up! They start to say, "Well this is permaculture," and they start to tell her about it!

Reaching Out Through Teaching

You get this ability to be able to talk and teach. Whether you like it or not you become some sort of teacher. Teaching is something that happens all the time every day, with everybody, in traditional community. Even explaining is a way of teaching. People from different backgrounds and different situations come to and respond to permaculture in different ways. Some are prepared to take on something that they would normally have not done because they've not felt qualified, and others have gone back into their qualified professional position, and changed the rules from within the camp because they have access. So instead of opting out, they've gone back into corporate banking or civil engineering or architecture or town planning, and influenced them from within. That's been very interesting and also quite subversive, in a very peaceful and meaningful way. Interestingly, we aren't much noticed for our subversion really. Apart from the environmentalists saying we use non-native plants. That's subversive but pretty lame. All the plants that we use for our needs anyway, are from global sources around the whole planet.

I've been trying hard to get information into students, to find simple ways so that anybody can understand it, even people who can't read and write. I've had to teach people who can't read and write who often learn permaculture as a system better than people who are academic. You can't talk in academic terms, so I've tried to make it understandable to anybody. As a teacher, I realised a little while ago that you need to help people find constancy in the universe that they can anchor their design on, and you can teach those as anchors in the design process. For example, evaporation always cools. Condensation always warms. Water only moves at right angles to the contour. Heat only moves by conduction, convection and radiation, just three ways. That's it. It doesn't matter whether they're designing a house or a system that needs heating or cooling, that's all they've got to think about. These are absolute constants, yet we all know the trendy sayings like, "The only constant in the universe is change." Well, that is true, but there are also constants out there. When I realised this, I started to see and to find them. I used them as tools and explained this to students and I could see them actually gaining in confidence around me. They'd say, "Oh, I understand that, no one ever told me that." It's about them being able to understand the science.

The emotions – pain or pleasure – anchor the information, that's the only way you can get people to remember things. We have our students bouncing one way then the other. You're laughing one moment and crying the next. You try not to be too painful but you touch buttons, and then you jump the other way and make them laugh; if you can. Or you talk about emotional things like love, hate, things that touch emotion and anchor the information. You can't really teach people much information but you can give them the infection of passion. They can get their information later. At the permaculture convergences I think it's getting a little bit better, because time has gone by and we realise that the decisions we make about this movement are not relevant; it's more or less out of our control; as it should be. Decentralised like a wild system. It's not controllable. It's like an epidemic that's infectiously spreading, like weeds. Weeds keep coming up in damaged country.

Patterns and Our Place in the Universe

Whether you're a vegetarian or whatever type of dietary scenario that you're in, you do understand the cycles of life and death. You understand as much about decomposition as you do about life creation, and extension. You understand that the decomposition cycles bring the soil alive and hold the fertility, and you know that cycle gives you the life-cycle. You're seeing cycles so that you can position yourself within them. You know where you are. That gives you a lot of understanding and an ability to approach the system with confidence. Everyone gets a bit of confidence and it's really nice to see what you can do for people.

There's actually a look that people get during a PDC when you get to section on pattern. It's the look of understanding. As you go through pattern the whole course changes. It's this sudden realisation that things are infinitely variable but very similar and limited in form. There are only a few limited forms in the patterns of the universe but there are infinite variations or slight imperfections. We're very interested in the slight imperfections more than the conformity of a few forms of pattern, no matter whether you look at the stellar galaxies or the tiny microbes, from the macro to the micro, the micro to the macro. What we're interested in is all these slight variations and imperfections that don't repeat.

Understanding the patterns of nature is about understanding the pattern forms. There are not that many forms of pattern. There are the branching patterns of rivers, which are also the branching patterns of your veins and arteries in your body, and the branching patterns of trees and plants. It's a dendritic pattern. And there are spirals which you find in the flows of water and the patterns of flowers. There are imploding and exploding patterns, and different tessellations and core patterns. There are the hexagon patterns of the beehive and the tortoise shells. All snowflakes have six-sided patterns. But the thing that's really interesting about pattern is that while all snowflakes are a six-sided pattern because water freezes in a six-sided pattern, no two snowflakes are the same. There never have been, there aren't any that are falling at this moment, there are none in the future that will be the same. Just like there are only two patterns of fingerprint on your fingers – the swirls and the spirals. Only two patterns and yet there's seven billion people on earth, but no two fingerprints the same. Identical twins have similar finger prints but not exactly the same.

There are very harmonic patterns and there are the destructive disharmony patterns which indicate chaos and break-down patterns. So we start to see a difference between harmony and disharmony. This is like the difference between something that is sustainable and productive, and something that's unsustainable and destructive. We start to see a difference between a positive and a negative. Mind you there's never a negative without a positive. We start to see the pattern of our behaviour. And there's something very interesting. As negative as we presently are, we could be equally positive. We're not inevitably negative. We're not inherently negative. We're just in a negative way of behaving. We could be equally positive. So we start to get hope. Permaculture design gives you hope.

This is very much anchored around the subject of understanding patterns in nature, because that glues our subject together. The other thing that's very relevant is the orders of size within elements in systems. There are five to nine orders of size of anything. It doesn't matter whether you're talking about the branching patterns of a river, or the branching patterns of a tree or the sizes of oceans or the sizes of animal form. Like how many different sizes of kangaroo or horse, how many sizes of river or planets or galaxies there are. It makes no difference what you choose, there are never less than five orders of size, and never more than nine. Things don't get infinitely bigger and things don't get infinitely smaller within specific elements. That immediately gives you a starting point and an end point of what we should be aiming to achieve, and a balance in the size of orders. That's very relevant.

Bill talks about when he was trying to conceive what it was he was trying to unravel with permaculture, what he was trying to find as a system that he couldn't quite put together in his mind. He said it was like he was looking up onto the side of a hill and there was a carpet there that was all rolled up and tied with rope. He had an epiphany one night, when he felt like he cut the rope so the carpet rolled out and down the hill. As he watched it rolled away, rolled out over the horizon, and into infinity. Forever. He felt like he jumped on the carpet and started to walk down it, and realised not very far down, that it was an eternal journey. It went forever.

But it's not infinite, not an endless swathe where you can be lost. It's almost a 'songline' or a road map to sustainability. It has its limitations and endless permutations. But there's an edge to it. We talk about it, and say, "How do you use this?" Well, then you start to design the edges. We can relate to people and say, "Remember when you had your first colouring book? You made a mess of it, and then what appeared to be an extremely wild adult, like Mum or Dad, came around and said 'Look, if you just carefully colour the edges first, you can colour the middle in really easily.' You remember that experience? Well nothing's changed. It's exactly the same. Concentration is on the form of the edges. Now that is pattern recognition at its simplest. In your childhood you can relate to those things."

What happens during a PDC is that students start to reformat the knowledge they already have in their brain. The life experiences and the knowledge they've accrued up until that point in time suddenly become relevant in a different way. It's like when you look at your desktop on your computer and you see all these files and folders in a bit of a mess, and you start to reformat and reassemble the information in different folders; in a relevant way, so that you can then access and use the information more effectively. I've noticed that when we get at least half way through a course towards the end nearly everybody comes in and says, "Oh, last night I was dreaming about planting food forests, or about contour gardens, or putting swales in." So I ask them, "If you dream about permaculture, can you tell us the next day?" And nearly everybody does. It's because their brains can't stop reformatting when they're asleep. They're reformatting it into an order so they can use it. Every time they come in I almost tick them off psychologically, "Okay I got that one, they're good, they're terminal. They'll germinate now or sooner or later; don't have to worry about that one too much."

After we've talked about all these logical ways of analysing designs with slope and orientation and zones and sectors and finding more than one use for every element and how many connections they have and what the inputs and outputs are and the intrinsic factors of every element – this big analysis process – you come to the subject of patterns. We see ourselves in this because we're not perfect. It positions you immediately in your own universe. You go, "Hang on a minute, I can see everything else and myself at the same time." People find their place in existence. They find their place in the universe.

Spreading Internationally

There is a high level of humility in the approach that was taken in how permaculture spread. Bill basically said to people, "Go out and do it." This is what permaculture teachers do. Quite often a movement like this can end up centring around one or two people. What's needed is some sort of humility, some level of willingness to delegate, willingness to have trust that people will just be able to take this on and go with it, once they've been given the basic ideas and the basic skills. We give people permission and confidence to act, starting at the back door and working out. We can give 100% infallible ways that will work if you go about the process in the right way. That builds confidence in people.

But then we also go right out on the edge talking to governments. In Brazil – the fifth biggest country on earth – we had good feedback from a questionnaire from the government workers who were our students. As a result we were invited to talk to the Minister of Agriculture. So we just walked in – we didn't even have suits on at the time – to talk about agricultural policy, and help to change the agricultural policy of the colonists in the Amazon. In that meeting permaculture was adopted as the 'family agriculture' of the Amazon. We were pretty arrogant and pig headed really, just three young lads, almost hippy types, trying to do something out there. But we were in a good position to get a government course going and influence the result. We get out there and we give people permission to act, to have a go, and to try and make change.

Now I've got fitted suits which I buy in Bangkok on the way through. I have to have the corporate camouflage to get in the office and I feel a lot more subversive than I ever did. I get in front of people and they don't even see me coming, I come under the radar. It's still me sitting in front of them, the same old person. But I've got a fitted suit on and a recycled silk tie from Reborn Wear, one of our aid projects in Japan, made from a 200-year-old kimono.

In international PDCs you see that people who come from an area where there is suffering and great need, immediately start work to help their local people. They move towards their local community and they act NOW. They don't think about it. They think, "If this is going to work, I'm going home and I'm going to dig a swale." Before the course finishes they've dug swales and banana circles and they're trying them out. What they do today feeds them tomorrow, and if it's going to work, they're going to try it. If it doesn't work they're going to tell you. It's immediate in areas of need. They don't have time to debate principles, that's a first world luxury. The principles are on the ground working and if they're not working, they're not worth even talking about. You don't get into a debate about whether we're starving or not. They are starving and they need to do something. If your kids are hungry, you move.

I was outraged when I went to Tanzania to see why one in three children were dying of cholera because the pit latrine toilets flooded in the wet season and it is so easy to teach people how they can build their own dry compost toilets, turning the problem into the solution as a fertilising soil amendment. Children dying of malaria just because poorly designed open storm water gutters full of modern plastic garbage mixing with greywater was creating an ideal breeding ground for mosquitoes and it is so easy to teach people to build their own greywater reed beds and create clean irrigation water and we can so easily teach people to recycle and turn waste products into an economy. Children with greatly depleted immune systems because the soil's fertility has been degraded through de-forestation for fire wood and cut over for the third or fourth time and it is so easy to teach people to plant and maintain diverse multi use fire wood forest systems which will help start the process of recovering soil fertility.

We can also easily teach people how to increase the organic matter returned to soils and increase the volume and diversity of the soil biology, and then plant a great diversity of crops and create an increased nutrient density. People say the African problem is not solvable. That's absolute rubbish. It's quite easily solvable and the people themselves can do it. It is our perception of wealth that moderates our population so when we use new indicators of wealth as clean air, clean water, clean food, sensible housing, community warmth and friendship we become healthy and truly wealthy. It's not about changing the world; it's about changing our minds. It's really that simple, and yet it's that difficult. But if we choose it we'll have no problem doing it. It's actually to our advantage. It's just that we're stuck on the other side of the line, so scared of making that change. That's what's so wonderful when you encourage people to be brave enough to step over the line. Step over that line and don't compromise with that system any more. We can welcome you in out of the cold. It's nice and warm on this side.

These things really matter, and as a teacher and activist it can really push you. You have to get very grounded and learn good anger management. You have to learn cultural sensitivity. I used to be angry about the state of the world, and that permaculture was not being adopted quickly and well enough. I was arrogant enough to think that it could be done better, but some of that was youthful arrogance. You have a lot of energy to spare and you feel like you could do a better job. Some of it comes from not really experiencing the level of pain and suffering that some people have to go through. You think you've suffered and you think you've been hungry, but you haven't, not really. And when you've actually been through that in its true form then you know what people are going through. We experience that ourselves when we work overseas. We live with the people we work with, sleep in the beds they sleep in, even if it is on the ground, eat the food they eat, use the toilet and the washing facilities that they use. We know what they experience.

I've already lost students. I've lost students to malaria, and some have been murdered. Recently I lost a student in Tanzania who was shot in a break-and-enter at the Arusha project. That's something I have to feel responsible for. With that level of commitment they died in the charge. As someone that actually taught them and inspired them to go, of course there's a level of responsibility, especially when you speak to their parents or brothers and sisters. We're all family in some way or another. But it's a sign of the commitment of people, the bravery of people, to help systems for the greater good.

If you get a result in emergency aid, there's only one place left that they can ask you to go to, and that is a war zone. You have to decide whether you want to go or not. If you do not feel psychologically stable enough, or realise that you are taking a risk with your psychological stability, consider that before you go. The ultimate negative in experiencing suffering is to live with people in a war zone where people are dying. Where the conflict is causing multiple deaths in almost every family. Where you've sat with people and listened to the stories of watching people die; watching their children tortured in front of them – such horrific things that it's hard to talk about them.

So you approach people hopefully with a level of cultural sensitivity; you sit with them and work with them to create a sustainable, abundant, peaceful future, help them find hope in the future. When they're telling you things they've seen that you cannot believe a human being can be telling you and still be sane, that knocks the anger out of you. You're either going to go insane yourself, or you're going to calm right down. You've only got edges there. If you go into a war zone, you either come out very psychologically damaged, or you come out very much calmed and in touch with yourself. You have to look at yourself clearly and seriously and examine yourself very honestly. Because if you're going to go and work in that situation, you have to think, "Am I doing this for my own ego? Or am I really doing it for the greater good? Am I willing to take this risk?" It doesn't have a lot of money value on it. You realise that you're being involved in the ultimate tragedy of human kind, the ultimate breakdown of communication – a war. As permaculturists, to go into an emergency aid scenario is to experience tragedy. Everybody has suffered. But to go into an actual war zone which is the ultimate suffering created by opposing people – that is the ultimate chaos.

So I tell my students, "Be careful. You must draw the line regarding where you want to go. Because if you are good at this, you will be asked back into more and more difficult situations. Inevitably you will be asked into aid, and then if you get a result, you'll be asked into emergency aid. You will see tragedy. It will touch you." As Bill has said, "You will have nightmares on the ceiling at night that you can't switch off."

Learning and Working with Permaculture Design

It's 'go home and garden', but then that's extended to gardening community, gardening the world. You agree to garden. You agree to experience the energy audit of something as simple as gardening, the energy you put in for the product you get out, the exchange of energy. If you can think it through, you can extend that to the largest scale you like. You can design continents if you like. No problem. Scale doesn't matter. It's scary, but once you get past a certain scale you realise there's no limit, you could just keep going. It's holographic. You can extend it from the smallest to the largest. So again, you're edge-thinking, going from the smallest to the largest edges. You can apply the same thinking to a nation as you can to your window box or to your house. You've got the same analysis of approach.

That's the wonderful thing about the *Permaculture Designers' Manual*. It's not like a normal book on ecology which is passively written as an observer, observing ecology and the environment from a distance, and writing it down. The manual gives you directives to act. Of all the names Bill could have called that book, as wonderful a book as it is and as definitive a book on design, he called it a 'manual'. Just like a workshop manual for a car. As a mechanical engineer, I work on cars and engines. If I've never changed the clutch on a particular type of car, give me a manual and I'll read it through to change the clutch. That's how I use the *Designers' Manual*. If I'm in a desert and I'm not sure how I'll harvest water in this particular landscape form and climate, I look at the manual. That's what makes it stand alone. In all my courses in Australia, students get a manual free of charge, and I have all the students sitting with the manual on their lap. Then I can reference the manual in relation to their questions or the lesson we're doing. Afterwards they thank me for it because they can actually use it. It's a daunting book.

It's important to be able to take that information and apply it in some immediate way, because then the information explodes and goes a lot further than you'd imagine. In that action there are numerous lessons. There may be twenty lessons in a paragraph, but when you apply it in action, there are a thousand lessons. In addition, if you apply each one of those, there are a million lessons. It becomes infinite. This is typical of permaculture. There are no simple systems, no such thing as a simple permaculture. You need to find the most concise truth of a design and the main frame needs to be very clear and concise. It will complicate itself in application and in evolution over time. Don't worry, the complexity will happen!

I don't write very well so I've tried to inspire people who can to write the book about the thousand or so absolute constants in the universe, for all the school children in the world. It would be a fun book; it would let the kids feel more confident about the world they live in. It's very simple and very deep at the same time. It's like a tool, which if you could 'get' it; it would form the foundation of your life. I can see it being something that would work as a kindergarten picture book right up to an academic manual for students at university, or a fireside book that Grandpa reads and feels good about his life in the final years and can say, "I've understood these things in action all my life." Someone will do it one day. It would be a really great toolkit. We'd be able to design sustainable systems with such a book.

The Future: Weaving New Community

There are a lot of barriers to us living in a sustainable, just way. Corruption, denial, greed, bureaucracy – these barriers might be decreasing, but they still exist. I think permaculture can help because it is a system and a movement that begins with an ethic, so we're anchored from the start in an ethic. It has three very simple parts. Earth Care, People Care and a Fair Share on return of surplus. Earth Care is about how all living and non-living things have an intrinsic worth. That in itself says that living things do not develop purely for evil. They have developed in relation to an opportunistic niche that needs to be filled in the ecology of Earth's natural systems.

As civilised people, we have achieved quite a lot of control of prejudice against others, as a society we accept the intrinsic worth of people. We're morally not supposed to be prejudiced against human beings because of their race, their religion, their beliefs or their culture. We should be working towards that in relation to all living things, not just people. People who are supposedly New Age and alternative and environmentally minded, who are often the most proud of not being prejudiced towards other people, are sometimes the most prejudiced towards other living things. Isabell Shipard (in my opinion one of the best herbalists in the world and a long-term friend and co-worker) said to me many years ago, "If you hate anything out there in the environment it will haunt you forever." You see the world through the framework of that hatred. We have to be very careful about that because if you hate a particular plant because it's not a native (it could be a hard-working immigrant) or you see some animal that's not a native, and you hate it, then that's the first thing you will see in the environment when you look. When you design you must look at the land with no prejudice, with a completely open mind. All you're doing if you have prejudice is putting a framework in front of the way you see the world. That actually doesn't give you a clear vision, but instead you're taking away some of your potential as a designer, to see and observe and design the world.

I use people as an analogy a lot these days. If people don't understand weeds, I talk about weeds as hard working immigrants, which works very well in Australia. When you talk about energy systems, if you put it into people terms, people suddenly feel personally affected by the analogy. It's because we immediately get emotional if we talk about people – ourselves and others, other cultures, other races, other religions, other belief systems. We need to work with that in the way we see the whole universe and to be able to switch off our emotions so we become better at designing. Then you can see the design or the design possibilities clearly. You always have options, and you need numerous options. With emotions, with prejudice, with anger, with love, with those emotions, you limit your options and your potential as a designer. It's an interesting skill. It's very enjoyable. People actually pay you to practice this enjoyable process as a designer!

When you have emotion, whether it's hatred, or love, you're not rational, you're not someone who can observe with a neutral mind. You have no emotion of unconditional love. You need to look at the world as an open space, almost a meditative space. You have to walk through the landscape. You are just someone who is inputting, taking things in. You're not outputting. If you're approaching it from emotion, you are outputting. As a designer you have to switch that off so that you're allowing yourself to observe everything without emotion. This includes people and culture. That sort of cultural sensitivity is hard because you are a human being and you're looking at other human beings and they will charge your emotions one way or another, up or down. Then you don't clearly see the culture.

In the modern, western colonist countries like Australia, America and Canada, the indigenous people mostly don't take part in government and the development of culture, so you have no traditional protocols left. You don't have the checks and balances that many traditional cultures have. These can be very longwinded, but they have worked wonderfully well for multiple generations. A while back we taught a course in a Maori community in New Zealand. The checks and balances and protocols of their culture are quite extensive but they work extremely well. We found during the course that if people agreed to fall into that process it worked for anybody, even those who weren't Maori. I'm not saying we should adopt Maori culture or protocols, but we probably do need to develop our own. But we've lost the ability to decide who oversees that. Who is the head person? Who is the person who over time has gained heritage and respect, who has that really strong sense of heart and ability, and carries that integrity forward through generations? It's difficult.

This is also relevant to community. In countries like Australia, people aren't as skilled in simply working together as those from other countries. A lot of people in urban areas in Australia are used to living quite isolated lives, isolated from their neighbours, driving wherever they've got to go. They don't actually have to meet and work with the people they live near. So when you bring people together in groups, they're not making a conscious choice about the particular people they're being involved with in the way that they would if they were selecting friends or people of like mind.

How can you possibly put your lifetime savings into an ecovillage when you don't have a process for developing and operating the protocols of community like traditional people have? It's going to turn into an 'ego-village', as these things so often do. What else would it do? Everybody's got an ego about their life savings, of course you have, it's your life's work. It's going to be very hard to tolerate everybody else's ego around their life savings because you're trying to share one common vision of an ecovillage. This is not going to calm down your ego. Only cooperation and tolerance within the traditional cultural protocol can do this. And this is something we don't have. We have to reinvent it but we don't have the fabric of community to enable us to reinvent it, our industrial society doesn't allow it.

With the Permaculture Noosa group that I started, I thought very carefully about this because I was terrified of it. How do I start a permaculture group and what are the themes? How do I theme this group to develop and be strong and demonstrate its own evolution and keep it going? The group is now 17 years old. Many other groups have modelled it, which is great. What I found was that after a few years of those people interacting together, through this design system, they were sharing happy little accidents and sad little failures. They were sharing things that worked well, that they didn't even realise would work. We didn't realise that dogbane helped citrus if you planted it as an understorey to citrus, and we still don't know why. Someone in Permaculture Noosa discovered it, we all tried it and it works. It's an insignificant little herb with a smelly pungent scent and succulent, furry grey leaves which grows easily from cuttings. In the subtropics young citrus often suffer and don't make it through to maturity, but the dogbane helped, so we wrote it down as a constant. Sad little failures happened and we shared those too.

After many different interactions with all these people, one night I went back to the group when I moved to Northern NSW. I couldn't get in! There were so many people in the hall that I was stuck outside the back looking through the door. Nobody saw me, so I could observe the group without being noticed. There were lots of new people, and lots of people I knew and I realised I loved everybody there, because they were there, a great big mixture of people, all sorts, all backgrounds, all professions, and different races. I also realised that I couldn't live with 95 percent of them. But that indicated to me the 5 percent I could live with. And I suddenly realised the process was one of re-tribalisation. I'd found my tribe. And within my tribe, I'd found my totem. And within the totem was my family.

I suddenly saw this vision of society like a tattered rag hanging from a hem with just a few threads. Some people will say, "Oh well if permaculture works you should be in a glorious community with everybody interacting and everybody working together and everybody getting on with each other." Yeah, sure! That would be a tapestry in the fabric of a community. We haven't even got a fabric. We've just got a tattered hem with a few threads. We need a re-weaving process. We need to re-weave the fabric so that we can then embroider it with the tapestry that is the personality of our local bioregion, the fingerprint.

What kinds of protocols are needed to re-weave the fabric of community? Good, meaningful sensible action, that supplies our basic needs in the local bioregion. This is permaculture. It's on page 510 in the manual in chapter 14, the bioregional resources map. The manual has a list which I've made into an interactive mindmap. I run my farm and all my projects and client consultancies on a mindjet manager mindmap. It's more or less infallible because it is a dendritic type pattern, like looking down on a tree from above. You can see all the branches and you can add to the branches up to about nine times.

Re-weaving the fabric of industrial society into a more traditional set of protocols will be the ultimate achievement of permaculture. That is the biggest thing you can achieve, the bravest thing you can do. Forget the bloody war zones and the aid projects, they're simple, it's not hard work. It's bloody damned obvious what you've got to do to stop people dying of cholera because of pit toilets or floods. Little puddles of mosquito larvae with malaria, it's damned obvious. It's re-weaving your local community in Melbourne or Sydney or Brisbane or some other town – that's hard. You stick your neck out where other people in your local community are ready to chop it off, and say you know a better way that we can live. You start to reformat and reprocess community. You start to set up traditional protocols so that we can be sustainable and work together. That's the bravest work you can do. And if you can do it, it leads to absolute abundance. That's where permaculture leads. We need to get towards something similar to tribal governance with bioregional links to traditional protocols where each bioregion has a distinctive personality, and identity.

For example, we could identify a bioregion occupied by people such as water catchment or gradations of water catchment within a region. We identify that bioregion, and then we look at all the primary production that is possible there. Not just what is there at the moment but what is possible with global species, plants, animals, fungi, and the way it can supply our needs. We make up a complete mindmap matrix of that, with all the livings and all the products directly from primary production. Then we look at what we can do to value-add and process the primary-produced elements of the bioregion. Next we look at the people and the services they provide, and the industries related to their trades. These will relate back to the processing of primary production and the primary production itself. Then we look at the arts and entertainment in that bioregion – the cultural variations that stem from primary production and processing the primary production – the services to the arts.

There are four professions of mankind. There are only four, it's quite simple. Primary production, processing of primary production, services and the arts. Those are your professions, your careers. All of us should have a right to at least five careers in a lifetime. It's boring just to have one. You might have one major one and three minor ones and two insignificant ones, but you need a portfolio of careers, not one. As a human being you have a right to clean air, clean water, clean food. You have a right to sensible, natural, long-lasting housing, and you have a right to community. These should be part of should be the bill of human rights. They are fundamental.

When I look back on my permaculture life, it's enormous. It's massive. If I look back on what I've done this week, it feels like a year. What I've done this year, almost feels like a lifetime. When I look back on ten years, I couldn't write it down. I don't know how Bill Mollison wrote that book *Travels in Dreams*. My farm manager said to me the other day, "Wow, I've only been here six weeks, but it feels like six months." Now that's priceless. Because what you've done is you've expanded your life into time experience. I think that's a great selling point. If we could explain to people, "This is longevity *within* the time cycle – you've extended your life experience – what would you pay for that?"

There's a lot I could still say. But I'd like to finish off with this – to quote Graham Bell, and his book *The Permaculture Way*: "If it's not fun, you've got the design wrong." You need to rethink it. No matter what it is, whether you're doing community action or gardening or farming or earth working, it should be fun.

Zaytuna Farm. Photo by Craig Mackintosh.

Russ Grayson

Russ Grayson is coordinator of community gardens and Landcare volunteer projects for the City of Sydney and media liaison for the Australian City Farms and Community Gardens Network. His background is in international development and journalism, and he has worked as a community food systems consultant, producing policy directions for local government to enable community food gardening and community food initiatives. Russ is also one of a group of four who started the food education and advocacy organisation, the Australian Food Sovereignty Alliance.

Russ taught in an urban Permaculture Design Certificate course through the 1990s and worked as a Development Education and Project Management Officer for a technology and food security-oriented NGO, APACE. APACE is involved in farming systems and training work in Solomon Islands and PNG. He later worked with TerraCircle Inc, which has further developed the work APACE started.

More recently, he wrote the NSW edition of the action guide for a local government community sustainability course, Living Smart, and was urban agriculture adviser to the Callan Park Masterplan. A keen advocate of social enterprise, Russ is a City 'Cousin' in the Sydney Food Connect, which connects country 'cousin' growers with consumers in the cities.

For Russ, the permaculture design system is about community development, design thinking and making our cities desirable, humane, and rewarding places to live ... about making them cities of opportunity.

Fiona Campbell

Fiona Campbell was inspired to become involved in permaculture when she attended the first International Permaculture Conference in 1984. She taught the Permaculture Design Certificate course with Russ Grayson through the 1990s as well as teaching permaculture at Ryde College of TAFE.

A one-time management board member of the then University of Technology based NGO, APACE, Fiona worked with people in the Solomon Islands NGO, Kastom Gaden Association. There she used her graphic design skills to produce a number of training manuals and an ethnobotanical manual for the Association and the remote communities it serves.

Fiona had trained as a facilitator and teacher, so becoming involved in sustainability education was natural, as was becoming Randwick City Council's first Sustainability Education Officer. She went on to instigate and project manage the energy and water efficiency retrofit of the community centre that became the Randwick Sustainability Hub. The Hub, with its associated PIG – the Permaculture Interpretive Garden, is a hybrid public open space-educational facility and the first of its kind in Sydney.

Fiona is also a keen gardener and has encouraged her fellow apartment owners to build the apartment block's shared food garden.

For Fiona, permaculture is a method of systems thinking for the design of resilient cities and cohesive, collaborative communities.

Fiona Campbell at a Community Garden Network meeting.

Chapter 9 | Gardening Community

Entry

Russ: Tasmania. Late 1970s. "There's this interesting man and he's talking about unusual things," said my friend with intrigue in his voice, while we stood among the vegetables in his organic garden that fine Sunday in Moonah – one of those suburbs that sprawl northwards from the Hobart CBD to follow the banks of the wide, grey Derwent river.

This 'interesting man', I learned, made sporadic appearances at the University of Tasmania, surprising for such a quiet, staid university in a provincial city at the far end of the Australian continent. Hardly the place to hear anything unusual! My only encounter with the university previously was visiting it with a friend and finding that one of the walls was covered with a huge, rambling banana passionfruit vine from which we ate well.

There was also another interesting man, I was told, this one somewhat younger, and a student in the environmental design course at the Tasmanian College of Advanced Education. I thought little more about these 'interesting men' until my friend showed me a book he had bought. It had just been published ... by not one, but by both of them. It was called *Permaculture One*. Exactly what that meant eluded us, intriguing though the content was. But now we knew the names of these two interesting men — Bill Mollison and David Holmgren.

A foothold in the Metropolis

Sydney. Early 1980s. Back in the metropolis, I was working at a radio station on an afternoon drive program; music mixed with recorded current affairs and live interviews. It was the unfortunate habit of the gnomes in the newsroom to send anyone they considered odd, anybody with an unusual idea, anything unorthodox or a bit out of the box, down the hall to find me. That's how this woman, recently arrived in the city from the NSW mid-north coast, tracked me down where I was hiding in the editing room. "Permaculture? Yes, I've heard of it and have a copy of *Permaculture One* and *Permaculture Two*. I left them in Tasmania. You want to set up a permaculture association in Sydney? Maybe a good idea. And run a course? Well, why not?"

So it was that I recorded a series of short permaculture talks with Robyn Francis for later broadcast. Thanks to the radio station I also managed to scrounge press passes for something called an 'International Permaculture Convergence' that was to be held, in 1984, in the hilly backblocks of the mid-north coast, a place called Pappinbarra. This I learned, was the first event of its type and although I was initially unclear about what would go on there, I was impressed that a small number of people had come a long way to attend.

Travels in Permaculture with Fiona

We drove up in the clunky white Kombi that belonged to Fiona Campbell, a smart civil engineering design draftsperson I had started keeping company with. "These people are doing really great things," Fiona told me with great emphasis, part way through the convergence … "and I've got to be involved!" Her statement would prove all-too prophetic.

Fiona: From my young days my father had influenced me about social justice and the 'people' side of things, and I had dreamed of ways to make our cities more liveable. I didn't know it was a profession – town planning – but had I known I would have gone into it. Instead, I went into maths teaching because I thought I could influence young people and open their eyes to new possibilities.

What caught my attention at the Pappinbarra conference was the urban renewal work of Colin Ball and his team in Adelaide, as described in *Sustainable Urban Renewal: Urban Permaculture*. I don't recall the names of the people Colin worked with, but their work inspired me and rekindled my interest in making our cities more liveable, and in the potential of permaculture design in doing this.

After hearing so many inspiring stories from around the world, Russ and I followed up back in Sydney by doing a Permaculture Design Course with Robyn Francis, her first-ever course.

A Buzz, Subdued Excitement and Inspiration

Russ: In the mid-1980s, Robyn reinvigorated Permaculture Sydney (there had been a small group using that name previously) and a flurry of activity followed, including the opening of the Permaculture Epicentre at 113 Enmore Road. The building now houses Alfalfa House Food Co-op.

In those days we were among permaculture's 'early adopters' and although permaculture was not very well known, you could see a growing interest in it. There was a buzz around the idea, an intensity and a subdued excitement. We knew that we were engaged in the early days of something with a significant but as-yet-unknown potential.

Of great interest and inspiration were the social innovators who came and went at the Epicentre, people like Damien Lynch, who had started Australia's first ethical investment initiative after doing a Permaculture Design Course, who had an office there. Then there was Max Lindegger who came to the city to promote a new venture in the backblocks of the Sunshine Coast, Crystal Waters Permaculture Village. The result of Max's visits was a small exodus of permaculture people from Sydney to his project up north. Those deserting the city included two Permaculture Sydney people, Francis Lang and Jeff Michaels, who went on to set up the successful mail order horticulture supply company, Green Harvest. One Saturday afternoon, Bill Mollison came to the Epicentre and cut the ribbon across the front door to officially open the place.

Late nights were spent cutting and pasting together the pages of the *International Permaculture Journal*, later known as *PIJ – Permaculture International Journal*, and the local Permaculture Sydney newsletter which changed its 'vine' name with every edition – Choko Vine, Passionfruit Vine and so on – before we settled on the name *Winds of Change*. These were the days before publishing was computerised.

Fiona: The Epicentre was an early permaculture innovation, back then in the mid-1980s. It had a large workspace where newsletters were assembled, a small kitchen and, out back, a small courtyard with beds of vegetables and a place to sit. Denise Sawyer, who later moved to Crystal Waters, worked in the Epicentre's little shopfront several days a week selling seedlings, seeds, books and other things. There was a flow-through of people, particularly on weekends, when workshops and courses took place in an upstairs room.

One thing I learned from the Epicentre years is the value of having a place where people can go to find permaculture. A physical location serves as a focus. This learning came in handy when I helped to set up the Australian City Farms and Community Gardens Network in the mid-1990s. Like the Epicentre before them, community gardens are places people can visit to see ideas in action, and to work together on little projects. In doing this, they can act out permaculture approaches such as 'cooperation not competition' and 'return food production to the city'.

Stepping Accidentally into Teaching

Russ: Having done Robyn's PDC, Rosemary Morrow approached us to teach a couple of subjects in her Permaculture Design Course at Pittwater, an affluent and somewhat reclusive waterfront enclave on Sydney's northern beaches. Rosemary was a Blue Mountains permaculture teacher, and after she went to Cambodia, people would phone us, "When are you running the next Permaculture Design Course?" they would ask. "We don't run design courses," I would respond. But after this had gone on for a while, Fiona and I thought, "Well, why not?"

The first design course we offered took place at the Coastal Environment Centre within earshot of the Pacific's swells breaking on Narrabeen beach. The next was on the Central Coast near Gosford, where an avid little permaculture group had set up. After that, we settled on the Randwick Community Centre as permanent venue. Sydney LETS (Local Energy Trading System), AID Watch and Permaculture Sydney had their offices there and the centre manager wanted to develop a 'green' centre in the Eastern Suburbs, complete with a community food garden. That she got when students in one of our Permaculture Design Courses designed a community garden for their course assessment and, after the course, decided to make Randwick Organic Community Garden a reality.

Fiona: For a while, we taught home garden landscaping at a community college at Menai in Sydney's far southern suburbs. People were moving into new houses there and the sites had been stripped bare of vegetation and good soil. Russ and I started teaching the 72 hour Permaculture Design Course, which had been developed by Bill Mollison. But Bill taught it as a storyteller and we soon realised that 72 hours was far too short a time to teach it in an interactive manner. So we extended the course, eventually to around 110 hours, and started to place greater emphasis on subjects more relevant to the urban environment.

Russ: Fiona was right as usual. We had been teaching city people a lot of course content that seemed, well, a little too rural. Sure, they learned about farm dams and soil degradation during our field day at Fairfield City Farm where Fiona and I were employed by Fairfield Council as Landcare Educators. But those people would never use that rural farming knowledge in the big city. What we needed was an urban-oriented design course suited to a metropolitan city.

Fiona: Bronwyn Rice, with whom we taught, was a qualified landscape designer and was one of a team that tried to gain access to a patch of land for a city farm in the then-new Sydney Park at Petersham. Had they succeeded,

Sydney would have had its own inner city community farm. Bronwyn became part of the Sydney education team I assembled to run our extended, urban-focussed permaculture design courses.

Russ: So the farming orientation gave way to lessons from creating small urban gardens and community gardens, to a focus on urban food systems, co-housing, food co-ops, urban bush regeneration, working in overseas development, appropriate technology (as defined by Fritz Schumacher of the Intermediate Technology Development Group in the UK), waste minimisation, community economics, project management and, importantly, people skills – decision-making, problem solving, working with groups – so that students could make their ideas a reality. We were motivated to educate our course participants in the skills that build social capital so as to develop creative thinking, social innovation and engaged citizenship.

Through the course we ran a cashless LETS trading system. Participants made low-waste lunches from local foods where possible, and we assembled a capable teaching team for what became a Permaculture Design Course in urban systems, suited to life in a major metropolitan city. The University of New South Wales approved some of its landscape architecture, architecture and urban planning students doing our PDC as a substitute for their general studies subject.

Critics said that a part-time PDC like ours would fail to achieve the student bonding that occurs during full-time design courses. Experience has proved otherwise and it's pleasing to see, all these years later, some of our past students using their permaculture knowledge in their work.

Fiona: The weekend, part-time Permaculture Design Courses were successful because they catered for working people or students who could not take time off to do the conventional two week, full-time courses. For the assessment, students set up teams and chose a project in which permaculture principles and ideas had a direct application. This did not have to be a landscape or garden-oriented project – even back then we were asserting that permaculture was not gardening, following Bill Mollison's statement of the same – but was about design thinking applied broadly in life. My idea was to encourage students to make use of permaculture in areas it was not then being applied to, such as working with people; in what Bill Mollison and David Holmgren called the 'invisible structures'. In the city, these are as important as the 'visible structures' of buildings and landscape. I had seen projects falter because they lacked inclusive structure and process, not because of difficulties stemming from any technical challenge.

You can think of invisible structures as the social and organisational glue that holds teams together and facilitates the success of projects. They can be envisioned in two ways. First – big picture – they are social constructs such as economic systems that are about value and exchange. They give rise to visible structures such as financial organisations like, say, Damien Lynch's ethical investment company, August Investments.

The other way to think about invisible structures is as the interpersonal bonds and working relationships that generate a sense of common endeavour. They stem from developing good relations with people and adopting open and fair processes within organisations. Without them, organisations become dysfunctional and fail. Positive, interpersonal, invisible structures are the embodiment of permaculture's second ethic of 'people care' which – dare I say it – has sometimes been the least successful area in the application of permaculture.

Realising the importance of getting the interpersonal thing right in permaculture, I did training in facilitation and group skills with Maria and Richard Maguire, who were members of the Institute for Cultural Affairs. The Institute had a three level, red brick, 1960s walk-up apartment block in Marrickville where some of the members lived. One of the apartments was set aside as community space. It was a 'vertical community' rather than one scattered across the landscape like an ecovillage or other intentional community, and it seemed somehow appropriate to medium density urban living. Later, the Maguires set up their own professional facilitation consultancy, Unfolding Futures, and offered two days of group skills and decision-making training as part of our Permaculture Design Courses and as the community contribution component of their work.

I also ran weekend workshops in facilitation, hoping to improve the skills of people in community associations and to make their operation more inclusive. In 1995 when students of our urban Permaculture Design Course undertook the design of a community garden at Randwick Community Centre and helped to build it, Russ and I were hired by the University of New South Wales (UNSW) Students' Guild to run a week-long permaculture course as the prelude to starting a community garden at UNSW. From this came the UNSW Permaculture Community Garden. Someone once said that a small spark can start a prairie fire; at UNSW the small spark of permaculture design ignited a project that ran for over a decade.

Like a Lush Vine, Permaculture Grows

Russ: This was the 1990s and permaculture was on a roll, pushed along by Bill Mollison's starring role in the four-part video production, *Global Gardener*, and his earlier appearance in the *Heartlands* series, also broadcast on ABC television.

Global Gardener had the effect of boosting the prominence of permaculture in the social marketplace for ideas, and a flurry of introductory permaculture courses were spawned. We taught introductory permaculture courses at St George and Sutherland Community College, Macquarie University, North Sydney Community Centre and Ryde TAFE, followed by the permaculture elective subject in the horticulture certificate at Ryde TAFE which we took over after Jill Beck left. Jill had in turn earlier taken over from Robyn Francis who got the subject started there.

Forgetting and Reinventing

Russ: Meanwhile permaculture in Sydney had moved into a time of lull, of quietude. Robyn Francis had gone overseas to teach and then moved to northern NSW to live, and others who were around the Epicentre scene in the 1980s had moved on to Crystal Waters ecovillage. Fiona had been away working in Albury, designing farm dams and irrigation systems.

One sunny Sunday afternoon we were sitting around in an upstairs living room in an old Victorian terrace house in Forest Lodge. Fiona and I were about to meet with two people who had contacted us about permaculture. Somehow, they had heard we had something to do with it in times before. One of these people, Brad Nott, was quite effusive about their intention: "We want to set up a permaculture association in Sydney," he told us. "I found out about permaculture from the ABC television documentary *Global Gardener* and thought it such a good idea that we had to do something about it." "Oh, you mean set up something like the old association?" I replied. "The old one? What old one?" asked Brad.

It was then that the realisation hit me that unless community-based enterprise is documented in text, image or video, knowledge of it is likely to be lost. Here were Brad (who did our Permaculture Design Course and now works in the organic food industry in South Australia and lives with his family at Aldinga Arts Ecovillage, south of Adelaide) and Ian Mason (also in South Australia and later to write a couple of books on community living), planning to reinvent what had been reality only a few years before. This they did and Permaculture Sydney version 2.0 went on to become a very large and active organisation.

So we set up a newsletter called *Permaculture Web*. Nothing to do with the Worldwide Web which was not in widespread public use then, but with permaculture as an interconnected web of nodes organised in a loose network. Phil Arundel established a permaculture farm, taking over a fruit orchard in Sydney's south west, at Rooty Hill. A feature of the permaculture movement in this part of NSW at this time was the Christmas gatherings at different places with Permaculture Newcastle, Permaculture Blue Mountains, Permaculture ACT and Permaculture Southern Highlands. We would meet at places such as Penrose Rural Co-op for a weekend (the Rural Co-op was an early, permaculture-inspired intentional community on the NSW Southern Highlands). There, we would celebrate, camp out, run workshops and have fun together.

Permaculture Sydney soon spawned Permaculture South, in the city's southern suburbs, and Permaculture Inner West. Earlier, Permaculture North materialised from the participation of its originators, David and Christine Leese, in Permaculture Sydney. Later, another permaculture association, Permaculture Hills to Hawkesbury, got started and flourished for a while after one of its leaders completed our Permaculture Design Course.

Apart from organising permaculture courses, my core involvement was to edit *Permaculture Web*, Permaculture Sydney's print journal. Fiona later led a participatory design process for the organisation's website after which Permaculture Sydney and Permaculture North developed a combined web presence. By the end of the decade Permaculture Sydney had faded for a second time, not long before *Permaculture International Journal (PIJ)* met a similar fate and left the geographically-dispersed permaculture milieu without their key link to news, ideas, opportunities and networking.

In June 2000, at a meeting at Djanbung Gardens, Robyn Francis' permaculture education centre at Nimbin, I joined the board of directors of Permaculture International Limited, publishers of *PIJ*. It was the year of the Journal's demise. Permaculture then was in a state of disorientation nationally, although groups and individuals still did great work in their local areas. Nobody was sure what lay ahead for the design system or what the effect of the loss of *PIJ* would have on the capacity of permaculture educators to attract students. It was hoped that *Green Connections* would fill the vacuum. *Green Connections* was a magazine that also reported on permaculture, though its content was broader and encompassed other segments of the sustainability movement. It was produced by Joy Finch, a rather organised woman from Castlemaine, Victoria, and I had been writing for it for some time.

But within six months of the collapse of *PIJ*, *Green Connections* had also ceased to exist. Permaculture was now truly without a voice in print. This was a dilemma because the networking of information and inspiration had proved vital to the development of the design system and was important to achieving coherence among its geographically-dispersed practitioners.

Permaculture now lay in a state of uncertainty and, at the two gatherings of the permaculture core at Djanbung Gardens in the opening years of the new century, a feeling of despondency lay close below the surface. To keep communication open, I undertook to get an email discussion list started (this became Permaculture Oceania for which Permaculture International Ltd later assumed responsibility), with the cooperation of Cameron Little from the UNSW Eco-living Centre, and to edit a print quarterly for Permaculture International Ltd called *The Planet*.

Community Gardens

Russ: In the middle of the 1990s, Dr. Graham Phillips created the Australian City Farms and Community Gardens Network and asked us to become NSW contact people. This we agreed to do and community gardens became a main permaculture focus for us and an avenue to apply permaculture principles in site and social design.

Fiona's main role was educating gardeners in basic horticulture, design and facilitation and meeting skills. We combined our complementary skills to provide training workshops and to assist new community gardens in getting started through a strategic planning process we developed. We also maintained the Network's website, something Fiona's work as a graphic designer and mine in journalism gave us the skills to do, and we continue to provide strategic planning services to community garden start-ups.

Permaculture – a Means to an End but not the End Itself

Russ: In working with community gardeners and with local governments interested in supporting community gardening, I realised that permaculture's role, as a design system, is to make things happen even though those things might not be branded 'permaculture'.

The prominence that should be given to permaculture was once, for a time, a controversial point among practitioners of the design system. After some discussion online, a consensus evolved that the main thing was the task at hand, that this might often come from groups outside of permaculture, and that we should keep that as the main focus. This means that where it is not initially a permaculture initiative, we make use of permaculture's principles but it is not important to brand the project as 'permaculture'. Permaculture design is a means to an end and not the end in itself.

For me, my role in the development of community gardening is not to use permaculture for permaculture's sake, but as a means of enacting the permaculture dictum of 'returning food production to the cities' and of making our cities more resilient places.

Teaching Organics

Russ: In the 1990s, as well as teaching our urban PDC, with its emphasis on developing people skills and on sustainable urban systems, we were active with Permaculture Sydney, providing workshops.

Another activity was our offering of an organic gardening course, through the Eastern Suburbs Community College, aimed at home and community gardeners. Evening meetings were held in a school and practical days in a community garden. In reality, it was more like a permaculture do-it-yourself food production course that included site analysis, design, water harvesting and storage, plant selection and so on. It was a very successful course that ran through that decade.

Fiona: The experience reinforced what I had learned while teaching permaculture, that most of these basic life skills, like growing food, have been lost. Fortunately, there is considerable interest among the public to acquire them again. I also realised that in medium density suburbs like those of Sydney's east, where around half of the population lives in apartments and other medium density dwellings, what we need to focus on as permaculture and sustainability educators is small, compact but intensively managed vegetable and herb gardens, because people do not have much space, and shading is often a constraint. That's why our course was geared to community as well as home gardeners. I focussed on teaching organic gardening because I knew that learning to grow food figured prominently among the reasons for people enrolling in permaculture courses.

Another thing we learned was that we need to keep things simple if they are to be copied and adopted by students. Replicability is an important consideration in permaculture and sustainability design, and is a principle deserving of greater consideration.

Not long before I started teaching these courses, Russ and I also worked as Landcare Educators at Fairfield City Farm in South Western Sydney. The students ranged from high school geography through to primary and, for the 'littlies', there was the Wanda Worm Club.

New Ideas on Scattered Islands

Russ: 1994. An airless, dingy, dungeon-like basement room at the University of Technology, Sydney, on Broadway. The bearded lecturer was talking to me as I prepared a shipment of equipment for posting to the Solomon Islands. "We can offer you pay if you can come in each week and manage the project," he said. I agreed, and found myself project manager of a food security/sustainable agriculture and livelihoods program for a little international development agency known as APACE (Appropriate Technology for Community and Environment). I was offered the role after agreeing to provide support to Tony Jansen, a young Sydney man, who APACE had just dispatched to the Solomon Islands to solve some hiccups with one of their projects out on the north of Malaita island.

My role morphed from straightforward project manager to something they entitled Projects and Development Education Officer, somewhat grandiose sounding for a little agency, I thought. I looked after the AusAID-funded program, while my counterpart, Lisa McMurray (who later worked with a church agency in Papua New Guinea – PNG), over-saw APACE's micro-hydro village electrification projects.

Fiona: Around the time I became involved with the community gardens network I also joined the management board of APACE. The agency was working in food security and other types of village energy development in the South West Pacific, mainly in the Solomon Islands. Russ was working with the Kastom Gaden Project (KGP) in the Solomons and with CanCare Lae, a metals recycling and small business project in Lae, PNG. Tony Jansen ran the project in the Solomons, and my work included field training, training staff in administration and project evaluation as well as membership of the management board.

Russ, Tony and I were three permaculture-trained people in what is now probably the longest-running international development project staffed by people with permaculture training. Here I learned again the value of simplicity in training people in basic food production skills so that the techniques of settled agriculture would be easy to adopt. Solomon Islanders, in some areas, are now finding limitations to traditional shifting agriculture and there is a need in those places to train people in basic gardening using slash and mulch instead of slash and burn, locally adapted seed-stock and planting for continuous production. The KGP didn't just import techniques into the Solomons but made use of local knowledge and crops, integrating them into an agricultural training package. New ideas and technologies, I learned, come as a package consisting of the idea/technology, plus training plus a maintenance program. Another thing I learned was the value of monitoring projects as you go and evaluating them periodically. This I brought back to my work with community gardens and later, as a local government sustainability education officer.

Russ: We didn't use the word 'permaculture' much in our work because few in the Solomons or PNG had heard of it. What we did, however, was to utilise principles from the design system and combine them with others from local traditions and from the sustainable agriculture toolkit employed by NGOs in international development, known as LEISA (Low External Input

Sustainable Agriculture). In this application, permaculture, as the means to an end, was about improving food security and establishing sustainable livelihoods.

We also used participatory practices which are common in development work but largely absent in permaculture, despite the work started by New Zealander, Robina McCurdy, to develop those processes on her return from working in South Africa. We made use of PLA – Participatory Learning and Action (originally PRA – Participatory Rural Appraisal), PTD – Participatory Technology Development with farmers and a process developed by KGP – Community Food Security Assessment. Yes, overseas development assistance is acronym heaven! Later, APACE pulled out of agricultural work so KGP set up a local organisation, Kastom Gaden Association (KGA) which incorporated the Planting Material Network, a national seed saving and distribution network modelled on Australia's Seed Savers Network.

In my work with CanCare Lae I worked with Tom Jumerii, a.k.a the 'Mr. Fixit' of PNG, who took me on a tour by ute along the Hilans Hiway, through police checkpoints where poorly-uniformed officers waved shotguns and an assortment of assault rifles, to the wonderfully wild west town of Mt Hagen. There I discovered just how dour and introverted American missionaries could be and quickly learned all about the 'rascal' menace threatening lawlessness in the country. Hagen is an edgy town. In Lae, Tom showed me his house, modest by Australian standards but with its food garden, fruit trees and ducks, a permaculturist's dream home.

During the Regional Assistance Mission to Solomon Islands (RAMSI) military intervention following the coup and inter-communal conflict, I returned to the islands on an AusAID-funded mission to work with a local illustrator to write a number of training manuals for students in a Kastom Gaden Association rural livelihoods training program.

It was thought that establishing viable, owner-operated small business enterprises in the villages might prove of greater attraction than joining local militias. The work took me to isolated locales sometimes involving what I thought to be somewhat chancy, long sea voyages in motor canoes, but which the locals thought of as routine. Interestingly some villages assumed that I was with the military intervention, only dressed as a civilian for some reason.

Here I met many motivated and innovative local people, including a retired Solomon Islander soil scientist who has converted his bush garden to a productive and sustainable alley cropping system, complete with legume interplantings, out on the shores of Malu Lagoon and who assisted our program by training village youth. There was a woman who, with her children, has created an agro-forest (agricultural forestry) in the hills on an island up near Bougainville. Then there was that young pig farmer out on that low atoll off the northern tip of Malaita Island, a woman with imagination and determination. And as for sitting on the verandah of the guest house on Gizo, looking out onto the volcanic cone of Kolombangara Island on the horizon as the cooling trade winds evaporated the sweat, that was heaven, no mistake about it.

Today, we continue to work in international development, now mainly in the production of training manuals and online communications, through the TerraCircle consultancy that was set up by ex-APACE people and others.

Permaculture Initiatives Long Forgotten

Inner Pod was the first attempt to set up a co-housing community in Sydney. It was the initiative of Nigel Shepherd, an associate of Permaculture Sydney then studying geography at Macquarie University. The issue was finding land near the inner suburbs where people worked or studied. But the process took too long and people moved on. I think that in future projects like this, the secret is to proceed with as much speed as is manageable so the thing can be got off the ground and happening, which is when people will be attracted. It's called the 'demonstration effect'.

Then there was Common Ground, the idea of a partnership of organisations that included Permaculture Sydney and the Alternative Technology Association. Our aim was to do a small-scale replication of CERES, the long-running sustainability education centre in Melbourne. That project foundered over access to a patch of land in Manly then being contested by different levels of government. In Sydney there is a history of unrealised permaculture projects like Inner Pod and Common Ground. There are others that started and vanished, including some that focussed on school food gardening.

Fiona: A barrier seems to be that many people treat permaculture primarily as a technical thing dealing with physical objects, the visible structures, and forget about the people side, the invisible structures that are the infrastructure underpinning the physical. There is a tendency towards doing, without careful thought preceding the action, and of rushing in to get things started when taking a little time might produce a better, more considered product. This is what I call the 'urgency addiction' and permaculture people are not immune to it. It afflicts much of our society.

Back in the City

Russ: By the late 1990s Fiona and I were working on a funded school grounds redesign project in inner Sydney and here we had the chance to adapt some of those participatory learning and action processes for use in the city. At that time schools were a new focus for permaculture, the idea having been seeded by Carolyn Nuttall (author of *A Children's Food Forest*) and Robina McCurdy, who offered a course on the permaculture design of school grounds in Adelaide in 1995, immediately prior to the permaculture convergence that year.

This was the time when the Australian City Farms and Community Gardens Network appeared, in which we continue to work. We provided training to new community gardeners for South Sydney, Canterbury and other councils, and were involved not only in other presentations and consultations, but also in maintaining the Network's website and publishing its *Urban Harvest* journal. It's hard to step out of that journalist role in which I was trained at the University of Technology, Sydney, after I left the adventure sports equipment industry that I originally entered when living in Tasmania.

Later, thanks to my work with community gardeners, I was contracted to produce policy directions documents to enable community gardening for a couple of Sydney local governments. On one of these local government projects, I worked with permaculture educator and designer, Faith Thomas, with her small business, Living Schools. We conducted a community consultation throughout the municipality and a participatory design process for a new community garden. It was gratifying to return to that garden which we had started, to see it under construction in late 2009.

For me, community gardens provide appropriate venues for permaculture people to practice their craft in ways that directly benefit communities and make them more resilient to changing social, environmental and resource trends. The community gardens networks and TerraCircle remain as part of my work associated with the permaculture design system today, and when asked to do presentations about them I take great pleasure in describing how the design system has informed what I do.

Sustainability Education

In 2005, Fiona was offered the position of Randwick City Council's first Sustainability Education Officer. This was a new area for council and for local government in general and it is one in which her permaculture experience is proving useful.

Fiona: I offer the Sustainable Gardening and the Living Smart courses, amongst other things. Sustainable Gardening includes the permaculture skills of site and needs analysis and small area food production.

We are piloting the Living Smart course in NSW, after bringing it over from Western Australia where it was developed by Murdoch University's behavioural psychology school and supported by the City of Fremantle and the Meeting Place community education centre. It's about behavioural change and goal setting for sustainable living and community involvement. Russ is writing the Sydney training manual for it.

I organise the annual Eco-living Fair that currently attracts 5000 to 6000 people. The 2009 event was like a permaculture fair with David Holmgren, Rosemary Morrow, David Arnold from Violet Town, Russ and others speaking. There was a successful full day workshop for sustainability educators with David Holmgren the next day. There's also council's Sustainable Schools Initiative.

Local government is still finding its way in sustainability education and, except when I'm tired from the long hours that are involved, I find it exciting to be playing a role in this new area. I have trained two people to deliver the sustainable gardening course I started – one is a horticulturist from a local community garden, and the other is a landscape architect who runs an organic garden design and construction company and who has a permaculture background. It's not nepotism, but he is a graduate of the Pacific-Edge Urban Permaculture Design Course.

Another outlet for our permaculture motivation has been participation in the Sydney Food Fairness Alliance (SFFA), an educational and advocacy organisation made up mainly of professionals from the health, nutrition and community work sectors with some city-fringe farmers, people with urban agriculture interests and some from the social sector such as the churches, as well as academia and Transition Sydney.

It's good to work with people who have a professional, action-based approach and who are used to working in teams. In 2009, SFFA partnered with five inner urban councils to run the successful Hungry for Change forum at Sydney Customs House to harvest ideas for the later Food Summit. Hopefully it's the start of developing a food policy for NSW. This we did through a participatory process known as 'World Café', a technique we had picked up from the Sydney Facilitators' Network. The speaker Michael Shuman, a US local economics advocate drew a full house.

The Future?

Russ: Where to now for permaculture in Australia? David Holmgren says that the design system has achieved its greatest acceptance as life education. I have no argument with this, however at times I feel a little twinge of disappointment that it is not more accepted by social decision-makers. There is some movement in this direction but all these years after my friend showed me his copy of *Permaculture One* and we stood together in his organic garden that sunny afternoon in Hobart, I realise that there is a great distance yet to travel.

Here are what I see as needed to progress permaculture into the future:

1. Greater rigour in permaculture education

First, permaculture education, and here I'm not talking about the accredited, certificate level permaculture training that emerged in the early 2000s, rather about the Permaculture Design Course. It could do with a little more rigour and with enlarging in scope to be more inclusive of invisible structures, interpersonal skills and project management. Permaculture education has to evolve to remain relevant and this will not be achieved by continuing to teach what we have in the past in ways that we have become comfortable with.

2. Greater collaboration and partnerships

Second, the permaculture movement would do well to develop collaborations and partnerships with other organisations in working towards sustainability. Working in the Sydney Food Fairness Alliance, I have seen how collaboration and partnerships have a multiplier effect on what you do and how it opens the door to achieving more than you can as a single organisation or individual. Permaculture, in general, is good at cooperation on projects within permaculture associations but is less effective at networked collaboration on a state or national basis. Developing a capacity for collaboration that is national in scale would be to take a leaf from the success of the Open Source movement.

In doing this, permaculture organisations would need to avoid dominating and branding the projects it becomes involved in with the 'permaculture' name because this would take due credit from those working towards similar goals from outside the permaculture movement.

3. Greater connection with potential allies

Third, permaculture is not the only sustainability game in town. There is out there a youthful milieu, some employed in sustainability industries, who are techno-savvy, digitally connected, smart, knowledgeable and in-touch with trends and events, and who have their own cultural milieu.

Presently, this amorphous milieu appears to have limited affinity with permaculture. Permaculture has values that are compatible with it but needs to learn to speak its language and to appear appealing if it is to connect with and attract today's youth.

4. Greater community application

Fourth, it's true that if we wait for government to act on sustainability it will be too late, that if we act as individuals alone it will be too little, but that if we act in collaboration with others then something might be achieved.

This is not an argument against individual initiatives in developing ways of living that are sustainable, but for doing so in cooperation with others on that same journey. Basically, this is a community-oriented approach.

As permaculture designers, or as permaculture or other types of sustainability educators, we need to ensure that our designs and the ideas we propagate are replicable. That is, that someone can take what we have done, adapt it and apply it in a different situation. This is a means of scaling-up what we do. Ensuring projects are replicable and scalable are criteria that could move permaculture beyond the life education role it has developed, valuable though that is. The capacity to scale-up is necessary at a time when we face problems of global scale.

Ideas Once Strange Now Mainstream

Fiona: Ideas that seemed strange only a few years ago are now in the mainstream. The important thing is to demonstrate these new ideas to show they work. In sustainability education, give people ideas to take home to try and get them to report back, help them to set goals and monitor progress and reward them. Give them the easy things at first because accomplishing these breeds motivation to do more and can eventually lead to sustained behavioural change. In sustainability education, and permaculture training which is one of its methodologies, providing credible, verifiable information is very important if we want to influence people and develop as an evidence-based practice.

Home gardening is important and it is something I like doing, but if we are to create the sustainable future we need, then we have to be out where the people are, discussing solutions and acting on contemporary issues from a permaculture perspective among friends, colleagues and communities. We have to work within the home base but we also have to step out of it if we are to scale-up permaculture initiatives. To paraphrase the great science fiction writer, Arthur C Clark: "… the home [garden] might be the birthplace [of permaculture] but sometime, we have to leave home."

I believe that permaculture remains a fine approach to motivating individuals and communities to move towards sustainability. This is why Russ and I did the accredited Transition Initiative training, with the trainers from Totnes in the UK in Bowral in early 2009. Transition Towns, or Transition Initiatives as it is now called, offers a new way to expand our influence beyond the people we usually access through permaculture. Interestingly, our membership and the people we attract through Transition Sydney are mainly from outside of permaculture. This is good. It shows that to use permaculture in the Transition movement we need to form partnerships with individuals and groups outside of the design system and move beyond the idea that Transition is just another permaculture initiative.

Influences

Russ: I'd like to recall those who have influenced us in permaculture. First are David Holmgren and Bill Mollison, two distinctly different personalities. David continues to make a contribution to the evolution of the design system. Then there's my friend standing in his organic garden in distant Moonah all that time ago and the good fortune of his showing me that first edition of *Permaculture One*.

There's Robyn Francis who gave Fiona and I our formal introduction to permaculture through her first-ever design course. Then there's Rosemary Morrow for our start in permaculture teaching. But then the names become legion, too numerous to list … Jill Finnane, Brad Nott, Morag Gamble, Max Lindegger and on and on.

In recalling Bill Mollison's description of permaculture as 'adventures in good design', I think that fits what the author, James Joyce, wrote about life:

> *All experience is an arch*
> *Through which shines that untravelled world*
> *Whose borders fade forever and ever*
> *As I move.*

Somehow, that seems to encapsulate the permaculture experience as one that has no fixed border…

Fiona: It seems a long trip when I look back on permaculture … a lot of trial and error, disappointment as well as triumph. It seems a long road from the present back to those rustic, ramshackle buildings in the hills of Pappinbarra, back in 1984.

Food Forest (overleaf) produce at a local farmer's market.

Annemarie & Graham Brookman

Graham Brookman is one of Australia's foremost permaculture designers and teachers. He specialised in horticulture at Roseworthy Agricultural College where he developed a healthy scepticism for exploitative land-use systems and marvelled at the effectiveness of geese in controlling grassy weeds.

He and his partner Annemarie have created a sustainable home on their 15-hectare permaculture property, 'The Food Forest', at Gawler, South Australia, which attracts thousands of visitors, including many students and WWOOFers (Willing Workers on Organic Farms). Graham is also founder and chair of the Adelaide Showground Farmers Market.

Graham and Annemarie have been demonstrating the practical results of well-designed homes, gardens, water systems and energy capture for 25 years. The Food Forest has an information-rich website and makes educational videos about permaculture which are available on the Food Forest TV channel on YouTube.

Photo by Kathleen Read.

Chapter 10 | Building and Living in a Food Forest

Whenever we are going to plant, build or arrange something on our property or within our lives we like to quickly run through the permaculture ethics – will what is proposed help care for the planet, care for community and avoid excessive consumption of non-renewables? If the idea passes those tests we can move on to some fine-tuning with principles and technologies.

Annemarie and Graham Brookman designed, built and live on one of Australia's most mature permaculture properties, The Food Forest. It was unambiguously designed in the mid 1980s as a demonstration and research property on which they would bring up their children and live for as long as it took for trees like oaks to bear fruit. The Food Forest has been recognised as one of Australia's best certified organic properties, best exhibitions of Landcare and sustainable business, as well as one of the best medium-scale permaculture properties in temperate Australia. It is a truly 'complete' permaculture design with everything from organic vegetable growing to composting toilets and tree cropping to solar power.

The enterprise offers permaculture design courses, educates school groups, and conducts human-scale research and development. It has possibly Australia's largest private collection of strawbale structures and is the focus for education and networking on strawbale construction. Thousands of people have visited The Food Forest or consumed its wine, nuts, vegetables, fruit, carobs and poultry products, available through organic shops or The Food Forest stall at the Adelaide Showground Farmers' Market.

Looking back on their 15-hectare property over the twenty years spanning the time while their children grew up, Graham summarises their journey with permaculture:

It has been an exhilarating rollercoaster ride, which has energised twenty years of living. It has introduced us to extraordinary people who have recognised many unsustainable aspects of conventional economics, agriculture, community and education with which we were brought up, and replaced them with human-centred models which work in developing countries as well as affluent nations. Permaculture allows you to let go of poorly-founded contemporary conventions and use wise ethics and principles to devise your own way of gardening, bringing up your children, designing an environmentally responsible home or doing business. Scientifically based, permaculture avoids dogmas, religions and leaps of faith. It is a framework for appropriate living in the new millennium.

Graham tells the story of The Food Forest....

The Food Forest Story

I was brought up in Adelaide (now a city of around a million people) and loved working in my parents' suburban garden. Dad was a railway engineer and a pragmatic and productive vegie gardener like so many people after World War II, and Mum was a social worker and a rose lover. Grandfather's garden was quite different, more fun and less work; he had chooks, almonds, a huge *Macrocarpa* hedge and a fishpond. Having demonstrated some persistence I was granted my own plot in the garden, from which I sold vegetables and strawberries to my mother, my grandmother and anyone else who was interested. I spent as many holidays on the farms of family friends as I could and became addicted to the space, plants, animals and the unquestionable productivity of agriculture.

Later, I majored in horticulture at Roseworthy Agricultural College in South Australia, a big teaching farm located in 450mm rainfall country near Gawler. It was a tough place, devoted to furthering the production of food and fibre for export through 'the education of young men', and plant breeding. I resolved that if I ever became involved in teaching, I would ensure a more creative, caring, socially and environmentally responsible atmosphere. (It wasn't going to be hard!) However, Roseworthy gave me a wide range of skills in everything from fencing to livestock management, and from public speaking to agronomy.

At that time, the Vietnam War was raging and before I had time to use any agricultural skills I was called up and found myself in Malaysia. I was purportedly protecting Malays from communist terrorists, but I was actually having a fascinating time hiking through jungle villages that were, from what

I could see, 100% self-sufficient in food and building materials. This was nothing like the export agriculture that was tearing Australia to pieces as rabbits, salinity, erosion and nutrient depletion all took their toll.

I returned to Australia when the Labor Party gained power in December 1972 and abolished conscription, withdrew the remaining troops from Vietnam and banned sporting teams from South Africa. It was a heady time but, having no farm to go to, I slotted quickly into Teachers College to follow the well-trodden path of city slickers who 'migrated' to the country as agriculture teachers, and sank every penny they earned into a farm. But the desire to see other forms of land use in faraway lands was strong, and in 1974 a gang of Aussie agricultural exchangees headed for Canadian prairie farms for six months. There was nothing sustainable there and we saw real ghost towns and bizarre religious communes that had opted out of mainstream society. On we went to the depressing barn-based dairying system of Denmark where cows simply stood or lay in concrete cubicles all winter, receiving an electric shock if they forgot to step back into the dunging race when urinating or defecating.

Running out of cash in Europe forced us to stay put in a little village in Holland, and it was in that diverse, well-watered farm-scape that I got to know a Dutch agricultural science student (rather well), whose family had left industrial Rotterdam some years earlier seeking quality of life. Annemarie was already keen on the idea of organic agriculture as a result of assorted chemical scares, whereas I'd hardly heard of it. I was seeking a system that was 'sustainable', a vague idea I had of land use that would enable the perpetual sustenance of biodiversity and farming communities, with the soil at its centre. Our journey continued via the extraordinarily sustainable farms of Yugoslavia, still practising systems common in medieval times; the classical systems of Greece with vegies on the fertile low ground, then vines, then olives and high on the rocky hills, the tough old carobs. We saw vines growing prostrate on mud walls inclined exactly to the winter sun in Iran and lived with the Mursi, an Ethiopian tribe who moved house every dry season, coming down to the river flats when the floods had finished, to plant their tobacco, gourds and sorghum with primitive implements, while the teenage boys went to the high ground where the goats they tended would be safe from sleeping sickness. We saw the tall rainforests of the Congo and danced with the pygmies while we could – their rich habitat was disappearing to be replaced by incredibly poor pastures.

After this extraordinary exposure to different methods of land use we needed time to articulate the lessons we'd learnt. Selling the trusty Kombi, I headed back to Australia, richer by three languages (all spoken badly), and feeling like a citizen of the world. I joined the staff at Roseworthy, with the task of introducing primary school children to food production. I realised through repeated experiences, that Grades 5 to 7 are key years in the acquisition of the complex concepts which dominate people's understanding of the world for the rest of their lives.

Annemarie came out from Holland and, thank goodness, liked Australia. Before long we were on the lookout for land and she thought it would be good to live by a river which was handy because I wanted deep soil and a useful aquifer. We bought 29 acres on the Gawler River in South Australia, not far from markets, schools and public transport. We borrowed most of the money from the man who sold us the land, and share-farmed the block with our neighbour, but we still had no clear design for our little farm.

My cousin was given a book called *Permaculture One* for Christmas in 1982 and happened to put it down next to me. I basically devoured the book as the party went on around me. In May the next year its co-author Bill Mollison delivered South Australia's first permaculture course, at Willunga. He lectured solidly for five days and changed our lives. He threw a massive responsibility upon every student to go out and actively show people how to design and live in sustainable systems. Bill's confronting style, seething distrust of multinationals, his critique of universities as followers rather than leaders and as production houses of nicely tamed graduates ready to perpetuate an occupation of the planet that was clearly heading for disaster, resonated with us. However, the clincher for his permaculture philosophy was its basis in science and its compelling logic. Having had more than my fill of Judeo-Christian dogma at school and having observed the gods of many other faiths in action around the world, I couldn't help thinking that it was time for personal observation, thought, planning and responsibility to take the place of faith in divine intervention. To me, Existentialism seemed a fair way of rationalising the simultaneous existence of the assorted gods – that way everyone who believes in a deity could have their own god in their head. This flexibility seemed important to enable the permaculture movement to be able to reach out to people of all faiths, but equally, it seemed vital that all humans adopt some basic secular human ethics and actually take personal responsibility for the way they live – perhaps the three permaculture ethics? Permaculture went further than providing three ethics (against which I subconsciously check out every decision I make) it gave principles and technologies for sustainable living that can be used anywhere on the planet.

With the short design course under our belt we rapidly planned our property. The aim was to show what ordinary people can do without significant government support. It was to be a demonstration and research farm, refining a land use system for the wheat/sheep country of southern Australia. We did not consider country that received much less than 400mm of annual rain, as we believed that such land would end up reverting to pastoral use when the energy being poured into it become too expensive to continue propping up such a marginal system. We wanted to show that an Australian family on an ordinary income could buy, develop and manage a piece of farmland in a sustainable way. Our aims were articulated in this old 'mission statement' which were all the rage in the '80s:

1. To demonstrate that land can be managed in an environmentally sustainable way, producing healthy food and a healthy income.
2. To share information and skills for land management, self-reliance, conservation and food production with other people.
3. To be a rich and beautiful place to live, to work and to raise children with balance, wisdom and skills.

Models for land use potentially existed anywhere on our latitude, north or south of the equator. The Mediterranean, Mexico and bits of Africa, South America and Asia all had plant species that might do well in the ecosystem we would build. Undertaking the design was one long adrenalin rush. To be establishing trees that may still be alive in 200 years, planting a tree that our children may climb in and forage from, designing a forest that may provide timber to build a barn in 2014 and assembling plant and animal guilds that had never been used on the planet before, was an immensely creative act.

Barriers, Hurdles and Helpers

We had our critics, especially amongst my agricultural scientist colleagues, but most people were, and have remained, overwhelmingly supportive. If we ever doubt ourselves or feel tired we have a squad of barrackers to whom we owe our persistence, and a planet to whom we owe our lives. We were scared out of our wits by the big floods of the Gawler River in 1992, wondering for two weeks whether the house was going to go under as the rest of the block had. We survived but the damage to trees and fences was considerable. Then, out of the blue, like some ANZAC godfather, Frank Payne rang up and said that he wanted to bring a group from the Australian Democrats political party to help clean up. Rather than clean up, we planted a whole lot of seedlings into the fresh wet silt the flood had left behind and congratulated ourselves. I think of him and his wonderful, idealistic mates every time I look at the jungle of native plants that they put in.

All normal humans suffer self-doubt, and pathfinders often look back to see whether they have missed an important point or are doing something really stupid. They have a strong need for people who have their respect to give regular, expert, honest feedback. I have relied heavily on colleagues at Roseworthy and the University of Adelaide, particularly professionals in agronomy, agroforestry, linguistics and architecture, all of whom have a commitment to sustainability. Annemarie has tended to develop professional friendships in the organic growing and retailing fields, in the Vocational Education and Training sector and in community organisations.

Education – the Ultimate Answer

Fostering the permaculture movement in South Australia during the mid '80s was a small band of committed people, including two of Bill Mollison's earliest students – Tim Marshall and John Fargher. But the most overt energy came from the community-based 'permies' who were associated with the Brompton City Farm, just west of Adelaide's CBD. They brought Bill Mollison over to run a full Permaculture Design Certificate course (PDC). He arrived a couple of days late, having been put in hospital with a dread disease, but when he was sufficiently rehydrated he'd wandered out of the hospital and made his way to Adelaide, still in rough shape but determined not to let the side down. It was during this course that I realised we could teach a PDC at our place. This plan was cemented through an Advanced Permaculture teaching course at Colin Endean's 'coach house' at Burra with the superb teacher training of team Robin Clayfield and Skye. We formed a teaching team to offer the first design course at our property and have offered it annually ever since.

Unlike Bill, David Holmgren, co-author of *Permaculture One*, had shunned the public spotlight, choosing to perfect the practicalities of permaculture design on his property at Hepburn, north-west of Melbourne, but after publishing a number of books he too began teaching. He joined our team in 2002 and has been a treasured presenter and houseguest, constantly providing new insights into the intellectual basis of permaculture as it links with other disciplines, yet bringing us back to earth with stories of containing goats, trapping blackbirds and soil management challenges on his own property.

Teaching has been the secret of our spiritual survival, bringing us together with a constantly growing and evolving group of enthusiasts and now the number of permaculture graduates worldwide is estimated at over half a million. To have simply lived on the farm, producing food within the constraints of organic certification and permaculture principles while the world around us gobbled up resources and became 'wealthy', would have shrivelled our spirit. Instead we have had the richest possible journey through the realms of ethical living, magnificent food, and caring, creative friendships. For me there has been the challenge of having a dual identity – one minute a student adviser at a conservative university, and the next, planting organic vegies on the river flat. If we lived in a developing country perhaps we would have had only the role of producers. But in Australia, with all the economic levers pulled back hard to maximise production and profit, there is still no 'right of place' for true environmental carers, artists, musicians and wise people, so most permaculturists 'fit in', earning a living by doing something as close to rational as possible and spending the rest of their lives on the real stuff of sustainability. Compromise is at every turn, for we are gregarious creatures who live in social and economic systems that are pervasive… No ecovillage can completely avoid the dollar, cheap imports or the odd McDonald's wrapper.

Identity

Giving our property and our ideas an identity was vital, as we had more and more produce to sell and courses to market. 'The Food Forest' said it all, borrowing the name of a well-established permaculture concept and signalling our intentions for the farm. It was to be dominated by perennials, reaching down through the deep river silt to the aquifer which had been sustaining river red gums along the river channel for millennia. It would be home to many native organisms as well as some outstandingly productive exotic species and would sustain humans through dint of their work. We drafted up a logo incorporating those ideas and have never been tempted to go beyond the strong black and white version. It is quite widely recognised in South Australia.

Children

Our son Tom spent many weekends as a baby in a backpack while we weeded and watered seedling trees early in the property's establishment. Children the world over have been with their parents as they work, quickly understanding the nature of work and the way humans naturally interact with the biosphere. He and Nikki have grown up at The Food Forest but have not been pressured to do more work on the place than they wanted to. They received excellent education at a nearby co-educational school and did all the things the average Aussie kid does in terms of sport, friends and responsibilities. We believed they needed to be able to function at a high level in society at large if they were to have a chance of becoming positive change agents in the years ahead. Since they have been at university they have taken progressively more interest and involvement in the education, production systems and marketing of The Food Forest. Both have done PDCs and majored in environmental studies at university. In terms of human succession in the design of The Food Forest, we will see what happens, but the place has been set up on two titles with three access points from public roads and two mains water connections. Additions to the house have been designed to make it possible for it to be divided into two small independent homes.

Going 'Certified Organic'

While we were restoring the old homestead, we employed and got to know a fantastic plasterer. He was interested in Annemarie's organic gardening techniques but admitted that he just didn't have time to grow enough vegies for the family. One day he proposed that he would like to be paid in vegetables as his wife had cancer and had enormous trouble getting organic food in Gawler. So began our small-scale commercial production.

As Annemarie eased out of TAFE teaching to spend more time on the property, the vegie supply grew and we developed a good relationship with two organic retailers. Eventually one of them said she was going to stock only certified organic produce so we went through the certification process with the National Association of Sustainable Agriculture Australia (NASAA), the nation's most stringent certifier, whose regulations have been significantly influenced by permaculture. The pressure was on me to swing the rest of the property over to certified organic but I felt very nervous about the creeping grasses, couch and kikuyu. We decided to get geese for grass control, but that would involve a 1.4km fox-proof fence around the bulk of the property if they were to simply free-range. However, the fence would also allow us to incorporate threatened native mammals into the farm ecosystem, something we had hoped to do in the original plan.

Building the Ecosystem

We had long been fascinated by John Walmsley's bold move in fox-proofing a dairying property, 'Warrawong', about the same size as our place, revegetating it and stocking it with platypus, potoroos, wallabies and many other native animals as an all-in-one ecosystem. He ran Warrawong very successfully as a wildlife park. We wanted to incorporate indigenous animals into a horticultural system and had already provided for a substantial area of native vegetation within the fenced area, so we held our breath and let loose Dama wallabies, brush-tailed bettongs and Cape Barren geese, hoping they'd be compatible with tree fruit and cereal crops. After some nervous times the ecosystem has bedded down perfectly, saving masses of work and non-renewable inputs and demonstrating that a bio-diverse system is generally more stable than a monoculture propped up by fossil fuels and their products. Who would have guessed that bettongs would eliminate the rat population and control sour sobs (*Oxalis*)!

The Truth About Trees

The average tree has a root system that can go at least two thirds as deep as the tree is tall. This gives it anchorage and the capacity to harvest water stored in the subsoil. But the very great bulk of the nutrient the tree needs is harvested by fine roots within a few centimetres of the soil surface. In the almost rainless South Australian summer, when most fruit grows, there is naturally very little nutrient-carrying moisture in that shallow soil, so the trees don't get enough nutrients to produce good crops unless you water them in. The most efficient way of doing that is by strategic, modest ('deficit') irrigation using water bolstered with nutrients from compost tea, composted animal manures or organic wastes. The treated waste water from a family of four will provide all the nutrients required by 20 fruit trees. The simple step of recycling water, through a reed bed or otherwise, is used by very few people, including permaculturists, who moan about their unproductive trees.

Chilling Consequences of Climate Change

It is increasingly important for permaculture proponents to master their design skills and demonstrate robust systems that can cope with global warming and offer sustainable living. Particular aspects of design need attention: design for catastrophe and design for succession. I truly thought the pistachio nut trees on our farm (our main crop) would be a resilient and relatively 'permanent' key component of our property design. However the warming of Australian autumn and winter temperatures recently has caused a failure of the trees to accumulate sufficient winter chill to set their physiological clocks for normal flowering in spring. This reduced our 2006 crop to 20% of the predicted yield. Cherry growers were similarly affected. Maybe some future years will be better, but the overall trend will be steadily downhill.

Consequently, at The Food Forest we are identifying and planting species with lower chill-requirements for the inevitable global warming ahead over the next half-century or more. A chill-requirement should be mentioned when describing plant varieties in permaculture resources and we should be reporting maximum temperatures that varieties can tolerate. The calculation of chill hours for sites should be taught in design courses.

Financing the Dream

Unlike some other permaculture teaching enterprises, we have not gone down the pathway of a foundation, trust or ecovillage as we see the place as, quite naturally, a family farm. But we would have been happy to be owners-in-common had any of our good friends been ready or able to participate in a jointly owned property when we started.

Starting in debt is a grand tradition in Australia so we borrowed to buy the original 29 acres and quickly got status as primary producers to be able to claim farm expenses and the interest payable on the loan as a tax deductible expense. This involved generating turnover to show that we were serious about the enterprise. Share-farming did the job for a while but eventually we were driven to planting gherkins to return a significant income while the tree crops developed. After picking and bagging-up fifty 60 metre rows twice a week for a couple of seasons, we never wanted to see another gherkin in our lives ... but it confirmed our seriousness. I continued to work for Roseworthy and the University of Adelaide during the development of the property and set the cost of improvements against my other income. Most farming families depend heavily on off-farm income to survive, with almost half of Australia's farm families raising at least 90% of their income off-farm. Perhaps the 35% of small farmers and 6% of large farmers who have left the land in the last nine years weren't told that this country is nothing like the USA where many farmers receive farm subsidies of $300,000 annually, or like European countries where rural communities and landscapes are valued and attract economic support to remain viable.

We were so keen to buy the adjoining property, with its old house and sheds, that we had even incorporated it into our original permaculture design. In 1986 our neighbours retired, we bought their block and we had the property of our dreams and a house (or rather, ruin) of our own! This meant that we could fully commit to the place and start implementing the design of the inner zones – the chicken tractor, stone fruits and vegetables – but it also provided a pit into which we could pour endless resources. We got the house liveable and put in a reticulation system before Nikki was born and resigned ourselves to some slow progress while Annemarie concentrated on being a mum and the cash-flow slowed. But it was a great time to plough some energy into the garden at the local pre-school and help the kids to become sociable little beings. The business had grown and Annemarie never really had time to go back teaching for TAFE seriously but she picked up part-time contracts teaching mainly in the area of native food production, which have led to some terrific long-term contacts. Education income at The Food Forest comfortably outstrips farm income; farming people is certainly more profitable than farming the land.

One of our perennial fears has been the possibility of losing the property. We lost two slices off the sides to the Gawler Bypass project in the mid '80s and survived the surgery. We are relieved to know Adelaide's billion dollar Northern Expressway will miss us, instead annihilating an organic vineyard and slamming through productive farmland. I'm not sure that I know how we'd cope with losing our place. Having planted every tree and watched them grow with our children, I can just begin to imagine what the dispossession of the Adelaide Plain may have meant to the Kaurna people, the original indigenous inhabitants. It certainly broke their hearts and their spirit. And our old homestead had seen it all happen.

Going broke is another way of losing your land. Being bed-ridden for a couple of months with a spine that refused to cooperate gave me plenty of time to think about that. We've gone for several forms of security in the end. Firstly there is our planting of a diverse, productive and tough range of species; they'll always produce food and fuel crops. We have also invested in on-site active and passive energy capture, solar hot water and photovoltaics (PVs) (and soon hopefully wind) for self-sufficiency, and we have good capacity to value-add our food products. Then there are the conventional things like superannuation, freehold ownership, the capacity to rent out half the homestead or even to sell one of the two titles if things got dire. Designing for catastrophe is one of permaculture's great strengths.

Value Adding

Self-sufficiency gardens are ideally super-diverse but when you go to farm-scale you need to have some clear winners that you can handle in reasonable bulk. We chose pistachios, carobs, walnuts, olives, apples and grapes as key crops amongst the 150-odd food producing varieties on the farm. This has led to a quest for small-scale and recycled equipment as well as the design and local construction of unique machines for processing and storing produce. Our deep tray dehydrator was designed and built up the road in Jamestown and has slashed the total energy involved in getting pistachios dehydrated by almost 90%. Our permacultural mania for energy conservation and for personal observation has led to some technologies of which we are intensely proud. Their designs will be of use to others who wish to reclaim the food chain for small/medium producers and the people who eat their food.

Perhaps the best thing we've done is to initiate Adelaide's first Farmers Market. It is in its early years still, but without it, many of the 120 stallholders would have gone broke at the hands of the supermarket duopoly. Instead they take home their part of the market's $8.5 million annual sales and they are flourishing. Most of the regular customers are formal members of the market and most shop there with the expressed intention of putting their cash into the producers' pockets.

An unexpected value adding bonus came in 2008 when the South Australian 'Feed-in Tariff' scheme was prepared for the SA Government by one of our PDC graduates (and an electrical engineer). As a result, green power exported from domestic PV systems into the grid doubled in value per kilowatt hour, reducing the payback time of the PV system by more than half, and is guaranteed for 20 years.

Permaculture, Energy and the Future for Humans

We are so bombarded by political spin, and energy is so central to permaculture, that I think it is worthwhile for me to share my understanding of where we currently are with respect to energy, greenhouse and a future for humans on earth. I believe that the path to a liveable future is still very much a permaculture one, but what I have always regarded as an 'energy sideshow', compared to the long-term need for a sustainably designed future is moving rapidly to centre-stage.

Having watched the Australian grape harvest fail and our own crop of apples roast on the trees due to climate change in 2008, the greenhouse issue presented itself as a very present threat. It is tempting to suggest that the fact that humans have used up most of the readily available oil and gas reserves of the planet would be 'good' if it meant that they had to reduce their use of fossil fuels. But in a fabulous trick on themselves, in 2007, governments seriously encouraged the production of biofuels from crops that essentially use almost as much non-renewable energy to produce as they make available and take up land that could have been used for biodiversity or food production purposes – all in the name of minimising climate change.

More recently the uranium lobby has become more vocal, claiming that 'virtually GHG (greenhouse gas) emission-free' nuclear energy is the way to solve peak oil and climate change while continuing to grow our economies. The alternative energy camp says that wind and solar are the way to continue living in the manner to which we are accustomed.

Having weaved and dodged our way around the main issues, humans must eventually confront the essential challenges facing them as a species and decide whether they can respond logically to their intellectual understanding of the world's limitations and thus overcome flaws in *Homo sapiens'* ancient DNA. Or will their primitive urges towards procreation, tribalism and power prevail, so that humans behave like other plague species and suffer a truly catastrophic situation because they have no control over their exponential population growth or urge to consume. World population more than trebled last century from 1.8 billion to 6 billion (in 2000) and has added a population of consumers the equivalent to that of North America since 1998; we are currently adding 3.2 humans per second. It is ironic that we should be agonising over this issue in Australia, the continent where wildlife species had adapted their breeding cycles to respond to resource availability long ago and where the indigenous human population had more or less reached a steady state with the environment over a period of some 60,000 years.

As an educator I believe that humans can be trained to step back from the abyss, but it will require a level of self-discipline and regulation that takes people to the edge of their genetic capabilities. Dominant religious, economic and political frameworks today still fail to take account of man's capacity to destroy the earth's life support systems. Often the concept of divine power is used to excuse believers from taking absolute responsibility for management of themselves and their earth, leaving an urgent need for an overarching set of global ethics and principles through which wise decisions can be made. Permaculture needs to be part of such a global agreement on the way forward.

Now that some leaders have accepted concepts of climate change and that oil is seen as a finite resource, it is time for the 'population reduction issue' to be addressed in public. Limiting the right of humans to reproduce has been a 'no-go' area even for most permaculturists but it is clearly the core issue as China effectively abandons its one child policy and Vietnam its two child policy. Our education, legal and medical systems must seriously address the matters of bioregional and national 'carrying capacities', mandatory family planning and euthanasia in a philosophical, humane and scientific manner. Every hour that we delay in developing a workable approach to population growth, over 11,000 extra people arrive on this overstretched planet. Without offending our increasingly 'conservative' students (I had one with nine children in a recent course) we need to seriously introduce this issue in our permaculture courses.

In order to spark rapid and intelligent change toward sustainable design, we need our universities to offer permaculture courses throughout their curricula so students can obtain recognised qualifications with a 'sustainability context'. The steady rise to influence of the trickle of young professionals with permaculture ethics is already enabling organisations and communities that are ready for sustainable change to do so confidently.

For the individual permaculturist the most effective contribution to change may be to demonstrate, through personal behaviour and work, that sustainable living can be both possible and prosperous. This contribution can often be amplified through membership of bioregional permaculture groups, Transition Town activities, community organisations, local government committee participation and support for green candidates. We have thoroughly enjoyed being involved in all of them. Happy designing!

Rosemary Morrow

Born in Perth, Rosemary Morrow (Rowe) was claimed early by the Earth; plants, animals, stones, weather. Some years in the Kimberleys as a young girl confirmed it.

Later she trained in agriculture science with which she was very disappointed, then moved to France where she lived in the L'Arche community. Later at Jordans Village in England she realised she would become a Quaker. Back in Australia in the 1980s Rowe's Permaculture Design Course provided the basis for a concern for Earth restoration. She considers permaculture to be 'sacred knowledge' to be carried and shared with others. Since then, when asked, she has travelled to teach the PDC to others who, due to circumstances, could not access it any other way. This took her to immediate post-war Vietnam as well as Cambodia, Uganda, Ethiopia and other countries.

Rowe's present concern is to make teaching sustainable and encourage others to succeed her as teachers.

Chapter 11 | Cherish the Earth

When I am asked who I am, my answer depends on the period we are speaking about. I think I have nearly lived my nine lives not because I am very old but because I am a risk-taker. However there is a thread through all of the lives.

I built a close relationship with the Australian bush very early. As a child in Perth's hot dry climate I escaped outdoors, played in the bush and lay on hot sand like a lizard. Later I was to bask on the hot Sydney sandstone. I wasn't conscious of it then but I felt centred when I lived as close as possible to earth despite my great love of water, especially rivers. I established the love and habit of interacting with the natural environment and finding solace there under difficult times.

Then I found gardening. My mother wanted to make a perennial border garden at our home in Castlecrag in Sydney. I dug and hoed and raked and felt marvellously elated and fulfilled in some way I couldn't put into words. I also found that I was strong and I could do whatever my mother could, in digging and hauling. Later she called me 'her little pit pony'. I loved working hard and that I think is due to my physical type; small and strong. I also loved long walks, tennis and swimming. My muscular body responded as did my senses. I don't know if other people need to do physical things as one way of feeling alive. For me there was quietness and satisfaction that came with hard work and being close to the earth.

Later, as a teenager I escaped to the Kimberleys and lived on Gordon Downs and Noonkanbah Stations. These were free, wonderful, open and vast spaces. I loved it and expanded in this environment. I discovered that big spaces were important to my wellbeing. Later, the sky and the ocean would fill me with such relief that I would breathe more easily. Living so remotely was a time of great ignorance, innocence and immense contentment. I've written about it in a memoir called *A Girl in the Outback*.

I felt whole – integrated in body, senses and the natural world. It was also a very safe world and one I could trust despite droughts, cyclones, duststorms, plagues and floods. By contrast, although I value people and love many, I must always be vigilant with them, but with the earth ... never. There is also a quietness and safety in my experience of big spaces. I think big spaces keep us small and that is very reassuring. I have probably always realised this but it took time to reach clear consciousness.

At 21 I was back in Sydney to study agriculture at Sydney University. But it began to feel like a paralysing mistake. I had thought that agriculture would be an interesting and vital study. I had loved what I lived in the Kimberleys, with animals, bush and gardening, and here I was in laboratories. It was so dry and removed from my experiences. It was so disappointing; all those techniques for cutting trees and planting in rows with chemicals and with tractors. Later I realised it was simply reductionist science, but I didn't see it at the time. The teaching was also very bad and I truly felt I would die of boredom in the lecture theatres learning about strains of food crops and how to apply fertilisers and biocides. From living in freedom on one million acres then being restricted to hours each week in lecture theatres and laboratories, often with no windows, was a huge culture shock for me. Writing interminable lecture notes I would fall asleep, my pen sloping off the page. I thought that learning by heart for exams was senseless when it was already in books. There was no genuine interaction with the natural world except through test tubes and microscopes. I became alienated from farming and gardening for a while because of it, but I finished my degree because I felt I had to. And many times since then I've been glad I did. Strangely, I enjoyed mycology, the study of fungi, which still thrills me.

I also enjoyed being in a male world and learned some really terrible vulgar sexist songs on hired buses. I liked the university vacations and getting experience on properties, and I went back to cattle stations in Queensland where they practiced cattle hunting and no farming. The field visits all over New South Wales to a wide range of farms were great, but I felt for many, many years, a great pain at having left the Kimberleys. I loved the natural world with such passion. I would breathe it, gaze into the night sky and smell it. Now I have the same love but there's much, much more awe in it. (And, at that time I was having a rollicking affair so this wasn't sexual sublimation!)

After graduation, I went on to work for the Department of Primary Industries in south-east Queensland. My job was to visit small farmers and provide them with technical advice on farming methods. Most were dairy farmers and I was their first woman technical adviser. The men would meet me at the gate and send me up to see the wife until I said, "It's you I've come to see." However invariably I'd be asked to stay for dinner or lunch. I would sit with the family who worked so hard, cared about their animals and demonstrated an old-fashioned sincerity, and I'd wonder about their future. I will never forget the farmer who invited me to dinner and then had his son take out his false glass eye and put it on the tablecloth where it watched me while we ate huge plates of chops and vegies.

Their lives seemed so complete on these small mixed dairy farms – intrinsically valuable I suppose. So I went back to my boss and asked about the government policy for them. No one would answer my question. Are we here to keep small farming viable and productive or, to shift them out? It is now evident that no government would ever actually say, "Well, we think we'll get rid of small-scale integrated farming because of lack of economies of scale." They were just allowed to die out and the land was amalgamated or sold for hobby farms. I suspect now that they either didn't have a policy, or if they did, it was to shut down small farmers, in which case I would have resigned. I really loved the small mixed farms and farmers with a few hundred acres living traditional lives; the fruit trees, dogs, pet lambs, vegie garden, old horse, two dogs and sheds. These farms had internal values to themselves, the farmers, to the land, and to Australia.

I didn't know then it would all disappear and be swallowed up by large companies, city suburbs or coastal development. But knowing it now, I am remorseful because I believe that every country needs a hinterland patterned with mixed small farms around its cities. From those times and since, studying in France and working in Vietnam and Cambodia where this pattern persists, I see how it structures a country, forming it with regional culture, produce and pride. Small mixed farmers give life to the hinterland. There is a social web created of trust and interdependence and a richness of landscape. The diversity gives strength; one farmer can only do so much damage. Farmers have pride in their regions; just think about the old agricultural shows and their special products. Small farmers give rise to localisation with its local wealth and long-term stability which holds the secrets of arts and handicrafts in farm living and working. For this reason, I see Victoria as rurally 'wealthier' than New South Wales.

After some more studies at the Sorbonne in Paris and Reading University in the UK, I worked in Lesotho in Southern Africa where my utter inability to use the agriculture from Sydney University was a mighty shock. After all, remember, I had never grown a seedling or a vegetable nor had any clue what to do about erosion or, more especially, hunger. I was agriculturally useless and lost in a country where malnutrition was dominant, so I worked in non-formal education which I loved, on a project which taught herd boys how to read and write through games printed on their traditional scarves. In the evenings they brought a candle to their rondavel (grass-roofed mud hut) to go to school. Later when they went to the mines in South Africa they could read their payslips and contracts.

It was here I was politicised by apartheid. I lived in the Basotho part of town, not with the white development agency people, all of whom had big fences and dogs. I took part in marches against apartheid in Maseru, the capital, and in fabulous singing and dancing parties in the shanty towns.

I started making the links between food and water security and poverty and development. A jigsaw of living and landscape, with some parts missing, was forming in my head. The Republic of South Africa was the huge consumer on the edges of Lesotho. It consumed the men in the mines, their money from gambling, their produce through cheap purchases, and emptied villages of workers, skills, knowledge and advocacy. When this happens there is rural implosion. The same pattern happens in Asia. (You can imagine how much worse this is to small, resource-poor countries in South East Asia, with Hong Kong, Singapore, Bangkok, and Tokyo not far away.) In such situations the water gets stolen, bad deals are made over trade and poverty becomes not just hunger and cash poor, but disabling through loss of meaningful lives.

So nearly a decade after I had left Australia, I returned and decided that horticulture offered possibilities. Horticulture was a strange study after a decade in developing countries. It was full of talk about 'the trade' and glasshouses, watering regimes, and woodland and cottage gardens. Another cultural shock, a worldview clash for me. Most of my cultural shocks came from Australia. Everything in horticulture was about either ornamentals, or profits, including royalties from plants.

But I was in TAFE as opposed to university, and so I did learn some skills. I learned how to grow seedlings, plant trees and was introduced to techniques and ideas in design. By the time I finished the course I was really aware that horticulture could not stop urban sprawl and the destruction of the Australian bush which I still loved so passionately that my heart spasmed when I saw its destruction (and still does). I later worked as a landscape designer in the Department of Environment and Planning in NSW. This was a rich and good experience. The salary was good and a certain amount of esteem and public approbation went with it. It was an enticing career path. It was while I was in this job that I enrolled in a Permaculture Design Certificate course (PDC).

The decision to leave the Department of Environment and Planning job was initially easy, but it became harder once I'd left. I had no income stream and was refused a mortgage. So I picked up Environmental Studies and taught it. I went to the local community college, which supplied the room and $25 per hour, and started my first class with eight people, two of them more that 85 years old. It was very good but I couldn't reconcile the conflict between the introduced and the indigenous.

Around this time, a good friend, Janice Haworth, said there was going to be a permaculture course facilitated by Robyn Frances at Newtown and that I would love it. Well, I immediately imagined everyone sitting around glass triangles with razor blades in them and chanting incantations and I was deeply suspicious of it. I'm suspicious of the lowest common denominator in group-think, such as 'permaculture is good'. I am a natural and confirmed sceptic. I love science for its mysteries but I also find evidence based thoughts and ideas the most exciting. However, I respected my friend's opinion and I felt that anything I was so prejudiced against I should try, and not submit to prejudice. So I did the course. When Bill Mollison later told me that permaculture was "about tangibles," I was particularly satisfied. I had imagined that permaculture would be about intangibles, which are not so satisfying.

Permaculture, that was it for me. There wasn't a lot that was new because agriculture, horticulture and environmental studies had supplied much of the natural science aspects. It was the approach which enchanted. It was interactive and overlaid and connected all disciplines. I was also intrigued by a course that began with ethics. None of my other studies had ever mentioned the word. I loved the statement, "What would happen if soil scientists had ethics?" As phosphate soil acidity and eutrophic waterways were increasing, this question was prescient. The need for ethics has become self-evident together with the need to integrate the study of all earth disciplines. I was coming home. I now had an approach to reconcile my need to grow food for food security, with the need to create habitat, preserve and restore indigenous land systems. Now courses I teach have these goals.

In the meantime, in 1978, I had become a Quaker and realised that their testimonies gave me a creative and meaningful way to live. I decided that an ethical life was the one which offered the most value and interest compared with lifestyle materialism. It seemed to me, and still does, that the ethical life is stimulating, challenging and is never fully achievable, yet it can be monitored. Everything we do and say has consequences, so it is vital to live life through ethics. Permaculture has its ethics, so there was a correspondence between Quakerism and permaculture. They have in common: care for people, simplicity, community, ethical use of money and right livelihood. They both render positive outcomes when practiced.

More and more I realised that my work with the earth and people could have ethics in every segment. For example, we could have ethics for our water use, and how we treat soils and other living species. The spiritual and the secular were coming together, and my appreciation, observation and awareness of life and its processes really developed. My body and breathing would almost stop when I considered the whole of life and my being able to reflect on it – to see, hear, touch, taste, tell, read about life. I couldn't believe that Life was so extraordinary and that I was conscious and part of it.

I became tender in my heart for all life but sometimes harder on humans, for their lack of appreciation and gratitude to be participating in the great miracle that is life. Later I was even further awed by cosmology and how remote the chances were that life could occur at all. I still can't think about it without swallowing hard and now I have moved to holding people compassionately.

Because permaculture is not an armchair study and I wasn't sure it would work, I bought a small house, impossibly sited in Blackheath in Sydney, and built my first garden based on permaculture design. And it worked. Later I moved to a couple of acres on the edge of Katoomba in the Blue Mountains west of Sydney, and there I really satisfied myself that permaculture does work. During that time I was offered work in Vietnam and Cambodia to teach their first permaculture design courses, and when I was in Australia, I taught locally.

Teaching is so humbling. I love the permaculture curriculum as a process which facilitates people's awareness and which leads to action. The content of the PDC is so good that participants do amazing things that I could never have thought of. And sometimes I find their lives have been transformed by its possibilities. Teaching PDCs is not a numbers game with set outcomes. I know that change comes from anywhere and there may be no one in a class of 40 who has a vision they can communicate. But in a class of three, there may be two people. This is my experience. I try to stay small, clear and personal in my teaching and keep the price available to most people. I believe it is important to teach locally, close to home and, over the years, build up a solid number of people who form a community. I have also done this in Vietnam and Cambodia and the results have been humbling. Every participant leaves the PDC course with a design they can implement at home.

I have taught in places in war and recovering from war, and I have been honoured to meet people who are so appreciative, willing and brave. I admired them then and still do when I remember how they had to trust me; they had no books or permaculture gardens to learn from. They set up gardens on trust – true permaculture pioneers. I have had close relationships with other cultures and people, and this is a privilege. Of course there's always the downside in being away from home for long periods with loneliness, sickness, misunderstandings, missing out on everything going on at home and difficulties in sustaining relationships. I spent nearly eight out of fourteen years away working with people in permaculture and related projects. Then the PDC spilt over into other projects such as food processing and project management. I taught the first PDCs in Albania, Vietnam, Cambodia, Ethiopia and Thailand, and other courses in India, Afghanistan, Ethiopia, Malawi, Uganda and China, as well as two each year at home, here in Katoomba, which have not been widely publicised.

My inner work is Quaker silence, the joy of community and the immeasurable treasure of long-time dear friends. My inner work is to understand that others have their reasons that I must trust are good even when incomprehensible to me, and when I ask myself "Why on earth would they do that?" I like to become very quiet and still. And keep a long view and not too subjective. After all, what do I know about the effects of what I do, or the future?

Having worked in war-torn countries, I have become a stronger opponent of all wars and all killing. Recent studies in cosmology have helped me understand a little better that we are an emotionally primitive species that has not yet found its niche or its wisdom. I feel sad that our brains and our intellect have grown so large, but without finding our niche on this lovely Earth. I believe we could become a peace species and it is important to live as if this is becoming so. I believe that everything each person does is important. Our participation is essential, every person and act matters.

I use a teaching approach from Alternative to Violence Program (AVP) which constantly draws on the resources of the participants and uses their knowledge, skills and questions to move further. It is mainly a Socratic approach, but is careful not to bring participants up against information they simply cannot know. This approach often elicits surprising answers. My postgraduate work in adult education had taught me, so it is now embedded that honesty, respect to and liking for students is critical for their learning. There are of course other factors such as relevance, but if the respect and liking is not there, their learning is compromised.

The hardest courses have been those with up to four nationalities all wanting translations in their own languages, as happened on the Thailand border with Thais, English speakers, Burmese and Khmer. It was very difficult. Other courses that require much careful preparation are those with illiterate people. In this case, I draw pictures and the interpreter translates. When I have a translator, I stand or sit to the side and let the native speaker dominate. This gives the respect for the knowledge to the speaker.

New courses can be very difficult and people are often so polite that they won't tell you when they disagree. On a recent trip to Vietnam, a former interpreter told me that after the first PDC the participants got together and discussed the course afterwards. They decided that I was mistaken because "climate can't change," and, they'd "never drink water from plastic bottles." So I had been suspect as a carrier of new information and ideas. The truth about some students' thinking emerges when I have developed good personal relationships with people and I am trusted. This can be slow.

My own culture, class and nationality have been barriers. I do not teach PDCs to people who suffer hunger for more than six months of the year. I teach the district people who, in turn, teach those who are hungry. They have all suffered hunger and they speak from experience. And they know what works and what is a priority. When I first started, I used to worry about relevance and the claims made in the PDC. It is no good telling people they will have food in six months if it is not possible. I have really worked at this. Kabul in Afghanistan was very hard indeed. There was the long cold winter, only melt-water and almost no rain and no biomass. I was careful what I claimed was possible because the people did not have strong garden culture background.

I know that I can never be poor because Australia has shaped me, it has offered me ideas, skills, abilities, health and contacts that other people cannot begin to think about. I am deeply grateful for this. Permaculture became my vocation and the more I worked with the content, the more interesting and the deeper it went. Then the links started to happen with special nodes around water, plants and soil. I saw design as philosophy and practice; the true subject of the course.

Even today I view permaculture as still a prototype. It is barely thirty years old and continues to grow and stretch out into people's lives and take forms of its own, especially if we think how David Holmgren has stretched the parameters. Personally I really like the focus it began with. I still remember Mollison saying to me, "Permaculture is about tangibles." Today I see the tangibles embedded in intangibles. My intangibles are the conversations, the solitude, the insights, reflections and feedback, and new findings in every part of the permaculture syllabus.

At times I have found it difficult to have to defend permaculture when faced with the same prejudice that I had carried before doing a course; that it was all about untidy gardens with strawbales, or it taught people to grow blackberries. It was a hard road to imbue permaculture and the doubters with the serious applied science which it is, and not simply, "Gee whiz, it's magic!" I have worked hard on interactive teaching but not only activity based, and to make it comprehensible for all levels, ages and nationalities.

I do yoga, and find huge rewards in working in my garden. I want to prolong forever and ever the days when the sun is warm, the breeze benign and the earth damp. They are the most satisfying. I don't ask for happiness but I do ask for contentment and meaning. I need both solitude and friends in different doses at different times. These practices and beliefs maintain my hopes for a better future. Quakers decided to become people of peace and refuse to carry arms and to kill during a time of terrible civil war, raiding militia and injustice. Permaculture emerged when we were beginning to threaten our existence on Earth through destruction of our habitat. I do not know that these can save us but they do give us hope. Now, taking a long hard look at the future makes despair likely, the human response is surely to embrace permaculture, practice it and teach others.

The world is in a dark space and I do not have hope for a wonderful future for our beautiful Earth and all species. However, I would be arrogant, overestimating my importance or ability to foretell the future, if I despaired or asserted that I have some special prescience. I am as blinkered as everyone else. But I have seen that change often comes from left field and may be waiting in the wings. I like the Catholic idea that despair is a sin. I do not have much hope but neither do I despair. The resilience of both Earth and people is miraculous. Perhaps running out of fossil fuels or financial recession will actually save us from ourselves. But if it is lack of water that reduces us to our proper population, size on Earth and behaviour, then it will probably be very painful.

I would love to participate in a low resource use world with everyone learning more about how to live peacefully. I would love us to specialise in the arts of peace rather than the arts of war. I want to live in a future with reverence for all living things; to grow in wisdom, dive deeper into the natural world and live more rightly and simply; to give opportunities to others to teach permaculture.

I imagine a gardener's world surrounded by indigenous vegetation and very friendly native animals living amongst us with confidence. For Australia I imagine a network of neighbourhoods working like ecovillages. Food, family, children all together and no fences. The neighbourhood is the 'ecovillage' of the future. It would be easy to transform the suburbs. My community is already diverse, engaged and aware.

For the world, I imagine that some countries will resuscitate traditional ways of living, as Cuba and Bougainville already have done. Countries like Sweden and Denmark are leading the way with technology for high latitude countries, and others like India and China have practices today which stretch over hundreds and sometimes thousands of years, all happening concurrently with modernity.

I am growing older, and faster than I think. I will soon be very old. I would like to grow wiser and more compassionate. I would like to be at home and have friends and family visit. I'd love to watch the indigenous plants grow and the animals come into this new garden to find a home. I'd like to read well and think more.

I'd like a more compassionate and fairer world. I'd like my community to develop more co-ops and collectives and work well together. I'd like this place with less traffic and more trees and more local food with complete and satisfying employment for the young, the elderly and people with disabilities. A slower place with time to sit on benches and let the sun warm us through. Less regard for money and more for our quality of life. But it will be hard for people to cope well with less water, less petrol, rising prices and dependence on all levels of government. I can imagine pockets of hope that model new 'permaculture' ways of evolving.

I see permaculture as becoming more and more important, and in a crisis, permaculturists being called up by governments to teach and work with communities. I also see areas such as my own – where people are practicing sustainability in many ways, in their lives and on the land – as pockets of sustainability to model how it can be done for the rest. I see the fortunate as people who have the knowledge, and privilege to begin now and to get their systems in place and then volunteer for wherever they are needed.

I have looked and researched many other sustainable ideas and systems, however only permaculture (I often spell it 'permaclutter') integrates so many disciplines. Without coercion, we need reverence for life. With reverence for life, we will practice permaculture.

Rowe Morrow with youngest PDC student at Rakai in Uganda.

Martha Hills

Martha Hills emigrated to Australia from the US in 1988 at age 40. She discovered *Permaculture One* at the public library in South Yarra and was so impressed that she took an Introduction to Permaculture course with Ian Batchelor in the early '90s. Then in 1993, she took her PDC with David Holmgren and Ian Lillington at Commonground, just outside Seymour, Victoria. Martha was highly active in Permaculture Melbourne during the 1990s and has taught with others on Introduction to Permaculture courses at the Council of Adult Education (CAE).

These days Martha doesn't really call what she does 'permaculture'; instead she 'acts with the ethics of permaculture' in how she lives her life. So when she's in her garden she focusses on that, but through the garden, she is conscious of the connections with community, food security and good relations with neighbours. For Martha, life is about creative optimism and enjoyment, and nurturing an instinct for cooperation rather than competition.

Photo by Kerry Dawborn.

Chapter 12

Experience is What You Get When You Don't Get What You Want

I'm a child of the 1960s from the USA. We were confident that we could save the world. Our parents were narrow-minded, materialistic and conformist. But we were a new creation and we would crash through those walls that held us back and spread our wings. Life was for experiencing and creating – drugs, sex, personal relationships. Bras restricted us, so we went without. Neckties symbolised the separation of head and heart. Our children would grow up in freedom. With our good will, there would soon be no war or poverty or pollution.

It took a while to realise that the world was a bit more complex than we had thought, that good intentions weren't enough, that we still carried around some bits from our parents anyway. Communes busted up. Druggies crashed and burned. Hippies became yuppies.

Although I did start wearing bras and shaving my legs, I still enjoyed growing some of my own food, marching against the evils of US imperialism and environmental dangers. I was straight but had lesbian and gay friends. My clothes came from the Op Shop (second hand shop), and food from the co-op. For a while I was a Marxist until they seemed as un-green as the capitalists.

Through a complex series of events, I arrived in Melbourne, Australia in September of 1988 with my new Australian husband, and we bought the house I've lived in for the last 20 years. Permaculture sneaked into my consciousness after I borrowed *Permaculture One* from a public library. Then I came across the organisation Permaculture Melbourne and a weekend Introduction to Permaculture course. The next step was completing a permaculture course at the Commonground community in Seymour, Victoria. Our tutors were David Holmgren, Ian Lillington and Glen Ochre.

AHA! So that's how things work! Patterns – see patterns – flow with patterns – participate in patterns – live in the moment and above it at the same time!!! Just as I had responded to Marx's Communist Manifesto clarifying and re-ordering my anti-capitalist thoughts, so permaculture now put together my disparate strategies for a self-created, non-mainstream life (that's a life, not a lifestyle).

My brain grew and changed shape during those two weeks, I swear, and it burst forth with enthusiasm for the next few years. There was learning more about the ideas, applying it at home, applying it at work (intriguing but totally unsuccessful), teaching others and participating in Permaculture Melbourne.

Although my engineer husband Simon had come along to part of the introductory course, our lives diverged. As I grew into permaculture, he became more focussed on work, and we parted. Still, I met and remained friends with two other women, who also were married to engineers, and they're still together. How would our lives have gone if we had found a way to combine our interests? Of course, there's no answer to that question, but I did become more adventuresome in ideas as a single person. There was no one else to consider, so I didn't self-censor ideas that arose from my musings.

I loved the way permaculture married the altruistic with the practical and how it assured me that the aesthetic grew out of the functioning pattern rather than out of an advertising campaign. "Caring for the earth, caring for people, and sharing your surplus," – this was the set of ethics I first learned about in permaculture. I also loved the idea that I could have little experiments before jumping in. At the same time Bill Mollison talked about avoiding major errors – that's one reason for planning.

Permaculture, along with some other programs and practices, has helped me to be present, receptive, creative and resilient. I could notice what was around me, rather than the task chosen by my mind. As I noticed 'things' I could add them to my 'plan'. Sometimes these were tangents, sometimes additions to a more complex plan, sometimes dead ends. Still, there was always learning. I made cards with messages like, "Experience is what you get when you don't get what you want," and taped them onto walls and mirrors around the house.

Permaculture has helped me avoid immediate judgments of good and bad. Instead I watch an element or combination of elements to see what the consequences might be. The next step is to try some possibilities of arranging them, what I could add to make a better system, how these bits fit into a bigger picture. Here are two examples of my permaculture way of thinking.

Example One

I am not good at cleaning up, at home or in the garden. Unharvested plants from an earlier season are going to seed.

Consequences:
- My garden looks different from the others on the street.
- There are always flowers to attract bees.
- Plants are shading the soil.
- The larger root systems will decay into organic matter after I cut the plant off at ground level.
- Seeds are available for harvest or for self-seeding. (I have permanent parsley, nasturtiums and mustard).

Only the first consequence would be negative for me, and so I am happy to continue with a messy garden.

Example Two

My house has its long sides facing east and west, which is perpendicular to the recommended solar orientation. The 'For Sale' sign said it was "deceptively spacious."

Consequences:
- A long rectangle is less efficient to heat in the winter than a square or round shape
- Only a short side faces north for winter sun
- The long west wall gathers heat in summer afternoons
- I had enough money to add lots of insulation
- I can move my activity areas during the year, spending more winter time in front rooms and summer time in the middle area.
- There is a north-facing roof area that has a solar hot water collector and another that could hold a photovoltaic system.
- The farthest back room (on the south) is used for storage, so lack of heat and light is less important. The less than optimal orientation can be lessened by modifying the building and by changing how I act within it.

As I mentioned earlier, the mental excitement of learning about permaculture carried me into new places. Also, other permaculturists have carried me, especially when I needed it. A few years ago, some permaculture teachers put together a course that would fit within the Australian TAFE system. When it was offered for the first time, I was invited to attend. I had been unwell and didn't trust that I had the energy for it, but was still invited. I would not have made it through a conventional program, but the combination of content, teaching strategies and other participants made it an uplifting experience for me. I appreciated that these people were living out the ethics of permaculture.

The permaculture that I learned about 15 years ago is not today's permaculture. There may be a more dialectic relationship between permaculture and myself; we dance together and, in doing so, influence each other.

Early writings on permaculture concentrated on gardening. Still today, I hear people say, "No, I can't do permaculture yet, because I don't have land." The term itself may be from an American book that has Permanent Agriculture in its subtitle. But the field itself has expanded immensely.

When we were told that the principles of permaculture were universal, that was permission to go out and see what we could do, wherever we went. Later writings included information on house construction, ecovillages, forestry, and many other subjects.

The boundaries of this field are loosely defined, although many think they have the right answer. While I was working in the Permaculture Melbourne office, I put onto our answering machine the information that someone was coming from another state to give a talk on keeping poultry and another on dowsing and ley lines. A long-time permaculturist, upon hearing this, left his message about leaving the fairies at the bottom of the garden.

Who knows what is coming? My buddy, Ian Lillington, has written a book and permaculture isn't even in the title. I have occasional flashes and continue to collect information. Maybe my local group will run a PDC. As I continue in my life, my understanding of permaculture evolves and I evolve along with it. I'm glad that permaculture and I are friends.

Martha's porch, the vineyard is out the back. Photo by Kerry Dawborn.

Janet Millington

Janet has developed a permaculture demonstration site in Eumundi, South East Queensland. She helped develop the Accredited Permaculture Training package and the Train the Trainer in Permaculture package. Janet has worked with Permaculture Noosa and Permaculture North Sydney to increase permaculture awareness and knowledge in those regions and to develop functioning group structures and community gardens.

In 2007 Janet started the Sunshine Coast Energy Action Centre on the Sunshine Coast, which led to the Sunshine Coast becoming the first Transition Town outside the UK. Janet and a colleague went on to develop the Time for an Oil Change course based on the work of David Holmgren, which led to the delivery of Australia's first Energy Descent Action Plan being delivered in March 2010.

Janet is the co-author of *Outdoor Classrooms: A Handbook for School Gardens*, with Carolyn Nuttall, which was launched in 2009. She is working with teachers and schools to link the garden to all aspects of the curriculum and sustainability. Janet is also part of the Edible Landscapes and Edible School Yards co-operative that is successful in establishing and maintaining home and school gardens in South East Queensland and is now extending into Cairns where she is currently working.

Chapter 13 | My Seat on the Train

I was born in Sydney not long after the end of World War II, one of the Baby Boomers that rode on the train of post war development. The war and the post war development impacted greatly on me and my generation, just as WWI and the Great Depression had impacted on our parents. Dad had been a signal man in the army and Mum was the local air-raid warden, and of course we always had the vegetable patch and the fruit trees in the yard. Mum had kept that going during the war and she used to tell me that the war was both the worst of times and the best of times. It had brought communities together, and it 'freed' women to take jobs before only given to men. It made everyone appreciate what they had and to celebrate the smallest of events, because no one knew when things would get worse.

We had a modest home in the suburb of Mortdale, about 11 miles from the centre of Sydney. The very next train station was a place where Sydney people took their annual holidays, on the shores of the beautiful Georges River. We also had strong connections to our rural cousins and holidays were planned around the annual cattle muster. We made rooms ready for country visitors around the time of the Royal Easter Show, where Dad and I would spend our time admiring the huge displays of fruit, grains, woolly fleeces, meats and abundant produce from 'the Lucky Country'.

We all knew Australia was 'riding on the sheep's back' and we were doing well. But one dark cloud lingered long in our household, or rather two large mushroom clouds. The effects of the atomic bombs that struck Japan affected my mother for the rest of her life. Mum never accepted that humans could destroy so much innocent humanity with such a small but devastating action. So she morphed her air-raid warden role into the local civil defense officer for our district. Our back room held all the white paint and tape to stop the windows blowing out in a nuclear attack on Sydney or on the Lucas

Heights atomic reactor. We had rations and iodine pills and all sorts of gear. There was only one Geiger counter for the Shire. I later learned it may have been more productive for her to have done yoga so that in the event of Sydney being a target, she could have more easily kissed her arse goodbye!

So from age 12 I knew the consequences of a direct hit on Sydney. I knew the fireball distance, the shock wave distance, the distance of being sick for days before death, the distance where people may linger in agony for weeks or months and the distance you needed to be to have some hope of survival but no hope of bearing normal children. Our prospects in Mortdale were rather bleak but mum assured me that Sydney would only be worthy of a small megaton bomb so we continued going to the meetings and performing the drills right through the Cold War.

I also learned you can't stop a plane with a couple of thermo-nuclear devices on board from killing hundreds of thousands or perhaps millions of people if they don't like you, or what you stand for, or if you are standing between them and something they need or want. I learned that making friends is our best defense and ensuring there is enough for all forever. If we ever have to start painting windows white and taping them up I'm going to do some extreme yoga!

Preparedness, that's what I learned through all that, and that there can be the best times in with the worst times. Dread, fear and concern never stopped Mum from doing something and working with people and celebrating life. She was a fun person to be around. She was also a beauty, a stunning looking woman and an absolute charmer, so she was the perfect networker and connector and doer. Wanting to get out of Sydney was another thing I learned in that period, which I achieved finally in 1992.

Theatre of the Absurd and the Permaculture Penny Drops

If I had been born a little earlier I would have been a Beatnik. As it was, in the early sixties I dressed like them and read all the same stuff. It made sense to me. I guess I was an early bloomer in abstract thinking because none of my friends seemed to know what the hell I was talking about and my poetry and bongo playing fell on deaf ears. (I can understand the bongos but my poetry seemed to me spot on and profound. Maybe I should have performed them separately…)

My literary heroes at the time were Ionesco and Pinter, and the theatre of the absurd was my solace. From that period to this very day, the dichotomy, the dualism and the bizarreness of the human condition continues to astound and amuse me. To see it played out on the stage and to be able to see it externally in action and to laugh, was simply wonderful. More wonderful was hearing others laugh, which told me they too realised just how bizarre the real world is and that they, like me, play the parts but know the underlying absurdity.

In the 1960s and '70s our local beach was Cronulla, where we used to go a few years before the film *Puberty Blues* came out. (I think they were talking about us.) Not having connected with the Beatniks, who were a fair bit older than me, I became a surfie. Yes, I lathered on the coconut oil and baked on the golden sands. Of course the ozone layer news didn't reach us till the eighties. Then it was fun in the sun with very little protection.

It was there at the beach I met my husband of 40 years, Mick. We travelled home on the 3.20 train together and have since travelled together all over the world. We've been through many adventures and some sadness but always totally united in our beliefs and our goals for the future. We have supported each other in our physical, mental, academic, emotional and spiritual journey and still share that love of nature and a belief in permaculture as a real positive solution to the problems that have arrived while too many others had their shoulders to the grindstone, not looking at the bigger picture.

Those were the formative years and looking back I can see the seeds of all that has happened since. Cronulla Beach taught us some lessons about the wider world. Underneath our tanned and blonded, salty exteriors, we were united in a love of the outdoors, the challenge of nature in the surf, the beauty of the landscape, our dependence on the weather and the sea conditions. That was our glue, that and our youth and our free time. The growth economy meant that we could stay on at school and even go to university without having to work really long hours to help support the family income. Our mums were off doing that and we were able to learn and laugh and play right through the Cold War and up to the Vietnam war.

At that time a new tribe emerged – the hippies. Perfect! They were my age group, they thought profoundly, they were critical of government and the status quo and they loved the outdoors. And they wanted to protect it from ... yep ... the growth economy and the post war boom times which had supplied us with the education to think widely and deeply, and the time to engage with and come to love the environment and the means to take action.

To be fair, we *were* in a position to learn and see what was happening, unlike those older than us who were working hard driving the boom. We were in schools and universities, on the beaches and in forests; we were travelling the world bringing back all sorts of ideas and taking full advantage of the good times. We joined the Peace Corps, lashed ourselves to trees, chained ourselves to bulldozers, turned on and dropped out. And the economy kept growing. Average earnings increased through the seventies but the quality of life began to decline. One by one I watched my hippie friends climb back into the mainstream.

By then we too had the conventional responsibilities – my teaching and Mick's engineering, home and two children, so we couldn't join the truly responsible who were still chained to trees. But we were able to lock up and protect some land. We bought 40 acres near Jervis Bay, and I took on extra work and Mick worked longer hours to pay for the property.

At that time I became very interested in health through nutrition, having read Adele Davis in 1973. Her *Let's Have Healthy Children*, first published in 1951, was the bible I used for raising the kids. She warned of chronic illness and extensive social problems if the American and Western diet continued to rely on processed foods grown with artificial fertilisers and sprayed with chemicals. Her other books explained that all disease arises through vitamin and mineral deficiencies, and how good healthy and preferably homegrown food, was the solution to human health issues as well as for the crooked thinking that was leading society into what both she and I thought was the fast track to disaster.

The day in 1978 when I saw *Permaculture One* in the local news agency, I was on my usual browse for any material on healthy living. I picked it up because of the cover ... the world as a garden and the promise of "A Perennial Agriculture for Human Settlements." It gave me hope and it gave me tools to think about the world and to make some sense of its absurdities.

As my understanding of both individual and social health required the intake of fresh and healthy food, I embraced the permaculture design system wholeheartedly, and Dad and I carried on with our small vegie patch in the back yard with new vigour. Permaculture worked with us because it offered a real explanation of why the things we were already doing worked as well as why we had some failures.

Mick and I worked on raising the kids and our careers into the mid 1980s when we had the opportunity to work on an overseas project. We packed up the house, farmed off the animals to set up house in Holland. We saw a lot of Europe by car and bicycle, and left six months after Chernobyl. Yes, the old nuclear demon was back. We were in France at the time and went back to Holland to our five-acre property to cut back the whole vegetable patch and burn the lot. My mother was frantic. She sent powdered milk and iodine tablets and begged us to come home. We promised to eat nothing that wasn't packaged before 26th April 1986, so we lived on food parcels, tinned goods, glasshouse vegetables, New Zealand lamb and Argentinian beef. I stripped the kids down every time they came home from school, washed their clothes and showered them. They were not allowed to play ball games, or to lie on the grass.

I took such strong precautions, but not because the Dutch Government warned us to. On the contrary, we were told that the cloud had completely missed Holland and had gone around the border. Being a mere seven kilometres from the border, we listened to what the German Government was telling parents to do there – that and the phone calls from my mother and all the books and manuals I had read about the long, lingering death zones and the deformed children zone. I vowed then to do everything I could do to avert such a tragedy in the future. Taking action to be ready and prepared is better than regret when there is no chance to do anything, and all that is left to you is acceptance of consequences.

On our return we sold our old faithful Kombi that had given us so many happy days and bought a zippy little red car more suited to getting us quickly through the city. I remember the day my mainstream friend said to me as I showed her the car, "from hippy to yuppy in one foul swoop!" Was I really there? Surely this was just skin deep and a passing phase? We had the house, the two cars, the two kids, the dog and the acreage property. It surely looked like it. But no. We might have had one foot in mainstream but the other was still in permaculture. We had a plan and we were milking mainstream to put resources into our permaculture property. We were sure we had it all sorted.

Arrows that Fell from the Sky

After my father's death I found it very hard to cope with his loss. He was always the steady influence in the family, a details man. Everything was well-considered and although he had generous heart he was a resource conserver. We had balls of string and elastic bands, we mended sheets till they were only good to turn into pillow cases. Old towels became new face washers. We grew food and bartered whatever we could, and there was always enough. He was an accountant, yet we shared a love of the outdoors and the garden and worked together in vegetable gardens right up until he had his major stroke which paralysed him and finally killed him three years later.

I was sure it wasn't fair and I was going to be angry and injured until something was done about it. My father deserved to live longer, to see his grandchildren grow. Less worthy men were living longer. He must be returned! I was mounting a protest to God and in the process just about killed myself. I reached a point where every appliance or light I touched blew up. Car batteries went dead. Starter motors would malfunction, and I became very used to finding alternative transport methods. This went on for nearly a year. My fellow teachers really didn't believe it was I who was having this effect on everything electrical until one lunch time I got a bit agitated over something and the fluorescent light above my head burst into flames. It was absolutely clear to me and everyone else that I was either out of balance and my energy was going haywire, or that I was not getting a very plain and simple message from the universe.

The final straw was when sparks flew out of the cook top. I marched very deliberately into the backyard, and I spread my feet apart so I was anchored to the ground. I threw my head upwards and showed my fists to the sky and yelled, "Go on, send me another one! Go on, I can take it! What is it you want from me? Send me anything. I can handle it. Just don't take my children or I'll be no bloody use to you! Go on, tell me what to do! You have my full attention!"

And it never ever happened again after that. But I was surely listening for every message and took note of the sequence of events. I read messages in broccoli heads and nasturtium leaves. When someone recommended a book I read it. When someone suggested I meet someone I met them … and there was peace.

My immune system had already been badly compromised by severe exposure to chemicals in 1979, and when we returned from Holland we found our neighbours had done a heavy chemical saturation of the soil beneath their house for white ants. This began to affect my health and then, with the shutting down of my immune system, I became allergic to the 20th century. But to show that I still had some control over my life as well as to increase the meaningfulness of my teaching career I completed a Master's Degree in Education in the Creative Arts. It has since proven a very handy thing to have.

Get Out and Get Going Into Permaculture

I managed for a while in Sydney but the traffic in the 'big smoke' made my quality of life very poor. After Mum died we decided to leave the city and head for Noosa, in South East Queensland. I took long service leave and I had a couple of years just getting over the previous 42. It was wonderful. I took long beach walks, went swimming and power walking. I read at least a book a week, all non-fiction. I had an insatiable appetite for knowledge. I needed things to make sense before I could heal my body. I had already begun my spiritual journey (between parental deaths) and I really began to make peace with God and His God too. It was all a plan. It just wasn't MY plan. So I agreed to go along with it as best I could. I learned to 'hear' the messages and instructions and my intuition was finally given back its rightful place in my decision-making tool kit.

At the same time we decided we needed to be productive and to re-establish our vision of acreage. We bought 60 acres in a beautiful rural village just 20 minutes from Noosa, called Eumundi. The difficulty here was that the subtropics has rampant growth and any advice about potential crops came with instructions about which particular chemicals to use at which particular times in order to produce them. We were having none of that as my family had seen just what chemicals can do to people and the planet. It was time to revisit *Permaculture One* and *Permaculture Two*. That was a system using no chemicals! We had done it in Sydney, so surely it would work up here, albeit more quickly. I had discovered crayfish would also die when exposed to chemicals and so we decided to farm them, but managing the rest of the property was still a challenge.

I think this is where my plan and the Universe's plan came together because I just happened to be in the same area as Geoff Lawton. We were both working out of the Cooroy Butter Factory. I knew I could do permaculture in the temperate zone as I had been growing there all my life, but I was not so sure about how to go about it in a place with a different climate and with species I didn't recognise.

So I used the shiniest new tool in my toolkit, my newly replaced intuition, and signed up to do the Permaculture Design Course with Geoff on his property in Cooroy. That was amazing. It gave me the same buzz as when I first read *Permaculture One*. So much commonsense! And expressed so well! It all seemed so right and so clear. I kept saying, "I always knew that," and, "I always thought that." Others in the course were saying the same things. I had found my tribe. From surfie to hippie, to yuppy and now permie.

In 1996 I went to the 6th International Permaculture Convergence in Perth. Quite a few of us from the Sunshine Coast were there and what a blast it was meeting people from the Rocky Mountain Institute, The Farm in the US, people from Brazil, Africa, Europe and all over Australia. It was amazing. We were part of something real and big and growing. It was a huge event for me and one of the big filing cabinet tip-ups of my adult life. It took me a while to put the drawers back, label them and sort the folders. I ditched some old files, put things into a new order, added a couple of drawers and began the most exciting learning curve of my life.

On Track and Fast Tracking

I took myself off to what was advertised at the time as the last PDC Bill Mollison would ever run. Bill has had more comebacks than Dame Nelly Melba (whom we are both old enough to remember), and is now teaching annually with Geoff Lawton in Melbourne. He surely impressed me.

Bang! Inner and outer worlds collided and fitted beautifully together. The gaps in one were filled by the input of the other. Some say that permaculture is a practical science and does not have a spiritual level. But this makes the hair on the back of my neck stand up every time I hear about it. No matter if it happens to me or if I hear it from others (and I hear it a lot), it confirms to me there is a plan and we are part of it. The fact is that permaculture works when done properly. To work, the system must fit with universal laws. Universal laws work on many different levels and in many dimensions. Many practitioners have found this and have taken their findings into some amazing techniques.

Bill Mollison calls this "woo woo." I believe he has concerns that this aspect may overshadow the practical design science and reduce its uptake into mainstream. I also think that coming to permaculture through the practical and grounded level is a good entry point because if, and only if, you do it well you see the magic and have the entry into other levels and dimensions. Just go and re-read chapters 1, 2 and 4 of the *Designers' Manual* if you are ever in any doubt or have been convinced by someone who heard from someone else who thought that permaculture was all just practical tips about land use! Who says permaculture lacks spirituality?

I couldn't get enough. I read everything even remotely linked with permaculture. I was wandering around amongst the undergrowth in a permaculture forest of ideas. I put a bit of this and connected it to that and put that over there and rearranged my thinking. I was joining the bits when suddenly two years later, I was thrown up above the forest canopy and looked down and saw the whole thing in place. It all made absolute sense. I saw the basic structure, the Ethics and Principles holding the branches, each branch an area of action and the finer branches the techniques and strategies. The leaves were all the tools and species names, the resources and all those other details.

It was then that I decided to teach permaculture with the idea of giving students the basic framework on which to hang the strategies and techniques, and then finally put on the leaves of species names and other details. I made a conscious effort to demystify the *Designers' Manual* so it could be used as a tool, and to instill in all the participants a sense that permaculture learning and understanding is achievable; that you don't need to know everything, rather you just need to know its place and how to find the information when it is needed.

It is this about permaculture that blows me away every time. No matter what you have done in your life, no matter if you thought it was useless or stupid or unrewarding, without fail when you take on permaculture challenges and you have to look into your bag of skills, lo and behold, you will need to use that very experience to take advantage of a situation or a resource.

I have derived great pleasure and satisfaction from teaching permaculture and seeing the excellent work of many of those who have studied with me. It is not something I want to do forever but I believe it is absolutely essential that a PDC is offered in every region in Australia and throughout the world.

When You Get the Call You Have to Go or Risk Arrows

Just when my permaculture teaching was ticking over nicely, and life was becoming more relaxed, Permaculture Education called for volunteers. The Vocational Education and Training (VET) sector of tertiary education was changing and we were losing students and funding to courses with recognised qualifications. As a movement, we needed to see if the dream since 1991 of having permaculture accredited within the National Training Framework of the VET sector could be a reality.

So Virginia Solomon, Robyn Francis, Ian Lillington, and I put up our hands. We worked with blind faith and a magic partnership with Guy Rischmueller from Hortus Australia for a year or so before we were sure what we were trying to do was even a possibility. Then we got the news – our application to have a nationally accredited permaculture course had been accepted. There was champagne in Virginia's front yard that day. All we had to do then was write it!

It took a couple more years for it to become Accredited Permaculture Training or APT, a course owned and managed by permaculture through its industry body, Permaculture International Limited (PIL). We achieved, for just a few thousand dollars in registration and administration costs, what other institutions were paying $160,000 to $200,000 for. We had also ensured that permaculture practitioners would dictate what was presented in permaculture courses offered in the Vocational Education sector.

How we managed this shows what permaculture is about. We had dedicated and committed people who valued an outcome more than payment in money (there was none!). In fact it cost us all not just in time and lost income, but in travel to meet, phone calls, office costs and so on. We all came with the skills needed to do the job – hairs on the back of the neck stuff. We were able to work together as a team to achieve the goals despite the fact we were all so very different and some of us had, and still do, butted heads on many issues. We all showed respect for each other and valued our differences.

We walked a tightrope. If we did it well it would support the PDC, attract a whole new group of people to permaculture and get us back into the funding pool. It would give us back the opportunity to work with indigenous people and communities looking to do land restoration, and line up against conventional horticulture and agricultural courses.

If we did it badly there was the potential to split the permaculture movement. This was Bill's main concern and became one of our core challenges. We ensured the PDC course was located in the Vocational Education and Training (VET) Certificate III and Certificate IV levels and the way these are balanced is still an ongoing challenge today. What we ended up with was a course that could be unraveled at any of the five VET levels, into the very same permaculture workshops or PDCs that were running at the time, yet could still meet all the national qualifications framework requirements.

This meant that we needed to train our trainers in packaging and unpackaging the units of work. We wanted our own Train the Trainer Course and the same volunteers, along with Naomi Coleman, once again said, "We can do that!" And we did. The course is known fondly as the COW and CALF. The COW was the Course Orientation Workshop and the CALF was the Creative Adult Learning Facilitation. And that went over really well with a lot of permies having the opportunity to not only deliver permaculture but to deliver any other of their trades or areas of special skill and knowledge in the VET sector.

Our COWs and CALFs became legendary (divine and bovine) and attracted people from the creative industries such as musicians, natural healers, artists and others. If some of them were not permies when they started the courses they were surely 'permified' by the time they finished. We also attracted attention from other Registered Training Organisations (RTOs) as well as the government, and our courses were held up as 'best practice'.

VET was due for its next big change. The government decided that workplace training needed improving (up to our level) and added another seven units to the eight of the earlier certificates. To stay in the game we had to re-write our COWs and CALFs to the new Training and Assessment level (TAA04). I took on writing with Robin Clayfield, doing all the stuff she does so brilliantly. But after delivering just a few my heart was not it any more. By this time I'd had nearly six years of APT. I handed my work over to Robyn Francis and Virginia Solomon. Virginia has since taken that work to a whole new level.

Back to the Little Ones

My heart was back in primary education. That had been my career, and I had never really fulfilled all I had wanted to do there. I had just walked out and started growing crayfish in Eumundi. Surely all that work and all those years had a little more fruit in them! I'd had 22 years in the classroom, lots of qualifications with two degrees under my belt as well as 28 years practicing permaculture. I was armed and dangerous!

I believed that if primary children had a permaculture course within their seven years of schooling then they would demand further learning when they reached secondary school. From my work with APT, I knew they could be offered the Certificate I at Years 8-10 and that perhaps the Year 11 and 12 students could do a Certificate II. I also remembered Bill saying at the many courses that I had sat in on or co-taught, that he thought that someone should write 'The 1,000 Things Every Child Should Know'.

I began to receive a lot of calls from schools to help them with school gardens. So I started to think I would write a manual ... after a bit of a break. Then the next call came. It was Carolyn Nuttall. She wanted to work with me to help schools establish gardens and told me she was thinking of writing a book. I had admired Carolyn's work in the *Permaculture International Journal*, I had used her books and I had seen her presentations at conferences. I was honoured to be asked and gladly accepted the offer. That was the beginning of another wonderful journey. A journey where we wrote a successful book together that supported school gardens and outdoor learning, where we worked as a team despite our diverse personalities and ways of working, and made many new friendships.

The Theatre of the Absurd Returns but No One is Laughing

When our then Queensland State treasurer Andrew Fraser, was asked at a regional conference in 2007, "How can we have infinite growth on a finite planet?" he replied, "Government does growth really, really well. We don't do the other." I gasped, and turned to look for others reacting similarly, only to see calm note-making and nodding. I was just a little concerned.

Sonya Wallace, a student of mine, was also concerned. Sonya had a background in Emergency Services and had been very impressed by a presentation by David Holmgren she'd attended. She decided to do something to prepare the Sunshine Coast for a time when we won't be able to count on governments who "don't do the other." I agreed to help her. The community must be made aware! We designed a course of workshops she called 'Time for an Oil Change', based on the work of permaculture teacher Rob Hopkins in the UK and in Kinsale, Ireland, as well as using David Holmgren's text *Permaculture: Principles and Pathways Beyond Sustainability*. We started our journey towards energy descent for our region and the move to make communities on the Sunshine Coast more self-reliant and resilient.

Rob Hopkins, then in Totnes in the UK where he had started the Transition Town Movement, noticed our work and invited us to join them in September 2007. The Transition Town movement is permaculture in action, facing as it does the twin challenges of Peak Oil and Climate Change. It returns to permaculture its full range of domains of action that have been eclipsed by our tremendous success in the garden and in land management.

Without using the skills and knowledge of permaculture, communities cannot reduce their dependence on fossil fuels whilst maintaining a good, or even improved, quality of life. We are certainly ready for the approaching wave of need as we now have three decades of functioning permaculture demonstration sites in all climates to show the way. We have plenty of practitioners, an excellent range of books and films, we have the education systems and we have the trained teachers.

My vision for the future is one of a gentle entry into a world with less access to fuel and overall energy. A future that has benefitted from the forward thinking that has used the last half of fossil fuel energy to create human-scale, integrated systems that provide for human settlements. I dream of a world where society makes decisions based on ethics and that has returned natural eco-systems to their rightful, fundamental place. This future has an increased quality of life for humans that is more tightly aligned to, and accepts immediate feedback from nature. I believe we need to finally properly fulfill our role as the only species on the planet who can protect all other species.

For more than half my life, permaculture has played an important part in how I think, what I do and how I work. I hope I have made the contribution that the philosophy and the science deserve. Our future will not come without effort and pain but I will remain forever grateful for the thinking that has made navigating the absurdities of this world easier and for the wonderful people it has allowed me to connect with.

If I am to live in the world I vision, I know it will be permaculture and its influences that will provide it.

Janet's granddaughter Molly-Belle Millington loves the family farm and joins in the chores.

Robin Clayfield

Robin Clayfield is a respected facilitator, trainer, author and musician with a passion for creative, interactive group work, permaculture, deep ecology, transition, empowerment work and ceremony. She taught over 30 PDCs from the late eighties through the nineties, created and taught Advanced Permaculture Creative Teachers Facilitation courses with Skye, led many Women's Wisdom weekends and has presented an amazing diversity of leading edge educational and transformative programs. She now focusses on her Dynamic Groups trainings, consultancy and specialist workshops and co-trains the Certificate IV in Training and Assessment course for permaculture teachers and others in creative and sustainability industries.

Robin is the founder of Dynamic Groups, Dynamic Learning, a holistic learning methodology and workshop program and runs her business Earthcare Education. She facilitates at conferences and festivals and offers her consultancy work and courses all around Australia and by invitation in other countries. Robin has birthed three books, including *You Can Have Your Permaculture and Eat It Too* and a CD of guided journeys set to music. She also finds time to play in a band, potter in the garden, be involved in community organisations, home educate her teenage son and be a Grandma.

Chapter 14 | An Edge Species

I've always been a bit of an 'edge species', like one of those persistent pioneer plants that grows anywhere, trying out new and challenging territory whilst striving to improve the surrounding soil and environment. I've never really lived in a city and spent my first four years growing up on a farm, while the rest of my childhood was spent in a variety of small to large towns. As the oldest child in a family that moved around quite often, I had to make the most of new situations and often felt a little out of place. Even as a teenager I didn't quite fit in, preferring to slouch around in my Dad's discarded khaki work trousers rather than strut my stuff in miniskirts, platform shoes and makeup. I was labelled a 'freak' and a 'hippy' and later, as a tertiary student in the late '70s, earned the titles 'radical', 'lunatic' and 'greenie'.

I began to feel marginalised and discriminated against. As a young person, I couldn't understand why the rest of the world didn't understand or agree with what seemed common sense to me. My extra-curricular activities at the time involved student rights, women's issues, anti-uranium mining marches, and well as encouraging awareness of ozone depletion, acid rain and forest loss along with other environmental and social issues. These seemed much more interesting and important than my Social Science degree subjects.

My experience of tertiary education in rural Victoria didn't prepare me for how 'the rest of the world' lived. At 22 when I dived into the big wide world, backpacking and exploring the treasures of Tasmania, I realised just how different I was from most people in our western consumerist and environmentally insensitive world. After a few months apple picking, 'working for the man', the reality of mainstream life seemed pretty scary and I felt even less like I fitted in. I was in a small minority, or so it seemed at the time, that cared about the environment and was alert to the impacts of pollution, forest logging, uranium mining and a host of other issues. I seemed to be repeatedly jumping up and down saying, "NO! STOP!"

Having lots of money, a career and a big house was not at all part of my agenda. I was pretty burnt out after spending my whole life at school and felt quite depressed about the state of the world. I also felt small and powerless, and had no idea what I wanted to do with my life other than make a difference in the world. So I quickly retreated, dropping out to the safety and tranquillity of the northern New South Wales rainforest fringes. I didn't really think about it at the time, but living quietly next to a sparkling stream, surrounded by birds, animals and majestic trees, breathing deeply of a less frenzied environment, was the best medicine I could have had. I began to dabble at growing my own food, building structures, collecting rainforest seed and dreaming up a healthier, more planet-friendly way of being in the world. I lived without power or running water for eight years and learnt to be thankful for the simple luxuries life offered. The occasional hot shower at a friend's place was greatly valued and appreciated.

I had started reading *Grass Roots* magazines for inspiration and support and was drawn to an advert about a women's Permaculture Design Course (PDC) to be held with Lea Harrison in Tyalgum, an hour's drive away. I had heard only a little about permaculture but it struck a chord and was clearly in tune with the way I felt about the world so I registered without really having a clear idea about why or how I might use the information. I did have a feeling that it could help me to grow my own food but other than that it just felt like a good idea. I guess, even in those days, I was following my intuition. The course proved to be a pivotal turning point in my life, changing my worldview substantially. It gave me tools to work for positive change in the world while offering very practical and achievable possibilities for daily living, community interaction and general involvement in the 'big wide world' again. I felt that I had been given a new set of eyes that could see the landscape in a completely different way. I now knew how to approach designing and growing a garden, had ideas for dealing with pests and the confidence to consider integrating chickens into my system. I was more confident working with groups and even started up a small trading system in my area. I embraced the attitude of 'turning problems into solutions', so that instead of saying, "NO!" all the time, I could offer alternatives that were viable, creative and proactive. Upon returning home from the course, and motivated by feeling isolated and alone in my newfound permaculture passion, I actively sought out others with similar interests. Quite quickly we formed Brunswick Valley Permaculture Group.

Local issues soon raised their head, begging for positive solutions. One morning I woke up to the sounds of blasting in the mountain behind where I lived. After several phone calls I managed to ascertain that we had one week until logging would begin in the mostly inaccessible and stunningly beautiful ridge-top forest. Instead of 'spiking' trees and sitting in front of bulldozers, our valley's response to the Forestry Commission 'invasion' was to connect and motivate local people, find common ground with the local family owned timber mills, and suggest alternatives to Government. The small mills were being bought out by a big sawmill, not because they wanted their business and infrastructure or to keep the staff employed, but to gain their 'quota' of the timber allocated for each mill to harvest. They'd then sell off all the plant and equipment and close down the mill, leaving a once family-managed business totally defunct. We talked with them about creating industry jobs by planting up some of the abandoned dairying country in the area or developing an industry from the rampant camphor laurel trees that were choking usable land and had even invaded towns. I became involved in State Government negotiations, made a personal complaint to the Ombudsman and took part in a number of creative campaign strategies, learning many valuable skills along the way. It took many years to save that particular area, but we did, and a unique forest system and all its inhabitants remain for future generations to enjoy.

Another example of success in 'turning the problem into the solution' was my response to a proposed second ocean outfall sewage system for Byron Bay. Previously my initial reaction would have been to write letters, say NO and jump into protest mode. But now, my first step was to gather together all the alternative models and ideas into a big file and thoroughly research the whole issue. I then felt confident enough to propose a ponding system as the most beneficial and environmental option. The Shire engineer eventually called me into his office, photocopied much of the information and went on to design and implement a system that later won State Government planning awards.

Permaculture had very quickly become the central theme in my life. The skills I learnt from the PDC were immediately used to grow an increasing diversity of food, design properties for my friends and work within grass roots community groups from a more holistic perspective. It was during the PDC that I first learnt about Deep Ecology. John Seed, long time environmentalist and rainforest activist, was a guest presenter in the course. His passion and work with Deep Ecology was a delight to experience and seemed a natural embellishment to the permaculture that we were learning about. Studying about forests and ecosystems from the perspective that we share a deep connection with all the beings in the forest helped me know how important and magical natural systems are. The roots of permaculture are in observing and emulating nature. Deep Ecology helps us to not be separate from that. To me they go hand in hand. The integration of this information changed the way I connected with the natural world and how I felt about the state of the planet. My motivation to act came now from a deep sense of being part of nature, rather than a fear that humans would perish and the planet would suffer irreparably.

My passion for permaculture and community living led me, in the late 1980s, to Crystal Waters Permaculture Village in Queensland. I bought into the concept plan and trusted that the designers of this innovative world first ecovillage were heading in the right direction. My two-year-old daughter would grow up with other children all around her and we'd thrive on clean air, pure mountain water and organic homegrown food, all in a context of growing community and living cooperatively. My teenage son, born into the fruition of this dream, now delights in the garden, his connection to nature, the commonsense of permaculture all around him and the many teachers in his life's education. I doubt he would be so confident in the bush if he were city bred. He has so many opportunities available to him and relishes exploring them. Within the community he takes part in weekly yoga classes, is mentored in cooking by one of his second mothers, walks, rides and swims in the river often, visits a neighbour regularly to learn Tai Chi, helps in the café, sometimes waters a bush tucker nursery, goes off to the movies at The Eco Centre on Friday nights, and has learnt bamboo flute making, weaving, archery, building and numerous other creative and interesting skills. His ability and confidence to communicate and interact with people of all ages is a gift I certainly did not possess at his age. He is excited to soon be learning from Les, our baker, how to make his famous sourdough organic, wood-fired bread.

Many children have grown up at Crystal Waters with the ethics and principles of permaculture evident all around them. Whether they absorb them from community and family members, or from understanding the examples, is not always easy to tell. Hopefully they have had an experience of common sense and environmental caring that grows with them into their future. So many other children do not have such an opportunity to live among kangaroos and wallabies, echidnas, goannas, geckos and an assortment of snakes, frogs, possums, bandicoots, parrots, glossy black cockatoos and numerous other birds ... the list goes on. A wildlife sanctuary is surely a very different playground from a town or city. Crystal Waters is also a sanctuary that a diversity of people call home, whether they retire here, raise their families, base their businesses here or go off to work from this home base.

When I return home after any trip away, I always take a deep sigh and say, "I live in paradise."

Being an ex-dairy farm, Crystal Waters wasn't always like this. Today there are many more trees and dams, 83 residential lots with architecturally interesting, environmentally friendly houses with established gardens, a visitors' area and small commercial village with shops, industries and a monthly market. We have licensed land used for agricultural, forestry, grazing and community use and even our own cemetery. There is a lot more to be done, but we have come a long way in 20 years of being a permaculture village community. Villages don't just happen overnight, they evolve over centuries. Permaculture villages grow from the back door out. First you have to build the house and work out from there. Income is still a priority so many residents have established business ventures or other ways of supporting themselves. Not everyone has moved into creating abundant food gardens though there are some excellent examples.

One area of improvement that some of us are working towards is community building. There was no social design done for the property in the initial Permaculture plan - the design was for a village, not a community. The one-acre parcels of private land are spread out in clusters over the whole 640 acres which has seemed, in hindsight, to hinder community building and the creation of a central village focus. With more awareness of this we are growing more together as a community, though some residents still choose to remain quite separate.

Crystal Waters seems to be a microcosm of the wider world. Slowly over time, we get better at communicating and living in a village together. The community provides many opportunities to learn how to work together more caringly and cooperatively and I'm sure that if we can work it out, so can the rest of the world. For me, if we can learn how to respect each other and work together, there is some hope left for the human species. When important community issues (such as land use planning, management restructuring or water supply issues) raise their head, we tend to hold forums, and more recently, 'sharing circles' where everyone can talk uninterrupted about their feelings, issues and ideas. The original plan proposed a system of community 'Elders' who could mediate if there were disputes or conflict between people or with our co-op or body corporate committees. This worked for some years but isn't used any longer. More recently communications workshops have been held for the committees.

Twenty years living in this experimental village has given me opportunities to do my own personal work. My early 30s saw me dive into a journey of self-exploration, healing and inner growth. I'd always considered myself reasonably politically and environmentally aware but now, a new relationship, several synchronistic friendships and the fledgling New Age movement helped me to be more aware of my spiritual and inner self. Re-birthing became a key transformational tool, supporting me to grow out of many old and restrictive patterns. I use affirmations as a daily practise to turn my often negative thoughts into colourful, positive statements and turned my mind and life around. Though I was never a totally negative sort of person, I seemed to grow up not feeling confident in myself. I was a product of a middle class, western, consumer society which was brainwashed to 'work for the man' and look like the model on the front cover of every glossy magazine. I had low self esteem, a poor body image and didn't feel at all creative or musical. I felt quite powerless to take full responsibility for myself and my sometimes rebellious actions. Permaculture had helped me begin to work in a positive way with my environmental, social and political endeavours. Now it was my time to work on my inner landscape, an evolving and never ending journey.

In the mid 1980s I had begun cooking for PDCs and advanced permaculture courses for my original teacher, Lea Harrison. Through her I realised the need for many more people to become involved in teaching as there was a growing demand for permaculture teachers but not enough people with the skills and confidence. Lea encouraged and mentored me in the process by inviting me to teach small parts of her course each time I was cooking. I remember doing my first guest 'spot' on 'Money and Local Economies', not as well as I would have liked, but each time I presented a topic I did a little better. By the time I felt ready to do the full teacher training course with her, I was preparing to teach a PDC at Crystal Waters with my partner at the time, Skye, and two other friends.

After completing the first PDC, even though we had excellent feedback from the participants, Skye and I began to seriously question our methods of teaching. We had been essentially lecturing at people based on the only model we had experienced, as well as being concerned about ethics of our pretending to be the 'expert'. We were teaching soils to soil scientists, building design to architects and plants and forest systems to ecologists. It didn't feel right. We wanted to find a way to draw out and weave together all the wisdom in the group while also working in a much more empowering way by supporting the participants to set their own curriculum, rather than having it dictated by us.

Our early experiments in this direction saw us dive into an intensively creative time with only our intuition, passion and trust to guide us. We promised to never lecture at people ever again and to develop and use creative, interactive learning methods which also facilitated the participants to learn what they wanted to know. We were not aware of any learning theories or the developments in educational psychology, accelerated learning or approaches like Neuro-Linguistic Programming. We had no formal teacher training so were not hampered with learning blinkers about what was OK and what wasn't; what worked and what didn't. This gave us the freedom to create and explore and led to one of our participants saying we were 'permaculturing' education. We had devised what became known as Creative Facilitation.

My journey as a facilitator of PDCs and later of Creative Facilitation Training for teachers, facilitators and group leaders has also been an exciting evolution. The shy teenager who had always been terrified of public speaking and who had never before considered the possibility of teaching anything in an official capacity was now running PDCs and teaching the 'Advanced Permaculture Creative Teachers Facilitation Training'. I could never have dreamed that I would go on to facilitate at least 30 PDCs and other permaculture courses in four Australian states and overseas, as well as numerous Creative Facilitation Trainings and 'Dynamic Groups' courses in every state and a diversity of amazing countries. I have been blessed to be able to visit and work in Mexico, Cuba, England, Germany, Kenya, South Africa, New Zealand and Thailand. Most of these courses were run using an intensive residential format which encouraged strong group bonding and greater connections and networking between all involved. I found that a strong feeling of 'family' permeated virtually every one of these courses and I continue to be involved in residential training whenever possible. Most of the TAFE Certificate IV in Training and Assessment courses that I am involved in these days are held as residential courses and use creative, interactive learning methods to bring the conventionally dry material to life. I find it very satisfying to blend in all my 'Dynamic Groups' work with a mainstream training package to help shake the foundations of education and training and support others to bring permaculture right into the mainstream.

I have learnt so much from all the participants in my courses and feel incredibly humbled by my connections and interactions with people both in Australia and in other cultures. We all have so much to learn from each other. Probably my greatest learning has been to be a 'Lifelong Learner'. Not to set myself up as an expert in anything, but to share openly and genuinely who I am and what I can offer, and be open and thankful for the gifts that others give of themselves. This love of learning, my respect for people and my passion for creative and dynamic group interaction has led me deeper into the landscape of facilitation. Rather than seeing myself as no longer being a permaculture facilitator because I have stopped teaching PDCs, I'm excited to weave the principles, ethics and attitudes of permaculture into the whole learning process and into sharing with trainers and facilitators the opportunity to 'permaculture' education. One of the greatest offerings that touched my heart was a physical gift from Fernandini in Cuba. He wanted to give a gift for the training he'd just been part of with Skye and me. He was giving us a tour around his abundant garden, offering us juicy sugar cane, homemade juice and other produce and finally scaling a coconut palm to cut us each a coconut. He said, "I want to give you so much but this is all I can give." To give of the food that sustains him and his family, to give from his heart, this is surely the greatest gift of all.

I've often been asked, "What is the highlight of your experience working in other cultures?" Of course there are so many that it is hard to pick just one, though probably the most profound experience occurred in a permaculture course we taught in Kenya. Skye and I had a reasonable briefing from our host during the two days' travel to the remote village of Saga (in Kenyan taxis, which are seriously life threatening). Nothing could have prepared me for the lack of joy in the people, the separation of men and women and the sense of timidness that emanated from most of our course participants, particularly the women. After our first teaching day I had a sense that this was a people whose spirit had been squashed by years of western religious colonisation. Men and women didn't talk to each other or publicly connect in any way. It was unacceptable for them to walk up the street together so consequently, the whole village stopped four times each day as Skye, our two hosts (all male) and I (female) walked to and from our training venue together. In our sessions it took a lot of encouragement to support the participants, especially the women, to speak up and share their knowledge or ask questions. On the very last day, as we were brainstorming and modelling around a large central image of their town, helping them dream up all the elements they wanted in their bioregion, I noticed that everyone was having equal input, all laughing and chatting, with the women and men all intermixed and not even aware of it. My heart burst open. This was how it was really meant to be. That afternoon two of the women walked up the street with us. A very symbolic and empowering gesture.

Based on our experiences, in the early '90s Skye and I were encouraged to write down some of our ideas and processes so that they could be captured and shared with other interested people. A couple of years of brainstorming, writing and typing eventually yielded the *Manual for Teaching Permaculture Creatively*, a 320 page ring-bound training manual which has now found a home in over 35 countries all around the globe. We never would have imagined that so many people would have an interest in teaching permaculture creatively.

My journey into publishing was not to end there. Possessed by a need to get twelve years of my life out of my head and into writing, and with the help of Marianne, a wonderful WWOOFer from England, I began a one-year process which culminated in the birthing of *You Can Have Your Permaculture and Eat It Too*. Many a day was spent testing out the numerous recipes in the book to check exact quantities, brainstorming species lists, checking spelling and editing the copious pages that were flowing out of the printer. Many a night was spent speaking sentences and instructions into a micro-cassette player. The next day Marianne would play back the tape and type my words into the computer while the sun was shining on the solar panels, allowing the computer to operate. She could always tell how tired I was or what time of night it was by how slow and drone-like my voice sounded, as most of this was done after the kids were in bed. The book seems to have not dated at all as I still receive many requests for copies, even twelve years later. Occasionally, in the depths of the long gestation, I did wonder if it was worth going to all the effort. Now, after much grateful feedback and continuing requests I feel blessed to have honoured the call to write it all down and for the support I received to self-publish my own book.

Robin Clayfield with her Earthcare Education market stall.

There was another important development still to come. I began to realise after numerous groups, workshops and years of 'inner work', that permaculture wasn't my whole life any more. Unless it could address my need to 'care for spirit' in addition to the accepted three ethics of permaculture, it was no longer providing me with a holistic framework for how I lived, worked and interacted in the world. Also, throughout the many years of facilitating PDCs from a needs-based perspective, I often received requests from participants for knowledge and experience beyond the standard course curriculum. These related to deeply connecting with the Earth and the recognition and importance of inner, spiritual work as well as the outer, practical manifestations of permaculture. Participants regularly called for the course to include 'spirit' topics such as 'Connecting with Nature', 'The Spirit of the Land', 'Song Lines', 'Self-care', 'The Spiritual Properties and Uses of Water', 'Nature Spirits', 'Earth Healing' and other similar themes over the 15 years that I ran these courses. I was mostly able to offer experiences and processes to cater to these needs, as I was by this stage very aligned with many of the requests.

I certainly noticed a pattern emerging within our culture relating to the need for spiritual connection in its many forms. Participants in two consecutive courses suggested a fourth ethic to complete the ethics of permaculture, one which reflected the spiritual side. The opportunity to do this finally presented itself at the 8th Australasian Permaculture Convergence (APC8) in Melbourne. The whole convergence joined me in weaving 'Spirit Care' into all the valuable work and explorations of permaculture that we'd engaged in during our time together. Permaculture had broadened and evolved to include and acknowledge themes such as the 'Spirit of Nature and All Beings', 'Our Inner Spirit and Self Nurturing', 'the Spirit of Enterprise and how we do Business', 'Spiritual Diversity and Renewal' and 'Caring for All That Is'. It is deeply satisfying to be part of this evolution.

If I didn't have this connection to nature and to my inner spirit I don't think I would be able to keep up the intensity of my life's work. I make sure I take time out to spend in nature and do things like attending annual 'deep inner work' events like 'The Joining' (a workshop gathering in honour of the Masculine and the Feminine) and 'Being Woman' Festival. I have a strong support network where I am able to share deeply when needed and I do my best to remember to give thanks for all the gifts in my life. Gardening also keeps me healthy and sane and I love to walk aerobically every morning around the ridges and dams of Crystal Waters. Seeing a mistbow at least a couple of times a year must surely keep me inspired.

Right now in my life I feel very humbled to be in a position to shake the foundations of 'mainstream' education (from the edge, of course), knowing that if our children grow up with a joy and passion for learning and with permaculture inherently woven into the fabric of their experiences, they'll naturally care about themselves, each other and the world. Likewise, as adults connect more with themselves and the Earth, as they increasingly feel a sense of connection and family with others through working and learning in groups, as they find creative and positive ways to work together, then the future of humans on the planet has a better chance.

Personally I don't feel the planet is in grave danger. She's hurting but I sense that she can just shrug her shoulders and we'll all fall off, giving her a chance to heal and regenerate. It is humans and the many other species who are in danger, and I am convinced that the key to sustainable humanity and survival of species is learning how to work together. If we learn that and get really good at it – quickly – we'll be able to make the decisions and implement the strategies that will support a positive future for the children who inherit a care-taking role for our beautiful blue green planet.

I feel that the Transition Towns Movement, initiated by Rob Hopkins in England, is a fantastic and creative example of a permaculture way forward. It emphasises groups, whole towns, entire bioregions and countries even, working together and using permaculture strategies to move to a lifestyle that is able to sustain the population in perpetuity and in harmony. I look forward to the evolving Transition Towns movement on the Sunshine Coast in Queensland where we already have permaculture groups and sustainability groups working in this direction with local towns, the local council and also within my home community of Crystal Waters.

The last couple of years have been an incredibly affirming time for me and many long-term environmentalists. Rather than saying to the world, "We told you so," mostly I sense a feeling of relief. Once the media really got the picture about the state of the planet and started giving it space, the domino effect began to happen. People are waking up *en masse* and the job now for permaculture people is to service the increasing demand for skills, knowledge, information, design ideas, technology, plants, cultures, animals, materials, and group processes. Add in a good dose of hope, passion, love, joy and wisdom to support the transition to a truly sustainable society and we have a great recipe for success that we can all share in together.

Alanna Moore

Alanna Moore is an Earth energy expert (geomancer), a dryland-temperate permaculture farmer (in Central Victoria) and international teacher. A master dowser with 30 years experience, she has created much educational material for her dowsing and geomancy students in the form of seven highly acclaimed books and nineteen films. Permaculture is woven throughout her productions and she has written for permaculture magazines worldwide. A firm believer in the sensitive approach to land use, she incorporates geobiology into her permaculture designing.

Alanna and her giant cabbage, 'Big Max'.

Chapter 15 | Permaculture for the Spirit

Early Influences

Born in Sydney in 1957, I was fortunate to have parents who had an ethos of environmental concern and cultural heritage conservation. My grandmother Nell Pillars was instrumental in setting up the district historical society in the year of my birth. She helped prevent the destruction of many of the district's often stately buildings and there is a plaque to commemorate her achievements in the High Street. We lived around the Randwick municipality in old historic mansions and even a fortress with a museum, on Bare Island in Botany Bay. That was an idyllic time, for I was able to roam the coastal wilderness, in what is now the Botany Bay National Park. There I developed my love of nature, wandering freely amidst beautiful bushland boasting banksias and flannel flowers. School life started and, together with the local Koori kids of La Perouse, I would often eat bush tucker roots that we dug up in the playground. I went on to become the only member of my family to really love gardening, particularly native and bush tucker plants.

In my rationalist upbringing there was no room for spirituality, so from that angle I was a 'blank slate'. When I was going through difficult times at age 14 and spirit helpers manifested themselves to me, I was taken by surprise. I didn't expect them to exist. So I kept it to myself. If I had been living in a traditional tribal society, I think things would have been very different and I might have undergone shamanic training. Later I was able to discover by myself the answer to the mystery of exactly why those paradigm shifting experiences had happened to me. But I had to travel the world first to find out.

Early school years were pretty awful and I was often in another world, writing novels under the desk. High school was just as tedious and my three best mates and I couldn't wait to get out into the world. The teachers were horrified when we finally dropped out after completing fourth form.

"You lot are all Hippies," one teacher pronounced in horror. Later those friends achieved high academic qualifications and became university lecturers, authors and consultants. Not me – I never bothered with institutionalised studies, except for occasional practical courses. I was more enthralled by the school of life. With one of those high school mates, Amanda Lissargue, I went and lived the good life in the country for a few years, exploring community living in northern New South Wales. I took odd jobs in Sydney factories and lived frugally so I could live for a while on the land-sharing community of Tuntable Falls Co-op. Here I learned valuable survival skills. If I wanted to eat a slice of bread, for instance, I might have had to grind the wheat and bake the loaf first!

Unfortunately the valley where that community is located was long regarded by the local Bundjalung people as highly sacred, and they considered it inappropriate for women to live there (as for much of the Nimbin Valley). I never see many women thrive there, I must say. I did meet a male Aboriginal spirit there, a guardian probably, which reinforced the sense of discomfort in me.

At 17 I was ready to move on and discover the bigger world. I worked to save most of my income for frugal overseas travel. Over nine months I was amazed and enriched by the Asian and Indian cultures that I experienced along the way. I discovered ancient animist traditions co-existing with Buddhism, Hinduism and Catholicism. Backyards in Bali and Thailand feature little shrines – 'spirit houses' – where offerings are made to the spirits of place. Bus drivers would stop at roadside shrines to pay their spiritual dues. The great beauty of the landscape, its people, their creations, and the joy and harmony I found there, was wonderfully palpable in such places.

I eventually ended up in the UK, in London – a city of excitement! I had new friends and I kept up my interest in spiritual studies, having probably borrowed every spiritual (mainly Theosophical) book in the Camden library system. I was introduced by a friend to the ancient art of dowsing and this just blew me away! I realised that it could be used as a tool for greater spiritual understandings of life. It opened up wide vistas of my mind and intuition and went on to become the basis for my life's career. I joined the British Society of Dowsers, a 50-year-old organisation which provided fascinating information to further inspire me.

Baby came along late 1980. My one and only child, Sky, was born at home. At the age of three months he was suffering from frequent fevers. One night I had a strange vision where I saw a little spirit being and a line of energy beneath his cot. I had only just heard about the problem that is now called 'geopathic stress', that Earth energies can bring discomfort and illness.

I even tried to pull up the floorboards to see what might be under there. But the important consequence was that I moved Sky's cot to the other side of the room and the fevers stopped immediately. Thus I was put firmly on my life path and went on to develop a talent for locating, by dowsing, energetic structures in the home, garden and greater landscape.

However life still seemed rather meaningless. That was until I went to see Helen Caldicott's film, *If You Love this Planet*. I was impressed by Caldicott's use of her emotional, feminine power to convey her concerns. If anything could save the planet from nuclear war or eco-catastrophe, surely it was love! I dedicated myself to planetary care and became involved in the Australian Aboriginal land rights and anti-uranium mining movement.

The mystery of my life-changing experiences with spiritual beings at age 14 was finally solved when I returned to the family home with my dowsing pendulum and discovered a whopping big Earth energy vortex where I had been sleeping in the family home before feeling compelled to leave home. My niece had slept there after I had left and she also had had disturbing experiences. The site could possibly have once been an Aboriginal sacred site, its energetic effects were re-manifesting in the form of some uncalled for consciousness raising and initiation into other dimensions of existence, a lifting of the veils.

Campaigning for the Environment

When Sydney Greenpeace members saw me in action protesting at the Commonwealth Games in Brisbane, they invited me to take a job with them. As a volunteer campaigner for a while, I dealt with Land Rights Support, Toxic Waste and other portfolios. In those days Greenpeace was a very radical organisation and we operated as a collective. I also trained as a volunteer bush regenerator, using the Bradley Method of minimal bush disturbance, and was active in weeding the bushland around parts of Sydney. I retained my great love of native plants and to this day I advocate to people to select a native plant first for their permaculture requirement, if one is available. Indigenous permaculture is the ideal.

As a respite to trying to save the world, I loved to visit Aboriginal sacred sites around Sydney and found it spiritually refreshing. I also started to do professional dowsing and geomancy work then too and in 1984 I was one of the founders of the Dowsers Society of NSW. It's still going strong, and the largest such organisation in the country. I also began studying natural therapies, and then in 1986 moved to the Blue Mountains, where I published my first book the next year *The Dowsing and Healing Manual*. I had just bought my first home there and lived near an Aboriginal artist, Burri Jerome and his family. I was fascinated to find out that Burri had been given the mission, from Aboriginal elders in Redfern, to 'renew the Dreaming' of the local sacred sites, including the famous Three Sisters site.

Discovering Permaculture

It was 1987 in the beautiful Blue Mountains, where, under the capable tutorage of Rowe Morrow, I was inspired by the principles of permaculture design, and took to it like the proverbial duck to water. If nuclear power was bad, here was a great alternative. A thoughtful means of responding to looming environmental breakdown, for making a lighter footprint on planet Earth, and of creating a future worth looking forward to. Gardeners, it said, could help save the planet, using sustainable and wholistic principles that put the Earth first. It was positive, it was life-enhancing, an antidote to the doom-and-gloom I had previously been immersed in. Permaculture spoke of harnessing the inherent energies of a site – water flows, sunlight, wind, movements of animals and people and so on – and I well knew about the subtle aspects of site energies, from my geomantic view.

Moving back to the Northern Rivers area of north-east New South Wales I took my full Permaculture Design Course (PDC) design course with Jude and Michel Fanton in 1991. It was their local knowledge of gardening that I wanted. I loved their enthusiasm for seed saving too – this got me even more fired up. I became a compulsive gardener and made many gardens at several homes in the region. Also a dedicated composter, I would even visit the circus and bring home elephant manure. Later I became a teacher of composting for Lismore City Council.

I also got involved with three Landcare groups and enjoyed being part of the vibrant communities of the region. Whatever I did was governed by my own rules for sustainability, being that food I grew was locally suited, inputs were sourced close to hand and that anything purchased was never brand-new and was preferably cheap or free. This is how local cultures have developed over eons, gaining their unique regional cuisines, their vernacular architecture and localised, low-energy-input food production systems. For instance, in the communities where I lived in northern NSW, many homes were made with walls in-filled with 3:2:1 sawdust/sand/cement, somewhat resembling Tudor style houses. This made for great insulation in walls. Plus, sawdust was pretty much free and readily available at the many sawmills of the region. The method of putting up formwork, then filling the sections between the noggins of the timber framing is cheap, but oh, so very slow and labour intensive. But there were plenty of time-rich people around to build in such a way, especially as it wasn't as hard to live long-term on the dole as it is today. I often used the services of WWOOFers to build my walls, as my days were spent mostly at the *PIJ (Permaculture International Journal)* office by then.

Since 1983 I had been teaching about dowsing and geomancy as well as writing articles for magazines, and through my own dowsing newsletters. So when I found out that people were needed to help put together the *PIJ* in Lismore I put my hand up. I worked as a full-time volunteer and was on the board of directors of Permaculture International for several years in the 1990s. It was great to be part of a team that was making a difference.

People around the world were lapping up the information, so hungry for permaculture. It was often a case of survival in many parts of the world where permaculture was being adopted. The quality of the *Journal* was very high, crammed as it was with great information. Unfortunately however, a dangerous notion crept in and took sway – that people should be making a living from permaculture. That was an aspect of the PDC that I was also uncomfortable with. It was a bit like multi-level marketing of selling a dream with unrealistic expectations. In the case of the PDC it was that graduates would be instantly qualified to be a permaculture consultant upon completing a 72 hour course.

Gradually the *PIJ* office filled up with people who were drawing wages and wanting more. The magazine's finances were much shakier than anyone realised. On analysis it would probably only have generated income for one full-time editor, if that. It was sad to see the volunteering ethic eroding. But good things can't last forever, the coffers were emptied and the magazine bankrupted. I had loved learning to be a journalist and magazine editor on-the-job and ended up achieving three permaculture diplomas – for implementation, media and teaching of permaculture.

Farming at 'The Channon'

For much of the 1990s I was fascinated by rare breeds of poultry, having bought my dream farm, a 5-acre plot at The Channon, inland from Byron Bay. I kept up to 10 different breeds at a time, including several colour variations within a breed too. So – lots of breeding pens and rarely a holiday! All part of the rhythm of life when you have a school aged child.

I wanted to create productive and edible landscapes wherever I could, preferably on previously degraded or wasted land. It seemed to me to be a criminal waste to have land and not grow something edible on it. At The Channon we became virtually self-sufficient for much of our diet – producing fruit, vegies, meat, eggs and spices. Bartering became a way of life too and we lived pretty well, despite not much cash. Yet capitalism inculcates the view that 'poverty', or living simply, is a bad thing.

Twenty years ago I had a friend who, as a 'Hamaker Coordinator', (people inspired by the Hamaker/Weaver book and wanting to help remineralise the Earth), was trying to get people to remineralise farm soil using crushed rock. But people in those days were not very responsive to this idea. Hamaker had argued that people's health will be totally undermined if they eat food grown on mineral deficient soils and that civilisations will fail, unless degraded farm soils are brought back to health and our mineral needs provided from our food. This still holds true today. Fortunately Professor Phil Callahan, author of several books (Acres USA) came along with his observation that the best rock dust to use for re-mineralising the soil is paramagnetic basalt, and that its energy gives biological processes a real boost.

I ended up with a botanical wonderland of edible landscaping and a menagerie of livestock. The not-very-fertile sandy soil evolved into rich, lavish soil in a few years, thanks to all the poultry manure and scratching, plus rock dust, Power Towers and weekly mulching with a big load of broom millet waste that came from the traditional broom factory in Lismore. A Power Tower is a paramagnetic antenna, located according to dowsing, which generates a spherical energy field that can benefit the health and fertility of plants and animals within its area of influence. American Professor Phil Callaghan came to Australia in 1993 to run a few workshops in Castlemaine and Lismore and he whet my appetite for these subtle energy installations, which were inspired by the Round Towers of Ireland. I have since created over 200 Power Towers across Australasia and even in Ireland and other parts of Europe.

With the rare breeds, as I often sold poultry breeding stock, I was able to flood the area with rare chook breeds, making them less rare. These were the breeds more suited to the outdoor, free ranging, permaculture lifestyle, so it was an antidote to the trend towards factory farming, which I was very much morally opposed to.

For me, the institutionalised cruelty inherent in factory farming of livestock is an abomination. How can we call ourselves civilised if it continues? Permaculture provides many of the answers. Again I was able to transform my anger into something positive. Poultry articles that I wrote for various magazines were collated to become the basis for my book *Backyard Poultry - Naturally*. I was also featured on an SBS TV documentary conveying the permaculture view of chook keeping amongst other methods. I'm still waiting for someone to market the 'permaculture egg'. For me this type of egg production would involve moving poultry through pastured alleyways along diverse food forest rows on a rotational basis. Fowl, being from the tropical rainforests of Asia, prefer to scratch a living in a shady, jungle type environment, very unlike the bare grassy paddocks that typify free range egg production today.

Geomancy and Permaculture

For some years I had harboured a longing to get back to doing more eco-spiritual work, which I had made no time for in my life. I had stopped producing my dowsing and geomancy magazines in 1989 and rarely practised or taught dowsing for a number of years. So life was really lacking, despite the excitement and promise of permaculture. I wanted people to realise that geomancy can be very relevant for today. So I started to present talks and workshops on practical dowsing and geomancy whenever and wherever I could, including the National Permaculture Convergence held at Crystal Waters in 1993. Many people responded enthusiastically to my approach to permaculture, however I was aware of the ire it could raise from the 'party faithful'. Bill Mollison had warned me when I first met him that, as the new member of the editorial team at *PIJ*, I should "not get too spiritual." But I was not one to listen.

If we look back at how cultures who were intimately connected to nature and the spiritual dimensions of Earth have sustained themselves over millennia, we can start to imagine what might be gained if we choose this approach today. On a small-scale, the benefits of maintaining good geomancy have been seen time and again. I knew that there was a general ignorance amongst Australians about Earth energies and their affects on the wellbeing of plants, animals and us. The inherent energetic qualities of Aboriginal sacred sites was also not generally known, nor what might happen if we find ourselves living over sites that have been razed. Geomancy had always seemed, to me, a perfectly obvious and totally legitimate tool for helping to analyse landscape qualities and for aspiring to live in harmony with the land. You might say that it offers a more feminine approach to permaculture design. Until then, permaculture came across to me as a very 'blokey' affair. Massive amounts of earthworks at sites, for instance, can greatly disturb things energetically and ideally need to be done with sensitivity. To honour the land where we practise our permaculture is to do it justice. My approach allows for the building of a bridge between European and indigenous cultural paradigms. In Australia we desperately need to find common ground here, so both cultures can work together for the sake of environmental integrity at least. However some people take a negative view of geomancy in permaculture. Why they just don't leave it be I don't know. Perhaps they have unrelated anger issues. Fortunately dowsers can become quite thick-skinned, as they suffer many insults.

At the Crystal Waters Convergence, a stall-holder who was a psychic reader told me that my path of permaculture was not really spiritual enough for me. I argued that it was a practical form of spirituality. I could see no difference anyway, between spirit and practice. If, for example, one was to bring together one's community through the creation of a community garden, this could well provide a great blessing for the people and the land on all levels. Not long afterwards, around the mid 1990s, I had a significant dream that left me feeling tragically lost and depressed. In this most lucid dream I was one of a bunch of maybe a dozen or so ragged children, out in some windswept desert, under dark skies. We were lamenting that we had lost the Sacred Tree. Would we ever find it again? There were no trees in sight, just sand and dust blowing in a howling wind. Onwards we searched, on and on and on we struggled along through that acutely surreal, devastated landscape. The feeling of desolation and loss was heavy. The land was wasted and there was no Sacred Tree to be found. I was very shaken by the dream and didn't know what to do. Generally I considered lucid dreams to be portents of change and this seemed to point to a bleak future on Earth. But I allowed the changes that I needed to make, to express the true essence of my being, to unfold in my life.

In 1996 I attended the International Permaculture Converge, a brilliant event held in Perth. There I presented talks and workshops on rare poultry breeds plus the integration of geomancy in permaculture. The Western Australians were very receptive, indeed, the convenors of PAWA (Permaculture Association of Western Australia) had already invited me out there in 1994 to teach dowsing and geomancy, which I have done on several occasions since. Like me, these women (the amazing Pat Dare in particular) had a balanced, or holistic approach to permaculture, and knew that we need sustenance for our spirits as well as our bodies.

By 1997 Sky had left the nest and I was free, so I began to actively seek work as a teacher of geomancy. Life really started to get interesting and things just flowed along – a sign that one is following one's heart. I began taking many long journeys, driving wherever people would ask me to come and teach. I would often find myself amongst groups of farmers in far-flung corners of the country discussing all manner of things esoteric that would benefit their farming practices. They were very open to trying out anything and there were some spectacular results. For instance, some farmers took up using dowsing to select minerals or remedies for their stock and went on to report a massive improvement in animal health. Farmers who installed Towers of Power were reporting lusher pastures, friskier bulls and heavier crop yields. Certainly there is the potential for a reduction in chemical use when these sorts of techniques are employed, so it can be win-win all round. Farmers have been poisoning themselves and their descendents in advance for too long, and the suicide rate for Australian chemical farmers is very high. I now hear great feedback from farmers whose passion for farming has been re-ignited by using energy techniques – that good old 'wow factor'.

Permaculture is about designing sustainable systems using the inherent qualities and energy flows at given site. Geomancy can provide a spiritual component to permaculture design, helping us determine the ideal placement of design elements, in order to maintain or enhance good *feng shui*. It recognises and attempts to work beneficially with Earth consciousness in its various forms. Geomancy and animism are closely related, but not all geomancers use an animistic approach. Permaculture and the geomantic paradigms are complementary; geomancy is another useful tool in the permaculturist's palette, and all are perfectly complementary for a holistic approach to Earth care.

But how can we gain an appreciation of the physical, spiritual and consciousness aspects of place? First, discover your locality. Check out your local region. Go slowly, cars are too fast and disconnect us. Find out the local history and discover where one can visit indigenous peoples' special sites. Develop a relationship with these places. If we visit the sacred sites with our senses and our hearts wide open, we can learn amazing things. One can absorb the Earth's wisdom firsthand, directly, at sacred sites. Especially at initiation grounds, such as in eastern Australia where they are often marked by circular earthworks called 'bora rings'. We can also visit sites where horrible things have happened and send some healing, loving thoughts to the place. Saying "sorry" to the land is long overdue in many cases. It all helps.

Walking the land in an open and respectful manner is also recommended at the very beginning of the permaculture design process. Where does the place feel special, or particularly energetic? These sites must be treated with care, ideally never to be built upon or disturbed.

Before major upheavals, such as earthworks, are begun, the respectful way is to give plenty of warning to the place about what is about to happen, well ahead of, and up to, the event. The same applies to tree cutting and branch lopping. Nature is intelligent, so talk to it! I have been respectfully developing my own land and there have been some beautiful results.

Mankind has been 'at war with topography' (as journalist John Pilger described the Vietnam War) for too long. I think that we owe it to the Earth to take a gentle, caring approach to our custodianship of Her. That can start in our backyard.

Creating a personal sacred site might be what is needed. The ancient Greeks would devote one wild untamed corner of their gardens to nature. This 'temenos' (wilderness patch) can be a great way of helping to conserve nature's biodiversity, but is usually kept out of bounds. Perhaps a circular oak grove? Or an artistic outdoor altar that could be a focus for the peaceful pursuit of creativity. A spiritual longing for harmony with nature has been building in the last few years in our communities. Its symbols are starting to pop up in unexpected places – labyrinths of stone in local council precincts, Aboriginal art captivating the world, and books about the fairies and Green Man, epitomising deep connection to wild nature, are now proliferating. The fairies seem to want to be acknowledged again.

Perhaps this is because mankind in the 21st century has inherited a spiritual desert, with orthodox religions giving justification for domination over nature. The denigration of indigenous wisdom worldwide paved the way for the cold hearted commodification of the planet, where a beautiful tree or rock outcrop becomes just another 'resource' to be plundered. Although ancient, the arts of dowsing and geomancy are still valuable tools today as they can reveal to us how we are all intimately connected together in nature.

Dowsing provides a means for intuitive decision-making and the detection of subtle energies. It can be applied to the selection of soil inputs, animal health remedies and the like, and it seems that with it we can tap into Universal Knowledge. Dowsing can thus empower us to be our own experts. Farmers in Poland that I taught were very happy to discover the dowsing pendulum. They told me that they knew of nowhere to offer them laboratory soil tests and now they had something they could do themselves, Similarly, permaculture encourages people to learn new skills and trust their ability to work with nature, meet their needs and create a better future for their community.

As for how dowsing and geomancy work with edible landscaping and permaculture, I like to use dowsing for geomantic land assessments when designing. It is a good starting point to find out whether any subtle energies might be impacting on a site before I can decide the best plans and strategies to use. The special energy points that might be discovered are then kept as free as possible from human intervention and ideally form part of one's 'zone five' wilderness area, the sacred 'temenos' corner for an eco-spiritual gardener. Here insects and other wildlife can proliferate as part of the healthy ecological balance. And some of the crop can also be tithed to nature as a feeding ground for insects to munch as much as they like. Plants for bees and butterflies should be included in the garden and a watering point for birds is a must as well. Plants also grow better when positioned at planting time according to dowsing, for there are small points all over the Earth that can give enhanced or detrimental effects on growth.

Design forms are ideally curving, if geomancy is considered. In pathways and pond sides there's nothing like a sine wave for extra 'edge effect'. In Permaculture this refers to the increased productive interface between two sectors such as land and water, and it is good for *feng shui*. In addition, if the gardener or farmer is an animist, one might find a high energy 'control point' to be used as a sacred site, with stone arrangements or 'Power Towers' and ritual to focus the power of positive, loving thoughts into the good Earth.

Moving to Victoria

A move to the Victorian 'deep south' in 2000 brought fresh perspectives. In Central Victoria's Castlemaine I worked on the excellent *Green Connections* magazine for a time. My editor friend Joy Finch had created a regional permaculture-inspired magazine that was not afraid to mix stories of the heart and spirit with practical how-to-do-it sustainable survival tips. Normally you would find 'alternative' magazines that had either an environmental or a spiritual focus. *Green Connections* appealed to both and was ahead of its time, for nowadays people are more open to non-mainstream forms of spirituality, 'greenies' included.

Towards its last days the *PIJ* was offered to *Green Connections* and merged with it. But this strained *Green Connection's* finances greatly. Profits had been meagre before anyway, and mid-2000 saw the introduction of the Good and Services Tax, which immediately caused the demise of many small businesses including magazines. *Green Connections* quickly went broke, despite our best and most feverish efforts to save it. Now, with the demise of the two Australian permaculture magazines, the only other environmental publications radical enough for me to want to write for were *Earth Garden*, *Permaculture UK* and *Acres USA*. But at least now I was a free spirit again and this time I took the plunge with my own writing ventures. I had raised enough money, through years of renovating several homes, to publish a book of my own. Three intense months of writing and editing in 2001 produced *Stone Age Farming: Eco-Agriculture for the 21st Century*. This book described the esoteric techniques that I had been introducing to Australasian farmers over the previous six years. It includes a chapter devoted to geomantic permaculture design.

Without many animals to tie me down I began an intensive teaching phase. Several journeys took me to Alice Springs to check out Aboriginal culture there. Soon I was producing books on geomancy and making films about the geomantic insights and esoteric techniques of Earth care. Everywhere I travelled I found out about amazing success stories in farming, where people have been thinking outside the box and solving all sorts of environmental problems by intuitive thinking and energy work. I also filmed Bill Mollison in Tasmania and other permaculture stalwarts as they showed off their gardens. But apart from that film, I kept my focus mainly on producing educational material designed to foster a sensitive approach to land stewardship. It was an approach that really seemed to be starting to resonate with greater numbers of people.

At the same time I was on my own journey in my own backyard as a custodian of an Aboriginal women's sacred site, comprising an ochre quarry at a spectacular rock outcrop, a place of great power and dreaming on 'my' land. The female spirits of place were the definite bosses there, I came to discover over the years. It remains a place of great enlightenment for me. And a hard slog in the drought to establish an edible landscape suited to the long-range forecast of lower rainfall levels.

The arrival of the internet took up the slack of permaculture information dissemination. What a goddess-send! For me too the world was opened up via my new website, and international speaking events, starting in 2002. In 2003 I presented a talk and workshop at the British Society of Dowsers' 70th anniversary congress held in Manchester, as an international celebration of dowsing. People in many places were as hungry as I was for intuitive approaches, eco-spirituality and a more caring attitude to our planet. Publishers elsewhere in the world were starting to take on my self-published books, and now *Stone Age Farming* is in Chinese and American versions.

The Future

With the passing years it has taken an ever gloomier, ever closer looming environmental crash, in terms of climate change, as well as severe drought in Australia, for people to wake up *en masse* to the un-sustainability of their lives. The green agenda has been elevated and the need for permaculture grows ever greater. We need a Climate of Change! People's psyches cannot be fed doom and gloom and survive for long. I think what is needed is for us to re-focus and instead use our powers of love and intuition to create a gentler, happier, healthier world. By nourishing our spirits in nature we can honour our connection to it and fill the spiritual vacuum that is so prevalent today. This approach may not be for everyone, but it is possible for anyone. We have the power and it comes from our hearts.

There are many people in the world who recognise the existence of elemental spirit beings, which personify the intelligence of nature, and they are actively working with them in their gardens and farms, using the 'co-creative' approach. Shining examples of the resulting phenomenal growth have been documented in the 1960s at Findhorn in Scotland and in more recent times by Machaelle Small-Wright, author of a number of books about her Perelandra garden in the United States. For many more people this can be a good approach to ensuring future sustainability in an unpredictable world. I use my dowsing faculty to locate where these devas reside in the land, so that we can preserve their homes, open lines of communication with them and to assist them with their work, in order to maintain Earth harmony. Just as traditional indigenous peoples might do. Once again, this work is not for everyone. But for those who take the plunge, it is a wondrous journey and never dull.

If we look back more than two thousand years ago around the world, it seems that people were thinking pretty much the same way and seeing the world through animist eyes. Nature was perceived to be alive, to be conscious, to have feelings. People honoured the high energy centres, the places where nature spirits congregated, and made these their sacred grounds, where ceremonies of consciousness raising were enacted. People carefully preserved such places and would never dream of damaging them, let alone building houses over the top of them. In Australia this has been ignored. The consequence of this is that such homes will have a high turnover of occupants. People will feel uncomfortable, unhealthy or spooked. Strong and tainted energies can irritate and disturb. Some people will go mad or fight with loved ones. Divorce, illness and accidents will be more common in such places.

Globally landscapes have suffered the greatest since the demise of animism. The rise of capitalism saw land become a commodity, so in our minds, it obviously had to be divorced from the sacred to allow this to happen. For me, geomancy and permaculture help return spirit and a sense of connectedness to our perception and ways of working with land, and with the Earth.

As for the future? Currently I am finalising my latest book *Sensitive Permaculture* and promoting and teaching permaculture in Central Victoria and also Ireland, with a website of my writings, including many archived articles. In it will shine the light of idealism and enthusiastic applications of the permaculture ethic. The workings of an eco-spiritual community will be pictured and love will rule the day. Heaven on Earth is out there waiting to be manifested by us. First it must be imagined, then comes the permaculture plan.

Naomi Coleman

Naomi Coleman is a primary teacher and permaculture educator, and with her husband Rick, runs the Southern Cross Permaculture Institute (SPCI). As well as actively promoting permaculture education, she co-convened the 2005 Australasian Permaculture Convergence in Melbourne, (APC8).

Naomi and Rick have developed their 10-acre property in South Gippsland in Victoria as a permaculture demonstration site. With their four children they regularly conduct Permaculture Design Certificate courses on site as well as hosting WWOOFers (Willing Workers on Organic Farms).

Naomi is currently working as a Teaching and Learning Coach with the Victorian Department of Education. She is passionate about catering for diverse learning styles, and brings this background to SPCI's permaculture courses. The courses are well known for their dynamic teaching approach, which enable participants to be actively engaged in learning, so that they gain the confidence and skills needed to turn the theory of permaculture into practice.

Naomi believes that the richness of permaculture has added to her life and that of her young family.

Chapter 16 | Finding the Inner Balance

From the first moment I heard the concept of permaculture explained, I knew I had come across something of immense importance, something that could offer me the hope I was looking for, as well as a direction and a vision of what could be.

When I introduce our Permaculture Design Certificate courses, I always start by asking people to reflect on what has brought them to this point in their lives, that they are ready to commit two weeks and a sum of money to learning about permaculture. I ask them to consider the journey that has brought them here, and what they hope to be able to achieve once they leave. I have heard incredible stories from people; of social isolation; of a gradual awakening that the world needs to be a different place if we are to be sustainable; of people knowing in their hearts that there must be a better way to live. My own story is similar to those I have heard so many times.

Starting Out

My partner of one year, Rick, and I had left Melbourne, the city we had both grown up in, seeking a different way to live. We weren't sure what we were looking for, but we knew that we wanted a different way of life, one that took environmental and social justice issues into account. We had spent a year in Queensland, and had come back to Victoria to try country living. We knew what we didn't like, but had not yet defined what we would do differently. I had a growing sense of social justice after living in Cairns, observing the living conditions of indigenous Australians, and the appalling treatment metered out to them by 'whitefellas'. Rick had befriended some community elders and we had been invited to Yarrabah, in far North Queensland, where we saw firsthand how Aboriginals who were able to live a traditional lifestyle were proud and strong in their heritage and their desire to pass on their culture to their young, and to share it with us, and any other white person who showed respect and interest. This differed markedly to the conditions we saw in Cairns, where racism was rife, and Aboriginals were treated as a subculture, to be kept out of places such as the pub where Rick worked.

I was a primary teacher, already using environmental education as a means of engaging my students in real world issues. I knew I didn't like consumerism, and felt strongly about the perceived status attached to material belongings such as houses, cars and clothing, and the superficiality of striving for that lifestyle. Rick was already experimenting with growing food and a variety of heritage and useful plants.

One night, back in Victoria, Rick and I were sitting in our little old rented farmhouse just out of Wonthaggi, in South Gippsland. We were watching the only channel we could receive on the television, the ABC, when on came an old fellow by the name of Bill Mollison in his first foray into television – *In Grave Danger of Falling Food*. It was 1988.

After watching Bill Mollison I clearly remember Rick's initial response… "This guy makes me feel like *I'm* not the dickhead!" This response was predicated on the bewilderment and often vocal comments people made on our decisions to live a different lifestyle. Many could not understand our ethics and beliefs, and thought our views on the impending global warming crisis unbelievable and alarmist.

So it was this sense of relief, this clear direction and sense of purpose that permaculture promoted which really attracted us, probably for different reasons, but with a similar goal. We felt an immediate connection to the concept of permaculture, because of its clear design guidelines, its practical commonsense approach, its integration of such diverse fields as farming and architecture, land care and overseas development, community development and vegetable gardening.

On a previous trip to Melbourne we had bought a pile of books from the ABC shop on organic gardening and when I mentioned to Rick that I thought one of them might have been about Permaculture, he literally raced me to the bookshelf. Sure enough, it was *Permaculture One*, by Bill Mollison and David Holmgren. Rick beat me to it, so got to read it first, but I'd get home from work and he'd fill me in on what he'd read that day. The book expanded a lot on the concepts we'd seen on the TV program, and our journey began in earnest.

By moving to the country we had inadvertently isolated ourselves from many of our friends, not by distance, but by lifestyle choice. Some of our city friends seemed to feel that our rejection of their lifestyle reflected on them personally. Others couldn't understand why we wanted to live in the country but respected our choices, and some told us how lucky we were! All seemed to feel that the drive out to us was far longer than the drive back to the city. My family was supportive and mostly wanted to learn more about why we were so interested in this way of life. So after reading everything we could find about permaculture, I finally felt that I could rationally explain our lifestyle choices to anyone who asked, and I had a framework for living in a more sustainable way. (It took many more years before I could explain in a sentence or less what permaculture actually is!) Knowing there were others out there working towards a common goal, who had already held convergences and were forging a way ahead to a sustainable future, alleviated the sense of social isolation that we were feeling. It also gave us a positive way forward. Rather than the negativity of railing against a system that we felt wasn't working, we could now focus on investigating a solution-based alternative.

Family Background

I grew up in an orthodox Jewish household, and attended a Jewish school, where I never felt that I fitted in. My parents had strong social justice values manifested in a commitment to work with the Jewish Community. My father was awarded the Member of the Order of Australia (AM) in 2008 for his outstanding commitment to community and interfaith relations.

My parents were both born in Australia, so were very different to the parents of many of my fellow students, who were children of Holocaust survivors. Those who had survived and had made it to Australia were often driven to provide a childhood for their children that they had never had. They were driven to succeed in business so that they could provide this lifestyle for their families, perhaps to prove in some way that their survival was meaningful. Looking from the outside, it appeared that status came from money, but I am certain for the parents it wasn't about money itself, but about security. However for the children, by the time we were in high school, status was associated with money, and a consumer lifestyle. Certainly religiousness wasn't considered important; those of us who were religious were often ridiculed. All of this meant that I felt like an outsider on many counts. I began looking for alternatives.

By the time I was 15 I had rejected the tenets of a religious existence, and I knew I didn't believe in a god, but my parents were activists of a kind, and it rubbed off. I felt I needed something to believe in; a passion, a meaningful existence. I always wanted more than a superficial consumer lifestyle. After 10 years in the wilderness (to borrow a biblical expression!), permaculture seemed to offer a way for me to contribute, to express myself, to work towards achieving my leftist socialist views that we could create a better, more just and equal society. It also gave me a sense of belonging to a community, and this has been an important part of my journey. Interestingly, my mother has now combined her work in the Jewish community with environmental work, and was a founding member of JECO (Jewish Ecological Coalition), which has held one of their Annual General Meetings on our property in conjunction with a tour of our site.

By 1990 Rick had enrolled in David Holmgren's first Introduction to Permaculture course, which was held over a weekend at Melliodora, his property in Hepburn Springs. Rick did the course with a friend and I was fortunate enough to sit in on many of the sessions. In 1991 I saw an advert in our local paper for a Permaculture Design Course (PDC) that was to be held over the school holidays and I knew that we should do this course together. Although it was Rick who really took on the practical side of permaculture, widening his repertoire of skills and knowledge, if he had completed it on his own, I would have been left behind; even in the short time since the introductory course I had seen the gap widen. So we enrolled in the PDC in Chiltern with Vries and Hugh Gravestein. The course was intense, and I learnt a great deal.

Finding Our Place

When I first came across permaculture I was young and naïve, idealistic, with a utopian view of bettering the world. In my first encounters I was disappointed that some of the permaculture people I came across seemed no different to the rest of humanity, despite an ideal set of ethics. I observed people in meetings, trying to spread the message of permaculture but getting bogged down in petty personal agendas that just seemed to waste energy and time. I watched people with strong personalities dominate the less assertive in setting the way forward, and I wondered how we could change the world if we couldn't change ourselves. I tried valiantly in one organisation to lead an educational agenda only to be told to leave it to those with experience. Following the permaculture adage of spending energy where it counts, I stepped out to form my own networks, rather than trying to change existing ones. I have since come to realise that most permaculture people are making a sincere effort to work together to lead a more sustainable lifestyle. Even if that effort isn't always all that it has the potential to be.

Over the next few years, Rick and I started teaching locally, beginning with Introduction courses at community houses, from which we formed a number of successful local permaculture groups.

We had heard that PDC graduates could undertake a second PDC for half price, to further develop their knowledge of the course material, but we could not find a provider who was willing to let us participate for a reduced fee. Turning the problem into the solution, we decided to conduct a PDC of our own, but to hire those teachers we wanted to learn from. By a stroke of luck, Andrew Jeeves, the illustrator of Mollison's *Permaculture: A Designers' Manual*, was featured in a front page article in our local paper at around this time. He had moved to Leongatha, where we owned a property but were not yet residing. Andrew was reported as being part of a 'cult' movement known as 'All One Voice'.

We contacted him, and asked him if he would be interested in teaching on our PDC. He was. So too was Ian Lillington, who had worked with David Holmgren for a number of years, and who has now published his own book on permaculture. After advertising amongst our local groups, we had 17 students, a venue (the huge lounge room in our rented house) and quality teachers. Our PDC career had started. It was September (school holidays again!), 1993. I was still teaching full-time at the local primary school, and filling my holidays with teaching permaculture courses, a practice which continues to this day. (I am writing some of this whilst in Tasmania, teaching a PDC over the summer break.)

Our aim was to develop our knowledge of the material, and to use my background in education to develop more interactive teaching techniques, that complemented our view of how adults learn best. Whilst direct lecture has its place, it is only one of many ways for people to learn, and when dealing with complex scientific concepts, not everyone learns readily this way.

Enhancing Permaculture Education

My journey into enhancing permaculture education had begun. I met Caroline Smith who was conducting research into how people learned permaculture and what the long-term outcomes were for PDC graduates. We put together a booklet of effective teaching strategies for permaculture educators, based on some of the educational research we had studied.

It was also in 1993 that Rick attended his first International Permaculture Convergence in Scandinavia (IPC3). Rick found it exhilarating to discover that there were so many like-minded people all over the world, involved in so many interesting and exciting permaculture projects. There were plenty of opportunities to get involved at the organisational level, and Rick also ran an education workshop with Robina McCurdy (NZ), Thomas Mack (USA) and Martha Mondragon (Ecuador). The thrust of the workshop was to share effective teaching methodologies. All four teachers went to IPC3 to learn more about permaculture education techniques; all realised the existing knowledge was threadbare and they had a lot to contribute. The workshop was well attended, and formed the basis of many presentations Rick and I have given over the years since, as we strive to improve the quality of how material is presented to PDC students so that they can turn the theory into practice and become confident activists.

We were also very interested in designing the PDC course to reflect the principles of permaculture. Our second PDC was taught at a rundown community house in Frankston, with full funding for participants from the Shire Council. Their aim was to secure a commitment from the graduates to improve and enhance the existing building and neglected garden area. Our aim was to have a venue for our expanding local group to meet and call home. The PDC ensured a win-win for all parties, and a relationship that lasted 15 years with the Two Bays permaculture group.

It was a whirlwind time! I started a new job in Leongatha, commuting an hour each way for the first 6 weeks until we were able to move to our property, working full-time all week and going back to teach the PDC on weekends. On the last weekend of the course we were about to celebrate with some blackberry wine that we had made as a group, when one of the participants told me I shouldn't be drinking wine as I was pregnant! I told her I doubted it, as we'd been trying for 5 years and had been told we were unlikely to conceive naturally. Our plan was to become more skilled in teaching permaculture, then to travel overseas to teach in developing countries, and possibly adopt children either overseas, or once we got home. But the next day I took the day off work as I was extremely tired, and went to the doctor, really just to get a certificate. I sang the whole way home after getting a positive pregnancy test result.

Our land in Leongatha has proved to be quite fertile actually; we now have 4 children, and at least two others that I know of have been conceived there. The timing was excellent. In the days of Jeff Kennett (Liberal Premier of Victoria), I was finding it increasingly difficult to teach in the State School system; my beliefs and philosophy of education were challenged daily, even though I still loved teaching the kids. I took a separation package, we paid off the farm, and opened the door to possibilities for furthering my permaculture career.

Back in 1995, while pregnant with our first child, we travelled to South Australia for APC5, the 5th National Permaculture Convergence and our first. We did so much networking, meeting people involved with permaculture from all over Australia. There was a feeling of intensity and energy in all our interactions, and it was an incredible experience to come together with so many passionate, intelligent and active protagonists. There we met Robin Clayfield, a permaculture teacher from the Crystal Waters Ecovillage in Maleny, Queensland. She was also, independently, on a quest to make the teaching of permaculture more engaging and interactive. Without an educational background, she had intuitively reached similar conclusions to us, as had Robina McCurdy, the PDC teacher from New Zealand whom Rick had worked with in Scandinavia. Comparing strategies and philosophies, we all learnt a great deal from each other, and by the end of the convergence we offered another workshop to other permaculture educators. It was quite well attended, and we sold out of the copies of our educators' booklet.

I also met Graham and Annemarie Brookman, permaculture teachers and owners of The Food Forest, in Gawler, South Australia. We visited them after the convergence, which saw the beginning of a fruitful friendship. We were warmly welcomed, and I found the property (in dryland South Australia) to be inspirational, but the biggest impact (apart from the delicious homemade pistachio ice-cream) was meeting their children, Tom and Nikki. Tom was about 10, and proudly showed me his design for a permaculture BMX track, where he could ride his bike whilst picking fruit to eat along the way. Now when I teach about 'snack tracks' to school teachers, I often recall Tom's enthusiasm and knowledge. When I recently met Tom again, as a young adult, I told him that he had had a big influence on me as a young mother-to-be. When people comment on our children's knowledge and love of our property, I do think back to those permaculture children I met, such as Oliver Holmgren, and the impact that their enthusiasm for permaculture had on me. I hope that I have also inspired others who have met our children and lived with us on courses and as WWOOFers (Willing Workers on Organic Farms), and that the passion for parenting in a permaculture environment grows through these experiences.

In 1996, pregnant again (I got quite a reputation at Permaculture gatherings over the years, I always seemed to be pregnant!), we travelled to IPC6, the 6th International Permaculture Convergence, in Perth, Western Australia. We arrived early and offered to help the Convergence team, and so got to know many dedicated 'permies' from Western Australia, contacts who have remained friends over the years.

We met Dave Coleman, no relation, but who was an older version of Rick in so many ways. Quite a few people were tricked by the two of them joking that Rick had found his long lost dad, they looked so alike. We spent a week with Dave and his partner Claire, helping on their PDC course, and forged a friendship that continues to this day. I remember doing a tour at IPC6 of a 10-acre permaculture property, and as we walked around I realised I no longer knew where the house was. Having just started implementing our own bare 10 acres, I wondered whether anyone would ever get lost on our property one day, not quite knowing where in the jungle they were. It happens regularly now.

The experience of meeting delegates from all over the world was humbling; hearing of the projects in countries such as Brazil, where Permaculture Institutes were being established; Guatemala, where activists were teaching permaculture to their Mayan communities; Africa, where permaculture training was providing programs for people in Malawi and Zimbabwe... Hearing people's stories in person helped me gain a deeper perspective of the global opportunity that permaculture represents. That opportunity is now more important than ever, in a world where the mainstream has finally acknowledged climate change and peak oil.

Learning and Growing Overseas

We had placed an advert in the *Permaculture International Journal* prior to IPC6; offering to teach PDC courses overseas free of charge, in exchange for food and accommodation. At the convergence we met some of the people who had responded to that advert, and made contacts that would form the basis of our overseas teaching trip. After meeting people and hearing their life experiences, I was actually quite worried about the whole trip, but not for the reasons that other people worried about on my behalf (how could we even think about dragging two children under three around the two-thirds world?). My concern was about what I would have to offer. My contribution to the permaculture community in Australia had been real and worthwhile, but all very comfortable and safe. I had no doubt that Rick would thrive and have a lot to contribute, as his ability to read landscapes and people-scapes, and to integrate the two together was already well developed and quite outstanding, but I wondered what I had to offer. I found it difficult to comprehend just what we would be doing, but I thought if I organised the whole trip and kept doggedly at it, that eventually I would come to terms with it all, and it would just happen. Even up to the moment we boarded the plane to Mexico, I was still trying to come to terms with it. But I am so glad that I overcame my fears and that we went.

The year we were away provided some of the greatest experiences in my life, and helped me to become wiser, and to gain more understanding of global issues such as poverty. It also helped me to see that mothers all over the world have similar needs, though those needs may be met differently. In many communities, where white people were rarely seen, the fact that I was breastfeeding was the icebreaker that enabled us to be welcomed with ease. In Guatemala they laughed at my expensive Macpack for carrying Matilda, and gave me a *rebosso* (a cloth sling) so I could carry my child over my back like they did. The next day as I walked two hours up the mountain to class, I wondered at how comfortable it was, and how versatile (though it didn't have a separate day pack for the nappies), and when I walked into the community classroom, there were cheers, songs and more laughter as I was warmly accepted into the community.

In Mexico, I was teaching about resources, and what a resource is, in a community that had no access to water, living in utter poverty with most of the men away in the US trying to earn a living. After the course, one of the women stood up and said (via a translator) that before the PDC they had thought they were poor, but now they could see that they were really rich! It was a very humbling moment, and extremely emotional.

We had asked my brother Tony to accompany us – talk about backing up your major functions! Tony helped with teaching courses, looked after kids, and even translated from English to Spanish on one course when the translator couldn't attend, and this was then translated to Qu'echi Indian. Tony has since gone on to work in the environmental movement through the Landcare network, and he and his partner and two children live in the next town to us. They are about to embark on building a sustainable seven-star rated house on a block just 15 minutes from us.

In Gaza, in Palestine, we actually did some paid work for World Vision, evaluating a funded project there. As we walked through the amazing food forest in the middle of a barren desert, Osama, the community leader, explained how the problem of open sewerage had been converted into a solution by treating the waste and feeding the abundant food plants. We ate guavas, and sat in the shade, whilst around us it was 40 degrees. The community was inspiring and the permaculture centre they had established was phenomenal. We were humbled by the five centre staff who applied for their Diplomas of Permaculture by presenting their portfolios to us. We had no hesitation in awarding them all with their diplomas. With their knowledge of permaculture they were able to transform their community. It says a lot for the power of the PDC, which has been the main dissemination tool of the permaculture movement.

In 1999, pregnant with our third child, we were invited to teach on a course with Bill Mollison, in Tasmania. We took Jarryd and Matilda, and worked with the great man, a rare privilege and opportunity. Bill was quite taken with four-year-old Jarryd, who was dressed as a superhero of his own making, 'the Townsaver', complete with specialised cape hand painted by WWOOFers! Bill later sent Jarryd an environmental children's book inscribed, "To the Townsaver … never doubt it! From his friend Bill Mollison, the World Saver. We will succeed."

Accredited Permaculture Training (APT)

After attending a number of permaculture gatherings, and working on accrediting permaculture courses through the national training system with good friend and colleague Virginia Solomon, we started conducting training across the country in the Certificate IV in Training and Assessment. We used permaculture principles and worked with other permaculture colleagues to develop a permaculture-friendly version of this usually dry course. Virginia and I travelled across the country delivering the training to interested permaculture trainers, as well as assessing permaculture practitioners for Recognition of Prior Learning. In the process we found ourselves at a gathering in Noosa, volunteering Permaculture Melbourne to host the next Australasian Permaculture Convergence, APC8. (Virginia volunteered – I just backed her up.) I thoroughly enjoyed the couple of years when Virginia and I travelled to Western Australia, South Australia and Queensland, delivering exciting training, and meeting exceptional people who were so proud and passionate about their permaculture properties. We'd visit places like Julie Firth's Dryland Institute in Geraldton, to complete an assessment of prior learning, and come away having learnt so much in the process, as well as making a new connection, developing a friendship, and being inspired. During many a plane trip we reflected on how lucky we were to be doing something we loved with a passion, and to be assessing people who had similar enthusiasm.

The uptake of APT has been slow but steady, and I passionately believe that it was a crucial step forward in the development of permaculture as a credible system. I also know that the training courses we developed as a team were outstanding in their approach to teaching, and have a lot to offer the mainstream way of delivering the traditional Training and Assessment package that all teachers in the Tertiary and Further Education (TAFE) system are required to be qualified in.

APC8 – The 8th Australasian Permaculture Convergence

Convening APC8 with Virginia was an exhausting but exhilarating experience. It was a team effort, and we used permaculture principles to design 'troikas', or teams of people responsible for each aspect of the event. We wanted to present the many facets and potential of permaculture to the public, to local government and to businesses, as well as provide an event that new and experienced permaculturists would benefit from. In particular I had an idea that rather than spending the first day of the convergence sharing in a large group, that we could make it more dynamic, and have people present their permaculture stories, introducing themselves to participants, but with a public audience. We called it 'Permaculture People Tell their Stories'. A new concept that had multiple benefits. It was a challenge to gather enough energetic people together to complete the work required, and many of our past students stepped up to run the expo troika, the tour troika, and the administrative role, with great support from the Permaculture Melbourne executive. It was a huge undertaking, but the buzz at the multiple events we hosted made it all worthwhile.

David Holmgren became the convergence patron, which enabled me to get to know him on a deeper level, through the many discussions we had around designing the event. David came to visit our property overnight in this period and we stayed up till all hours talking. His intellectual rigour and depth of thought continues to impress me. He liked our property too, which was validating.

Permaculture – Gifts and Challenges

Through the permaculture community I have been privileged to meet people I would otherwise never have met. There are many passionate, highly skilled and articulate permaculturists, all of whom have strengths in a diverse range of fields. I've seen properties that have been restored from infertile and degraded land into paradises. I've met teachers who, despite lacking the benefit of formal training, search for better ways to convey the content of permaculture courses so as to engage their students more actively. I know people who have put their careers on hold to travel the world to share their permaculture expertise. I've visited intentional communities trying to invest in a sustainable future. I have become friends with activists who have inspired and motivated me, who have stoked the fires of my sense of purpose when, feeling burnt out, all that was left were dying embers. I have been sustained by the diversity and integrity of colleagues and students alike. I have been to places I may never have had the opportunity to visit, and worked with impoverished communities where people live in conditions we could hardly imagine, yet do so with optimism and an enthusiasm for life and family and culture that we Westerners seem to have lost.

Permaculture has given Rick and me a shared vision which has excited both of us. For the last 20 years we have developed that shared vision and grown our relationship, our family, our permaculture business and our property. Together we have grown as people, respected within the permaculture movement for our commitment to educating others, both in Australia and overseas, and for our time as volunteers, helping develop structures that enabled the message of permaculture to spread.

The journey has been amazing. Rick and I have discovered that many of our skills are complementary. The vision we have shared has often been the glue that has held us together. But there have been other times when it has also felt like the grout rotting away between old tiles. Permaculture has shown us a way forward that is positive and offers hope, but it has also exposed our differences. It has made us examine our ethics, our beliefs, our daily actions, and it hasn't always been easy to find the common ground between us.

Being an activist can be very powerful in terms of feeling that you are actually making a difference in the world. For me that has been an important part of my permaculture journey. Yet the problems of the planet can be overwhelming, and our individual actions can at times feel so inadequate that it gets depressing. Though I would describe myself as a positive person, there are times when I have struggled with depression. I am sure that being conscious of the woes of the world has had a major impact on my mental health. Unfortunately I have also seen this in other dedicated permaculturists, who have so much to offer, but suffer from burnout, or a feeling of being overwhelmed by the task at hand. I wish I had the answers. Sometimes it feels like it would be so much easier to just live the way most people do, not worrying about tomorrow.

Sometimes I struggle with the lifestyle choices we have made, particularly the untidiness, but I don't feel at ease using fossil fuels to mow the grass, and our constant travel makes it difficult to use biological resources such as animals to do the work. Rick likes to 'chop and drop' when he mulches trees. After years of nagging, he now makes the concession and chops and chops and chops before he drops. All these internal conflicts, and daily grappling with ethics and beliefs, can take their toll. When I take a long, hard look at myself, I think I need things to be orderly and neat because that means I am in control and systems are working well. When things around me are in chaos, I feel like I am also in chaos. Part of that makes sense to me: things should be well designed and flow easily, that is what good permaculture design is about after all. There are aspects of untidiness that I would like to be better able to accept. But I don't and I can't; I am not yet able to just love the untidiness. Believe me, I have tried! Rick has trouble seeing my need for tidiness as valid and it has been a real source of tension between us. Even when you passionately believe in something, which I do, it can still be hard to shed your skin easily. Maybe I need a good moult.

I have always needed to be more than a wife and a mother. Though I love both those roles, I need other stimulation and purpose in my life, a residue from an upbringing where my parents were always doing something for the community. This has created conflict for me and for my relationship. For me because I vowed I would not be a mother who was always working during the day or at meetings four nights a week. I would be there for my children and quality time would be my focus. It is hard for Rick to deal with my energy and desire to take on new projects, when it means less time for him, quality time that is needed to keep a relationship healthy. Having four children under 14 and working full-time in a fairly intensive job, running a business, convening a national permaculture conference, running residential courses, hosting international students; it's a juggling act that has often had me dropping the balls. And ultimately I know it is not sustainable.

Yet I know that the permaculture lifestyle has afforded our family a high quality of life. Our children are well adjusted, knowledgeable about the state of the world without being down about it, able to converse with the many different people who come through our place. They have met a diversity of people that most country children would not have access to. They know the farm better than I know it myself, they pride themselves on being able to take people on tours, and the names of plants roll effortlessly off their tongues. They know when to pick fruits and vegetables, and how to prepare them, they look after the chooks and know how to compost. They can catch yabbies and cook them for a snack. They happily swim in our dam each summer and have picnics in their treehouses, foraging from the many productive trees, vines and bushes. They go on adventures in the woodlots and tunnels Rick has created from the foliage. They have friends who love to come to our place for the freedom and the adventure it offers them. Our children have a rich and diverse lifestyle that we, through permaculture, have provided for them.

Our children have all attended the local primary school, where I returned to teach. Our permaculture students have often asked why we don't home school our children. Many people have questioned us, especially given my teaching background. I give a multifaceted response: if all of the alternative people in our society send their children to alternative schools, or school them at home, how are we contributing to the diversity of mainstream culture? Are we not contributing to more of the same, by isolating the things that make us different, and keeping the knowledge to ourselves? How can we hope that our children can make a difference if they are not able to make their way in the mainstream world? Are we not protecting them from the things we ourselves dislike and despise, and thereby sheltering them from the truth? If we really want to effect change, I believe we can most effectively do it from within, rather than from the outside. So our choice has been to educate our children in the ways of the world by exposing them to all its foibles, and discussing them rather than avoiding them.

Yes, this means they are 'different' at school. Each of them has made his or her way. Our oldest, Jarryd wants to teach permaculture when he is older and completed a full PDC once he completed Primary School, after sitting in on components of them for many years. He has probably felt the most different from his peers, but not necessarily because of permaculture. My daughter Matilda understands why I don't wear makeup but sometimes she chooses to and I let her. She understands that her friends who live in huge houses and have four wheel drives don't necessarily need those things and that there is an environmental price to pay, but she likes visiting them. We discuss various products and which ones meet our ethical criteria (yes you can have tea tree deodorant but you can't have Mum!). Our youngest two boys, Riley and Sean, take our lifestyle for granted, because for as long as they have been around our place has been an abundant source of foraging material and an excellent adventure playground. They prune out little tunnels and playgrounds for themselves and create places for fantasy adventures.

I know I have developed a sense of wisdom that is valued by my colleagues, but I do sometimes feel that I lack technical knowledge. It is Rick who can answer the technical questions about applying permaculture on a site, where the swales should go, what species to plant where and why; he's the one who can accurately read the landscape and apply the patterns thinking required to make permaculture an effective approach to sustainable land use. Whilst I know a lot about permaculture, my skills are more in the social context. My strengths lie in working with people to galvanise action. People leave our PDC courses with the knowledge needed to apply permaculture, and the action plan to become activists in their chosen field. I help people see where they want to head and how to get there. On a larger scale I use these skills to contribute to national discussions about permaculture education. On a smaller scale I run our business and organise the work. I value these skills and know that I have used them well and have made a significant contribution to the development of permaculture in this country.

Permaculture is as much about the social as the technical and we all contribute what we are best able to give – because permaculture IS holistic it needs development in all these spheres. I do know this, but the very breadth of permaculture means we can know a little about a lot of things. Rick often says, "I'm an expert in nothing; I know a little bit about everything and not enough about anything!" Luckily I know a little bit about some of the things he is less strong in, like managing a business, so we complement each other.

2008 – 2010

At the time of writing I am working as a Teaching and Learning Coach after returning to primary teaching a few years ago. I decided to go back to working off-site because I was spending so much time and energy running our permaculture business for little financial gain. I was burnt out after the conference, and felt that if I was going to work this hard, I should at least have a little financial security. I also felt that I should go back to something I know I'm good at – running a business requires such diversity of skills, and I'm not trained in business management. I often feel I have missed opportunities in promoting our permaculture business, and I have also grappled with the ethical dilemma of what should earn money and what should be done for the greater good of all. It all became very hard, and returning to teaching seemed an easier option.

Even now I find it difficult to acknowledge that I can't do everything. The State Government recently announced that the kitchen garden program is going statewide with the Stephanie Alexander Foundation. I feel that I have missed an opportunity to promote the possible contribution permaculture could make to this program; other organisations without the permaculture banner are taking it on. I would have liked to have been at the forefront of this school initiative, but it hasn't been possible at this point in my life. I want to write a series of children's books with a teachers' guide that could be used in school programs. I'd like to establish a bioregional network of permaculture groups in the southeast of Australia. There's so much to do, so little space and time to achieve what could be done.

I would like to introduce more permaculture into the schools I work in, but it's not part of the coaching job description! However I look for opportunities to promote ways we can educate kids differently, especially kids who are disengaged from mainstream education. Hopefully over the next few years we will work towards developing more hands on projects with permaculture techniques, to re-engage these kids with community and encourage their own self-development.

A new school is being built in Leongatha in the next three years, and the design phase has included a kitchen garden to supply the canteen. Once the school is on site, I hope to develop more links between my school teaching and my permaculture teaching. Rick is already working in the school garden program as a volunteer, taking groups of children to teach planting and propagating skills. Permaculture has an integrated approach that can be fully utilised in developing meaningful and engaging school programs that will equip the nation's children for leading more sustainable lives. It ties in well with the Victorian Essential Learning Standards (VELS), which promotes deeper and reflective thinking, and developing active citizens who can make informed choices. Sustainability issues underpin the VELS documents, and I believe permaculture is an outstanding vehicle for integrating Primary School curriculum. I hope to contribute more to that discussion in the future.

I have been involved in the permaculture movement for 20 years. I still do not really enjoy getting my hands dirty, though I'll do it when it needs to be done. I love preserving the harvest, but not necessarily the harvest itself. I am good at jobs that require a concerted effort, but not good at maintenance. I have become far more honest with myself about my wishes and lifestyle desires. I took time out of the business and hired one of our interns to work for us, which gave me the space to choose whether I wanted to continue working in the business. Now that she has returned to England, I find that I am again enjoying organising PDC courses, answering email queries, negotiating overseas contracts for Rick (although I have at times had a secretary once a week to help with administration, I currently am doing it all myself).

Our last four PDCs have been fully booked with three months to go, and our presence at the Sustainable Living Festival over the last few years has given us excellent exposure in Victoria. I still love seeing a new group of students go through the empowering experience of completing a Permaculture Design Course, and the bonding that occurs as the community builds on our site. I love knowing that our skilled facilitation helps so many people make changes to their lifestyles that impact positively on the planet. While I still love it, I'll keep doing it! For me that's the best part of being a permie. Making a difference and working with inspirational people – there's no better way to spend my time!

It's hard to know what the future now holds. Due to back problems I am limited in movement and in my capacity to work on the farm. For both Rick and myself, our physical health has significantly impacted on our ability to run our farm the way we would have liked to.

I find it difficult to leave jobs undone, so I picked up the slack for many years, but after our fourth child was born I learnt the hard way that I can't do everything, as I ruptured a disk in my lower back. I was out of action for three months, living with my parents and the baby in Melbourne, while Rick and the other three kids stayed on the farm. Determined to maintain our lifestyle we battled on, but then I ruptured a disk in my neck in 2008 requiring us to move off the farm and into town after surgery. This latest injury made me really question how sustainable it is to live the way we do. However we pushed on determined to get back to the farm and use the opportunity of being away from it to renovate, and build a certified kitchen for our courses. Four weeks after we knocked the inside of the house out, Rick was in Tasmania teaching a PDC when a freak storm hit, and a massive tree branch fell on him, crushing one vertebrae completely and breaking 5 others, leaving him in a critical condition. He was evacuated to the Austin Hospital where he underwent surgery, and over 6 months made an incredible recovery. In a way it was lucky we were already living in a house in Leongatha, as now it was Rick who could not walk up steps, drive or do any work. Being in town made life a lot easier. We are now happily back on the farm, in our renovated house which is now warm and more functional.

This has been very challenging for us as a couple and for our kids too, but they have shown incredible resilience through our periods of injury. Our family, friends and past students have rallied around us, helping to implement raised garden beds and move us back into our house when we couldn't lift a box between us. Sometimes I would love to see some of our active past students come and manage the property for us for a couple of years, and give us a rest from it all, still running courses, and Rick still supervising, but not doing as much manual work. But Rick continues to inspire and amaze me, and has managed to keep us in fruit and some vegetables, with the garden fairly productive.

Rick has even started teaching overseas again, and just returned from a 7-week consultancy in Mongolia. I know I want to be involved in keeping permaculture prominently exposed to mainstream culture, though I have come to understand that I need to let go and delegate to others. I want to stay on our property now that we have put 18 years into making it productive, but I wish it was easier to manage, and tidier. (That 'lazy gardening' idea propagated by Bill Mollison is a bit of a myth in my experience). I would have loved to build a sustainable house but I know that our budget and our bodies weren't up to it, and so I am content with my new look kitchen and office. I want to watch my children grow up to enjoy the fruits of our labour, and hope that we can safely guide them through teenage-hood, and all its associated risks.

I also said earlier that I knew the juggling act was ultimately not sustainable, and this time I think I have really learned the lesson. Last time I learned it but I didn't master it! As soon as I was better I just picked up the pace again. It is no good running yourself into the ground – if you can't nurture yourself, it's pretty difficult to nurture anybody else, whether that be family or students. So I feel comfortable with taking time out now, beyond the initial recovery period, and cutting myself some slack, instead of picking up the slack as I used to. For that reason, apart from running the business, I have not been actively involved in the Permaculture movement or the last few years. One day I may be able to once again dedicate passion and energy, as I am still a firm believer.

I am still committed to permaculture as a way forward, as an answer to many global issues. I know that many of our past students are activists, and for now I have to be content knowing that we have sown the seeds in many people to deal with the changes our society will face, that we have established a safe and productive home that we can escape to when the going gets tough, and that until that time, I deserve to nurture myself and ensure that I make it for the long haul. It's not the direction that I had expected but it seems my body is telling me it's what I need. Zone 0, the inner self often scoffed at, is the central part of permaculture that needs nurturing and care. Without personal energy you can't create or maintain the other zones, and so the next little while is going to be devoted to restoring our health.

As I put it to a friend the other day, I'm putting the 'me' back in Nao-mi!

Southern Cross Permaculture Institute. Note the dense planting and high bio-mass compared to the surrounding land, suggesting much higher productivity/acre, and higher carbon capture, helping reduce carbon in the atmosphere. When Naomi and Rick Coleman bought this property it was as bare as the surrounding farms.

Virginia Solomon

Virginia has been teaching permaculture for the past ten years, half of them in secondary schools and the other half at the tertiary level. She has a background in landscape design, conference management and training and was one of the key drivers in the development of Accredited Permaculture Training™ (APT). She is a teacher and a trainer and worked for five years delivering Certificate I in Permaculture to more than 500 students in Year 9 at Eltham College, Melbourne. She currently co-ordinates and teaches a Diploma of Permaculture in the post-secondary division of the school (Eltham College Training Services).

Working with Robin Clayfield and others, Virginia developed a version of the Certificate IV in Training and Assessment (lovingly known as the COW – Course Orientation Workshop, and CALF – Creative Adult Learning and Facilitation) which has been used to train permaculture people all over Australia in the mysteries of the Vocational Education and Training (VET) system. She continues to work to support trainers in this system and is currently developing course resources to support APT™.

Virginia lives in Melbourne in a sprawling extended family home with her husband, three adult children, her parents and an assortment of pets and livestock. Her passions (aside from teaching and her students) are food (especially cheese-making and preserving), crafts (especially patchwork, felting and basket-making) and gardening in her riotous colourful chaos.

Chapter 17 | Ethical Decision-Making for Secondary School Students

Sustainability is rather like motherhood – nobody much disagrees with it.

Hilary Whitehouse

The planet is in urgent need of repair and my generation of baby boomers is on the wane. It will be up to the current and future generations of young people to retrieve us all from the predicament we have got ourselves into. To me it seems vital to model sustainable practices and to find ways of teaching ethical decision-making through permaculture, at every available opportunity. Every seed planted has potential to bear fruit, and sometimes it is the most unlikely of seedlings that prosper if they just happen to get water and nutrients at the right time.

As a child I had always loved spending weekends with my grandmother who would take me out into the garden and talk to me about the good insects and how to imitate the birds (I can still do a convincing thrush whistle). At night we would lie on the grass and look up at the stars and she would name all the planets and constellations as well as many of the main stars. My parents were in the diplomatic service and, from the age of five, I had lived with them in some interesting but confronting places – Israel and South America among them. The shocking slums around Lima, the misery and constant cold in La Paz, the sight of dead dogs left to rot on the streets – these things appalled and sickened me but I felt powerless to make a difference. In Israel I had been intrigued by the alternative settlements – the Kibutzes and Moshavs. I particularly liked the Moshav system with its private ownership of houses and land, but collective marketing and production systems, factories, transport, banking and community facilities. This model is very permacultural in many ways – it addresses individual needs while providing a fair share for all.

I first met permaculture when I borrowed an interesting looking book from the local library. It was a big black book and I took it with me on a family holiday. During that holiday I read Mollison's *Permaculture: A Designers' Manual* twice, from cover to cover. It was full of the most amazing detail about all the things I had wondered about but never been taught ... things like soils and nutrient uptake, dynamic accumulation of minerals by plants, trophic levels and root zones, but also practical building ideas, ideas about social structures and ethical communities. It seemed to me to be both scientific (and anything scientific had always fascinated me) and creative. It was a revelation, but too much to take in, which is why I closed the book and then opened it again and started reading it a second time. The second time I made lots of notes, my imagination took off, I started planning my future, how and where I might live, what sorts of priorities I would have and why. Permaculture as a philosophical framework had me hooked.

Permaculture seemed to be like a balanced meal; there were all these options to choose from and all were complete. Thinking back, I realise that permaculture came into my life at a watershed. My children were young and I was finding my demanding life as a conference organiser too much. I was looking for a new career and had decided to be a garden designer. The *Designers' Manual* also plumbed the depths of my life experience and reminded me of what was important; reminded me of my ethics and the resourcefulness I had learned from my parents and grandparents. These ideas had been dormant for so long as I tried to conform to the booming society of the 1980s.

I now teach permaculture at a co-educational independent non-selective school in outer North Eastern Melbourne, one of the few examples of permaculture being embedded within mainstream education. The students are generally from relatively affluent middle-class families. I could have directed my energies at poor or remote schools or used my foreign languages to work in the majority world on aid projects, but I feel very strongly that one should start in one's own community. I also believe that working with these young people is as important as working with the disadvantaged, since it is likely to be they who will have the means and opportunity to become the future leaders. It is also likely that these young people, whose comfortable consumer lifestyle has the furthest to go to a conserver ideal, will be the ones whose conscious choices will drive the most profound challenges and changes to first world consumer culture.

For me, teaching is not so much about the dispensing of knowledge. In fact, today young people have access to more information and sources of knowledge than ever before in the history of our species and it is impossible for teachers to be the fount of all knowledge. Rather, I see my role as an inspirer or a guide towards wisdom. I am a seed-planter. I open up windows. I hold up mirrors and I offer new viewpoints. I always try to keep my teaching positive, even though it would be all too easy to make it dire and desperate. I don't believe teenagers respond all that favourably to warnings, and anyway, as they all seem so jaded and worldly, I doubt that my flappings would ruffle their sleek feathers. Instead I prepare them for possibilities, encourage them to see themselves as fortunate, and challenge them to view themselves as members of one species in a finite and limited world. For these young people, resourcefulness now needs to compensate for the coming lack of resources. Skills in developing connections, lateral thinking and reducing energy dependency will come in very useful in the not too distant future. Deep down many of them know this. They just don't choose to acknowledge it at the moment the seed is sown.

As with any area of study, sustainability does not appeal to every student, and meeting the needs of adolescents is a vexed question. My own experience, however, tells me that one can only inspire a will to learn if one can make something meaningful, if one can make a connection to the young person's world. I had one student who was going to be a fashion designer and for whom talk of a future with less oil was irrelevant. What had any of this to do with her? People have to wear clothes, don't they? I asked her what kinds of fabrics she wanted to have her designs made in. Would she be encouraging sheep farming by using wool? Would she be part of the water-wasting cotton industry? She responded, "Oh no, it's OK, nothing like that! I will be using modern fabrics like two-way stretch and soft synthetics." I asked her to research the fabrics she liked and to tell the class how they are made and whether they use any water or energy in their manufacture. She brought in her research the next day. Not only did she discover that her fabrics required water and energy to manufacture, but they were actually made from hydrocarbons; her beloved silky soft fabrics came from the petroleum industry! She was completely nonplussed. I don't know whether she continued with her dream, but at least she became aware of her choices.

On another occasion I caught two students surfing the net in their lunch hour (students were not supposed to use the computers for non-study purposes at lunchtime). They were looking at runners and basketball boots. They tried to tell me that they were doing school work as well, which gave me an idea. I asked them to find out about the company whose shoes they liked. How much did the workers on the production line get paid? How many pairs of shoes did they have to make in a day? How much did a pair of those shoes cost? What portion was profit to the company? By the end of the lunchtime they had discovered that a well-known shoe brand paid their workers the equivalent of 8c US per hour, that the workers had to produce 11 pairs of shoes per hour, and that the company had made millions in profits for its share holders. On top of that they had discovered that the same company had recently closed down its operations in one South East Asian country and moved lock stock and barrel to another as soon as the workers had organised to ask for higher pay. The boys were disgusted and decided to look at another brand which did not advertise so heavily, and which promoted itself as 'Fair Trade'. The shoes, they said, were nearly the same and cost half the price.

As Ron Miller tells us, every child is more than a future employee so every person's intelligence and abilities are far more complex than any scores on standardised tests can reveal. Schools are strange places in many ways, and education is now so politicised and rationalised that it is often difficult to see the process of learning in amongst the funding arguments, potential duty of care litigation, rankings and results data, and curriculum and assessment policies. People spend 13 years attending schools and many spend several of these years hating the experience. How strange that a society so devoted to pleasure and instant gratification should condone a schooling system that alienates so many at the peak of their potential for creativity and imagination.

I firmly believe that all students have something that can and should be released to shine. It concerns me greatly when I hear teachers and others labelling, or pigeon-holing, a student (or anyone!) as being dumb, disruptive or difficult. I had an experience of teaching horticulture at a high school where 50% of the students could not read and write English. They could read and write Arabic or Vietnamese, but that didn't help them. In the staff room the teachers spent much of their time making sarcastic remarks to each other about the students. If a student came to the door, they would be gruffly acknowledged, as if they were intruding or wasting the teachers' time, and a pantomime of sign language, grunts and eye-rolling would extort the sought-for teacher to slouch over to hear the student's request or accept their work. The teachers laughed at my techniques of taking students for a walk down the road to look for plants they knew or recognised, counting the occurrence of a particular weed in a square metre of the grounds, or playing games to learn Occupational Health and Safety, and writing them off as "that will never work here." I was warned about students, but not once did I find their judgement of a particular student to be accurate. Frequently the most 'disruptive' students were actually the bright but bored ones who wanted to *do* something and couldn't sit still.

I do not believe that achievement for its own sake is a measure to be valued. Academic achievement, exam results, facts known and repeated, do not of themselves indicate a good school, or for that matter, a good education.

Many schools use these as yardsticks, but are they an accurate measure? I believe that students will learn things that are useful or useable – meaningful things – and *they* will be the judge. A student who cannot see the point will 'tune out'. It is the teacher's job to make the material as meaningful as possible for as many students as possible. No use? No point? Forget it ... literally!

Developing a sense of meaning, of purpose, of positiveness involves making connections between what is taught and what is important in life. It needs to be personally relevant and accessible, but it also needs to be reinforced and reasserted and, where possible, demonstrated. If students leave Maths to go to Geography without the sense that the two are linked and, even more importantly, that they both have a continuing meaning or usefulness in the world as a whole, they may well consider the study of Maths or Geography to be pointless.

As Bentley (1998) puts it:

> *The capacity to integrate, to make knowledge coherent and consistent across the range of our experience and actions, is not sufficiently developed by mainstream education, constrained as it is by curriculum and time demands, limited resources and a reliance on text-based assessment. The contexts in which curriculum-based knowledge are used and displayed become bound-up with the situations and routines of school. The connections between different opportunities, both to learn and to demonstrate understanding, are too weak to become meaningful or useful.*

Seeing connections is central to the study of permaculture. The sustainability program at my school is predicated on the philosophy of permaculture, and students receive a Certificate I in Permaculture through Accredited Permaculture Training (APT™) on completion. They quickly adopt the ethical statement *Earth Care – People Care – Fair Share* and have constant opportunities to practice and reinforce it throughout the program. Links and connections are made between elements in the system, between each other, between subject areas, between parts of the school grounds and between themselves and their world. They develop a sense of care without ownership, and a notion of giving that is unusual in this world where even 'charity' has to come with a prize. Why do we need a chocolate bar if we support the football team?

Something we did early on in our sustainability program is to invite a snake expert to give a presentation on snake behaviour. Many students are afraid of snakes when they arrive for the presentation, but by the end of the class, they are fascinated and want to know more. The presenter makes his information useful and meaningful and links it to ecology. He doesn't only demystify snakes and give a salutary warning about the hazards of not understanding them, but he places them in context and enables the students to see the world from the snake's viewpoint. Along the way he explains keystone species, broadly-based and narrowly-based gene pools, the value of biodiversity and the worrisome uncertainties about genetic engineering.

To come to see oneself as just one species in an ecological system is both confronting and comforting to our students. There is a lot of stress associated with expectations and the petty concerns of their worlds. It puts it all into perspective when you realise that you are both small and insignificant *and* powerful. In understanding the concept of stewardship rather than ownership, and then starting with small steps, students discover that they can effect change incrementally. They can see the difference they make and it gives them a sense of success.

In a world of change and instant everything, this overall grounding wisdom will be more useful than facts and figures. Frank Zappa reminds us that, "Data is not information. Information is not knowledge. Knowledge is not wisdom." Where connections can be found, meaning can be made of the myriad facts, data and information. Every student's information needs in life will be different. To remember that we are only one element in the system of our world, that the resources of the Earth are finite and that we must share them with each other and all the other species and elements in the system, is not just an ideal, it is an imperative. No individual element can survive on its own, and no community can exist without links between the parts of the whole. It is the connections, the integration, which forges the wholeness. This is a fundamental teaching of permaculture.

The getting of wisdom involves reinforcing the connections and developing in our young people a 'can do' attitude to problem solving, an ethical approach to decision-making, and a sense of stewardship. Schools are uniquely placed to take up this challenge for the 21st century. If schooling involves the getting of wisdom, the concept of 'school' can be the environment, a church, an experience of déjà vu or a word of advice from a stranger. Schools of one kind or another have existed since the first woman warned her children about poisonous mushrooms, since fathers and grandfathers initiated young men into the rituals of the clan, or since minstrels sang the myths of creation. The schools of today are both more and less than that. The primary school classroom is often organised around themes. Students might spend a term studying the sea, or space, or the Olympics, and all their topic areas relate to the theme in some way. The information is integrated so that there are links and hooks to hold it all together. When students get to secondary school, however, the classes usually become subject-based with the boundaries jealously guarded by teachers and timetable makers.

David Orr (1991) notes that:

> *In the modern curriculum we have fragmented the world into bits and pieces called disciplines and sub-disciplines. As a result, after 12 or 16 or 20 years of education, most students graduate without any broad integrated idea of the unity of things.*

This packaging and pigeon-holing is limiting the scope for students to see their education as greater than sum of the parts: the whole. Have we begun to lose the connections to such a degree that students can no longer see the point of school?

Something seems to happen when students arrive in secondary school out of the more integrated and nurturing atmosphere of primary school. I watched a group of students from my son's primary school arrive for orientation at their new school nervous, excited, full of anticipation and apparently eager to learn. By the time I saw them in Term 4 of their first year of secondary school, many seemed to have lost that excitement. By Year 8 they had found strategies for subverting the flow, avoiding 'work', making excuses and beginning to 'opt out'. How is it that students can go from excited, engaged and enthusiastic about trying new things, to wary, weary and cynical about trying new things? Is it just their hormones changing them from angels to devils, or is there something about their environment, the school where they spend most of their time, that is not meeting their needs?

It is not uncommon to talk to students entering Year 9 and discover that they are depressed, even morbid about the state of the planet. They feel that there is no hope and that they have no future, so why try? One student I remember particularly well. He was a slightly 'odd-ball' sort, determined to be original, restless and a bit prickly. He was not one of the 'jocks' (his word); despite his tall, athletic appearance, he was not interested in sport. His music interests were extreme and he was into shock value, practical jokes and hunting. But he also had a love of nature. He had a detailed knowledge of snakes and other reptiles, amphibians and fish, and a deep love of rainforests. On one of the first days of the program he was talking to my daughter who was in the same year level and, half-laughing, said, "What is the point of having plans? Human beings are going to self-destruct, none of us will have a choice." But by the end of the semester spent studying permaculture, something had changed. Permaculture seemed to have shown him that he did have a choice, that it was possible to effect change even in a small way, and that all sorts of interesting options were available with a little ingenuity and lateral thinking. The same boy went on to do a Permaculture Design Course after Year 12 and is now planning to make a yurt to live in out of old woollen blankets, which he will felt in an old washing machine.

By Year 9, students need to be 'kept in the system', and many schools have special programs to do this. Some involve Outdoor Education, or a separate campus. My school has a City Campus where students are exposed to the 'real' world of work, challenged to be leaders and to ready themselves for senior school. One part of this program deals with sustainability and used to occur at a special 'bush' campus with a domestic house and garden, a large shed, dam and windmill. The orientation to Year 9 sustainability involved a tour of the campus: At the 'meeting point', a totem pole designed and decorated by students from an earlier group, the 70 students are divided into 'mentor groups' and embark on three separate tours. First they see the staff areas with their desks and the little kitchenette with its compost bin and clip of recycled paper by the phone. They move on to the yoga room where they see a pile of cushions and are invited to bring their own in from home. They enter the first classroom, a sunken sitting room beside a big kitchen. This is the student kitchen, with complete cooking facilities and an urn. Student mugs and pigeonholes line the bench, there is a fridge, an oven, a stove and a pantry with various labelled boxes. The next room is the computer room, cosy and carpeted, den-like. It's another sunken 'classroom' but this one is carpeted and has tables and chairs around the sides. There is a silent scrolling noticeboard, which on that day reads, "Happy Birthday Amy." The tour proceeds outside, where the students see projects undertaken by the last group. There is a guinea pig called Pedro and three chooks (Henny, Penny and Jenny). Everything has an explanatory sign, but they don't read them. Up at the shed there are benches with electronics gear, more computers, tables and chairs, more pigeonholes and a whole lot of neatly organised tools and boxes of screws and nails. There is a professional drawing board and a half-made scarecrow, a plastic folder labelled 'Zone Games', coloured paper, pens and other art materials.

Mentor Group A is now assembled in the shed. An activity begins. "Who is the oldest person in this room?" The teacher is identified. "Who is the youngest?" A student identifies himself. Right. "Arrange yourselves in order of birthday from oldest to youngest *without talking*." A few attempts at sign language, and whispering, but the shed echoes and soon the noise is deafening again. The teacher raises one hand and covers her mouth when she thinks the line is organised. The students quickly get the hang of the sign language. The teacher explains that, if there were time, we could do several things with this data (you, your ages), but we are just going to do one thing: what is a 'mean' number? Students answer the question. Who is the 'mean'? Laughter, students step out of line to count, get confused, some try to organise others, one takes charge and the job is done. The 'mean' steps forward. The activity continues, students make pairs with their neighbour and the 'mean' gets to choose her group. They are charged with reading every sign in the garden and developing a theory as to what it is all about. They are told there are two key signs and that there are 13 in total. Ten minutes, work together and pool resources. They reassemble after 10 minutes. It's a game. Is it about the environment? Is it about sharing? Several theories are put forward. A quiet and thoughtful sort puts up his hand, "I think it is about connections between us and animals and plants and things." Nods of recognition. Time is almost up, but the teacher has time to explain to the students that what they have just experienced is an integrated Maths, English, Studies of Society and Environment and Science lesson, and the signs in the garden were about 'permaculture', a way of thinking to do with sustainability.

Sustainability, as I have said earlier, does not immediately recommend itself to the average teen, so activities need to be flexible and unpredictable or unexpected. Learning needs to occur in spite of prejudice and suspicion. We ask students to calculate how much energy their bedrooms use; some of our students have an amazing array of electronic gadgets. They add up the power rating in Watts of their electrical equipment and appliances, multiply it by the number of hours they have used them over the last week, and then by the $ rate per kilowatt-hour divided by 1000. A simple but relevant maths problem.

Another activity involves looking at lunch from the perspective of food miles, of embodied energy or of embodied water. Students choose a 'lunch' from a set of cards, and add up the information on the back, then research other aspects of the products, or other lunches to expand their knowledge. The activity does not take more than a lesson, but it gives students a clear perspective on their ability to choose what they eat and buy, and that they are able to control the energy and water consumed on their behalf by making different choices. Often students have never thought of noodles being made from wheat which is grown in Australia, shipped to Fiji to turn into noodles, packed in Thailand with flavourings from somewhere else and dried vegetables from yet another country, shipped back to Australia to a central port, on-sent by semi-trailer to the supermarket warehouses around the country, then to the individual stores where they are purchased by consumers, transported home by car and eaten in five minutes. They have never thought about what happens to the packaging thrown in the bin after use, to be transported by council truck to the tip and turned over by bulldozers for years. How much additional energy is used in all those processes? The challenge then becomes: "What could I have in my lunch that does NOT cost the Earth?" The conversation has begun, the connections have been made and the student has the tools for an ethical decision.

Many students have a fascination with bodily functions, so a visit to the 'poo farm' creates both interest and disgust. Shock value is always good for teaching! The Western Treatment Plant at Werribee, west of Melbourne, is an amazing place for all the reasons students do not expect. It powers itself on methane which is trapped under huge tarpaulins and pumped into the gas power station, complete with flame. But it is also one of the country's foremost bird sanctuaries, and no chemicals are used in the sewage treatment, rather the processes are 100% natural using only bacterial digestion. The final treated water which goes out into Port Phillip Bay at the end of the cycle is clean enough to swim in.

The first cohort of students who completed the Certificate I in permaculture has just completed Year 12. In their Year 12 year a group of them spontaneously started a grass-roots action group at school, which saw the enacting of many changes in a few short months. They had all available lights converted to compact fluorescent globes, they applied for and received an Arbor Week grant from the local council to revegetate part of the school's Environmental Reserve, and completed the project with the help of students from grades 4-12, with the help of some teachers. They drove a change process at the highest level, the School Board, which resulted in the establishment of a strategic planning committee for the adoption of sustainable practices across the whole school over the next five years: to include curriculum design, business practices, resource use, waste management, community action and student action. All these students had completed the Certificate I in Year 9. By Year 12 it had become a natural part of their experience to make ethical choices every day.

My experiences teaching permaculture to teenagers have made me optimistic. In spite of their apparent disinterest, their lip and attitude, our 15-year-olds will grow up to lead, inspire and make ethical decisions into the future. The unlikeliest seedlings may yet become the tallest trees.

Virginia's son Guy mixing pollard mash at the course completion workshop.

Ross Mars

Dr. Ross Mars has had a long involvement with wastewater treatment and reuse, and has presented papers at a number of conferences, both in Australia and internationally. He has had several papers published in scientific journals. Ross has developed a number of greywater reuse strategies and technologies, and a range of approved products, and has installed these in homes and commercial premises throughout WA.

Besides his work in greywater and rainwater tank installations, Ross is one of the leading permaculture teachers, designers and consultants in Australia, and author of four books on the subject. He has also produced two videos on energy efficient housing design and renewable energy systems for power generation, and is recognised in Who's Who in the World for his contribution to education.

Besides designing, building and installing greywater systems, rainwater tank systems, gardens and water-sensible irrigation systems, Ross is currently setting up a wholesale nursery, Red Planet Plants, in Mundaring.

Chapter 18 | Straw and Greywater

My wife Jennifer, is to blame. It's all her fault! Jennifer did her Permaculture Design Course (PDC) in 1990 and soon I was getting dragged around field days where I saw straw, tyres and vegetables.

In early 1991 Bill Mollison came to Perth and did his usual thing – outraged bureaucrats and government officials, and talked in riddles. People would ask, "I've got blackberries in my cattle paddock. What should I do?" His reply was, "Grow an apple tree." Then he added, "The apple tree will grow up and the cattle would step on the blackberries to get to the apples."

I was a science teacher at the time (I taught high school students for about thirty years) and promoted passive solar design, energy efficiency and waste recycling. Eventually, I began to realise that permaculture was more than straw and tyres, and I undertook my own PDC in 1991. By the end of '91 I was teaching permaculture, initially with a couple of weekend Introduction courses and then my first PDC in January 1992. I have taught at least one PDC every year since 1992 and in some years up to four.

Since 1992, I have also taught advanced courses in design and teaching (some with well-known Western Australian (WA) permaculture teacher Dave Coleman), and in between school teaching, Jenny and I started building our house and setting up our property.

Like many permaculturists, we decided to buy a few acres and set up a demonstration site. This was the beginning of the original Candlelight Farm. I decided in late 1992 to take some time off from school teaching to focus on our permaculture activities. Over the next seven years, mainly by trial and error, gardens and trees were planted, a shadehouse, hothouse and chicken pens built, and water movement through the landscape was managed.

Candlelight Farm was to become a showpiece for sustainable living. It boasted composting toilets, a complete greywater reuse system, propagation facilities, organic food production areas, a variety of aquaculture endeavours and a very large number of different types of fruit and nut trees, bamboos and fodder species. The website still has photos from those days and many articles were written about the property.

Candlelight Farm, the property, was sold in 1999, and after two new owners it still has many of the original trees and gardens, and the very energy efficient house and outbuilding is still functioning well. Sadly, some of the trees that were planted have been removed, and there have been changes, but in other ways there have been improvements, which have been satisfying to see. Candlelight Farm, the business, still operates in its capacity in offering permaculture courses, selling books, and design and consultancy work.

Candlelight Farm was a labour of love. There was no topsoil to speak of, only clay, and this presented problems when you wanted to grow lots of different plants. Many lessons were learnt and it became obvious that you really had to consider soil type, climate, water demands and other needs for cold, sun or appropriate pollinators, to grow certain plants. When Jenny and I both started out, even though we did have reasonable knowledge of particular plants, we wanted to grow all of those amazing plants that you read about in permaculture books.

We spent a lot of money, and wasted some of it, lots of time and effort, and lost lots of plants. We soon learned which plants suited our local area, and we began to focus on them. While it would have been nice to grow kiwi fruit, jakfruit, jaboticaba and many other food plants that preferred a warmer growing environment, it wasn't to be.

Developing Candlelight Farm also caused other problems. In the early days some of our neighbours, not so much immediate neighbours but those who lived way down the road and passed our place, complained and presented petitions to have us stop our endeavours. We were seen to be degrading the amenity of the area, and there were fears of devaluing their land.

Looking back, it all seems so trivial now, and their concerns unwarranted. At the time, however, we were devastated to think that people could be so petty, but it just highlights the difficulties many people face and the prejudices good people experience when new ideas are proposed which other people may not understand.

The '90s were exiting times. Perth held the 6th International Permaculture Conference, and Candlelight Farm was one of the tours. I wrote a couple of books about my experiences at Candlelight Farm during my design and consultancy work. Jennifer and I co-wrote *Getting Started in Permaculture* to help people develop practical skills to recycle materials, make compost and build gardens. It still is sold today and has versions in Australia, Europe and the Americas.

I personally found Mollison's *Designers' Manual* great for an overview of permaculture strategies for different landscapes throughout the world. However, I wanted to write a book that not only showed things I knew from experience worked well, but also to discuss other things like building soil, permaculture in schools, harvesting and using water, and how designs were undertaken.

Jenny and I also spent time in the UK in December 1994 and January 95, talking to well-known permaculturists such as Graham Bell, Maddy and Tim Harland, Chris Dixon, Patrick Whitefield and the great Robert Hart who developed Forest Gardening. I remember walking through the forest in Wales with Chris Dixon observing nature at work, and Robert Hart's garden in Shropshire, England, and both explaining how the land can teach us. It was Graham Bell who said to me, "Permaculture is not a destination, it is a direction." You don't suddenly arrive at some point and declare, "I'm a permaculturist." It is a lifelong journey of change and growth.

There were many others who had stories to tell, but unfortunately their names escape me. As I get older I am finding it harder to remember all the details. I wish I could recall names and events, as many people have shaped my worldview, and helped me better understand our complex world.

While Jenny and I travelled throughout England, Scotland and Wales I observed a number of strategies that work in cold climates. This was important as many permaculture teachers in Eastern Australia talk of rich volcanic soils and tropical fruit. I grew up and still live in Western Australia and all I knew about was sand and clay, my only experience in those early days of common fruit was of citrus and apples. I needed to understand about other systems in nature and I soaked up knowledge like a sponge. All of these experiences, both in Australia and overseas, helped formulate the book, first published in 1996, entitled *The Basics of Permaculture Design*.

In the late 1980s and early 1990s it was clear that as more permaculture teachers and designers were being generated there was a great need for a professional association which could act as a support and networking group and in 1988, the Permaculture Institute of WA (PIWA) was born. This group was a separate identity from that of the Permaculture Association (PAWA), which comprised like-minded people with an interest in permaculture per se. This body of professional permaculture teachers and designers was always a loose association, and even though it had a short hiatus in the early years of this last decade, it re-emerged to introduce Accredited Permaculture Training to Western Australia.

I became immersed in permaculture and began to be known as a permaculture teacher, designer and consultant. I was soon asked to do talks at garden clubs, for environmental groups, at teacher in-services and community groups. I became the longest serving editor of the Permaculture Association of WA Newsletter, which in its heyday was usually 24-28 pages long (sometimes longer) and was sent to over 800 families. We even had a few editions with colour covers and two colour spreads in some pages. I have also been interviewed and appeared on television and in state and national newspapers, and written for local newspapers, permaculture magazines, the solar energy information centre, and a number of newsletters for nursery and garden clubs.

Besides editor and committee member, I have been convenor of PIWA and co-convenor for PAWA. I'm not sure how, with Jenny and I having eight children between us, and 13 grandchildren, I found time to also become the inaugural manager of the nursery and resource centre, The Permaculture and Environment Centre of WA (PECWA), which evolved into Gaia Nursery, Swan Garden Centre, and is currently the Green Life Soil Company. This nursery in Midvale has operated as a permaculture nursery since 1993 when I first set it up.

Having been a school teacher I had a particular interest in getting permaculture into schools. Dave Coleman and I offered a PDC during the Christmas holidays in early 1994 and targeted teachers and other school staff. We ended up with 72 participants, about half of them teachers. We designed and built gardens in four schools and soon after worked with teachers to do the same in other schools. It was the beginning of introducing permaculture in schools in Western Australia, and it was the time that I developed the 'Outside Classroom' concept which I first heard about during a trip to the UK in the early '90s. I still work in schools. These days I am helping young people build gardens, install rainwater tanks and greywater systems. Permaculture in schools is ongoing work and at times increases in waves.

There is an upsurge of interest in better utilising school grounds, installing raised vegie gardens, water tanks and growing food plants as part of a greater awareness of healthy living. This, coupled with former government programs such as Community Water Grants and Green Vouchers which have enabled schools to install rainwater tanks, undertake water efficiency upgrades in toilets and for hand basins, and greywater reuse for garden irrigation, have all helped the community better appreciate the role of permaculture in our daily lives. To quote a popular cliché, 'kids are our future', it is clear that students today are far more aware of their environment and not as 'innocent' as we seemed to be at their age.

In 1995, I decided to undertake some further studies. I enrolled in a PhD program at Murdoch University and my research focussed on the use of wetland plants to strip nutrients from domestic greywater. I was fortunate enough to be able to undertake my research at Candlelight Farm as I had a composting toilet and had designed and built a greywater system to deal with kitchen, laundry and bathroom water. So I had easy access to a continual source of greywater to be passed through tanks planted with vegetation.

These studies opened doors for me. I was able to present scientific papers at a number of conferences both in Australia and overseas, as well as having papers published in scientific journals. More work came in for permaculture projects and greywater systems.

I finished my studies in 2000 and obtained my PhD in Environmental Science. As I soon became a leader in greywater design and systems, I was often jokingly referred to as 'Doctor Greywater' in the local and state newspapers. I rarely use that title except when I give talks in the public forum or when I meet with corporate business and government agencies. Sometimes it is good to have that recognition. Even though I am not now involved in the academic scene, I have presented papers and seminars on greywater recycling at international conferences, and at the Australian Permaculture Conference (APC8) in Melbourne in 2005.

Besides my academic qualifications, I have proudly obtained other qualifications in permaculture: Certificate in Permaculture Design (PDC), Advanced Teaching Certificate, Advanced Design Certificate and Diploma of Permaculture Design (from Mollison's Permaculture Institute) in the categories of Site Development, Site Design, Education, System Establishment and Implementation, Media and Communications, and Administration. In 2004, I became the first person in Western Australia to receive the new accredited training qualification Diploma in Permaculture.

Having had a greywater system approved in the early '90s, I was approached by the Health Department to examine if I had wanted to make any changes to my approval and if I had other comments to make on the new Greywater Guidelines they were developing. I was fortunate enough to proofread the draft guidelines and make suggestions about the regulations, which were adopted. At the same time, around 2000-2001, I thought this was an ideal opportunity to develop new greywater systems. I went crazy and spent untold hours building and trialling a number of different systems, which were added to the new approval list.

By 2004 I decided to leave the teaching profession and jump head first into developing my greywater business. I started off small, with the occasional help of a friend who I employed as a subcontractor. I would go out and install systems during the day, and then come home to answer telephone enquiries and write applications to Councils for the installations of greywater and rainwater systems for houses. It was hard work and long days.

Ross installing a greywater system.

Unfortunately, as the business grew, and I put on extra workers, I seemed to have less time to make every thing work well. The days were still long and every night was spent doing invoicing and replying to an increasing number of requests for information. I once heard some business advice that if you made it past three years you were going to make it and survive in the business world. I think this is good advice, and from my own experience rang true. After about three years of struggle the business picked up. I actually started making money and started quoting more realistically. I always felt guilty about quoting too high and making my systems financially prohibitive, but I quickly learnt that you have to charge enough to make the business viable and profitable.

I had to get out of the home office syndrome. Having a home business is all right to a point, but you can't separate your home time from work time. I leased a factory office and retail outlet/workshop in the light industrial area of Mundaring and set up shop with my family and workmates where we could store stock and manufacture greywater systems and rainwater tanks. To make the whole project worthwhile I diversified in true permaculture fashion and set up another business – making steel rainwater tanks. I also set up the business structure differently. Part of it deals with greywater jobs, part looks after all rainwater tank and irrigation installations, and a third makes rainwater tanks and raised garden beds. Candlelight Farm, the business, still sells permaculture books and DVDs, undertakes permaculture consultancy and designs, and offers courses, both PDCs and Accredited Permaculture Training (APT).

In the late '90s I wrote the scripts and produced two short videos on *Passive Solar Design of Buildings* and *RAPS – Remote Area Power Supply*. I was lucky to obtain funding from the Alternative Energy Development Board to pay for the majority of these. Both were finalists in the WA Energy Efficiency Awards in 1998 and 1999 respectively. I was very proud of the presentation of the videos, their content and their acceptance by libraries, schools and TAFE colleges throughout Australia.

Permaculture, as a business and livelihood has been a struggle. I don't think there are many who are solely dependent on their permaculture designs and teaching for their income. It is my observation that active permaculturists rely on associated enterprises as their primary income. For me, it is the supply, design and installation of greywater and rainwater tank systems. These are, of course, closely related to and an integral part of permaculture thinking. And to be honest, I would not have ventured into these fields had it not been for permaculture.

The Permaculture Conference in Sydney (APC9) was a highlight for me. It was great to finally meet people I had only heard about and to renew acquaintances with permaculturists that I hadn't seen for years. APC8 and APC9 were the only two conferences that I have attended. I would have liked to have attended others but as I was a high school teacher until 2004 it was difficult to get time off during the normal school term. I felt honoured to have presented talks at both of these events, and I am sure that I will be involved in other conferences in years to come.

While we were in Melbourne and Sydney, Jenny and I took some time to visit permaculture properties and people. I really valued this time and these experiences. You can read all about Mollison's property at Tyalgum in northern NSW and Crystal Waters Ecovillage in southern Queensland, but nothing beats firsthand observation and experience. We all need to see that permaculture strategies do work, but we also need to acknowledge that we have to maintain, change and strive to improve, our designs, work and endeavours.

This has always been my own personal belief, and when I first started in permaculture I was disillusioned with 'armchair permaculturists'. I came across many people who even taught permaculture courses but didn't live that life, people who didn't 'walk the talk'. And might I quickly add, with humble confession, that I too sometimes have, and still do fall short.

I believe that people are only true teachers when they can teach from experience. It is not sufficient to have the knowledge, you must have the practice. You must be able to deal with design problems drawing on your own worldview and using skills and understandings you have developed. It was with this in mind that I wrote my permaculture books and articles on permaculture-related topics.

During the last few years, when I left teaching and took the plunge into working for myself and developing a new business, my family and I began to become more informed about the Peak Oil, Climate Change and Food Scarcity issues confronting us all as we head towards a new decade. In hindsight, it seemed fortuitous that these things fell into place: I left teaching to pursue working in the environmental field, with my permaculture bent, in a changing world where permaculture will be able to offer people knowledge, skills, and hope, to meet life's challenges head on. More and more people are beginning to grow their own food, undertake permaculture and organic growing courses, find ways to make their house and lifestyle more sustainable, and prepare themselves for what is to come.

Peak oil is reasonably well-known, but little understood, with its consequences for products we take for granted. When petrol goes up in price as it will do, as less oil is found and available for us to use, then almost everything else increases as well, either directly or indirectly. So far, we have witnessed huge increases in the cost of food, steel, plastics and manufactured goods.

I feel privileged in some ways to live during these times in our history. I believe I have much to offer in teaching others how to grow food, how to design our houses to be energy efficient, how to tread more lightly on the Earth, how to reduce our consumption, how to capture and reuse water, and how to repair damaged soils. As our climate changes, as our seasons change and natural cycles become disrupted and out of sequence, people will be looking for hope. They will be looking for help, and they will be seeking others who can teach how to live, and how to live well.

It is ironic that the country where permaculture originated seems to have the least to show. After thirty years there aren't many fully developed properties to see. In other countries, maybe because of desperate need, permaculture has had far more impact. I think this will change as we move into a post-oil future.

I have always realised that permaculture was hard work. It has been portrayed as 'be a designer then a recliner' but this is far from the truth. Jenny and I would imagine how great some permaculture properties would be, and then we would visit them. Sometimes we would feel sad about the lack of maintenance a property had suffered, yet at the same time we understood the amount of effort required to devote to caring for gardens, animals and households. I am not pointing the finger; sometimes it just works out this way. But we need to learn the lesson.

Permaculture-designed systems only function well when they are managed. They are, in fact, cultivated ecosystems, which left to themselves, turn feral. Gardens can become a jungle, ponds become meadows, trees struggle to find light and so fruit is not evident. It is just the normal succession of natural ecosystems.

Some would argue that this is good – that an uncared-for property manages to survive and thrive. However, if our aim is to maximise food production in order to survive, whether that be as fruit and nut trees or in aquaculture, or from any other endeavour, then we do need to care for and maintain our cultivated ecologies.

If we don't prune, mulch, feed and water the many gardens we have built, then we start to battle those wonder pioneers we call weeds. Mulching was great for the first year or two of a new garden, but if we didn't continue to perform simple maintenance tasks, we ended up with a crop of weeds. That is their role in nature. They love to grow, they love to cover bare ground and they love to make our life difficult at times. Not everyone can have animals to control and eat weeds, and some weeds even animals don't eat, such as cape tulip and Patterson's curse. So don't ever think that permaculturists are all lazy gardeners. You will get a shock!

We seem to be going in a circle, revisiting the ways of life of our parents and grandparents where everyone had a rainwater tank, fruit trees, chickens and a vegetable patch in the backyard. And maybe that's not a bad thing. It is a return to a simpler lifestyle where a person took more responsibility for their food supply and to deal with household wastes. Many of the highlights for both Jenny and myself have been the simpler things, like walking out the back door and picking a broccoli head or plucking lettuce leaves, watching chooks fossick around or picking a plum or apple and eating it straight off the tree.

Jenny and I are now in the process of setting up Candlelight Farm II in Mundaring. We have two and a half acres (1 Ha) and we are building two strawbale units as an eco-stay, putting up tunnels as part of Red Planet Plants, a commercial wholesale nursery we first started in Midvale a few years ago, And of course planting our stock plants, windbreaks and food forest. We are starting all over again, but the development of Candlelight Farm II will be another story.

I have visited and talked to many well-known permaculturists in Australia, some of whom you have been reading about in this book. I count myself fortunate to belong to this group of people, all of whom are leaders and shining examples of those who endeavour to work towards sustainable living. My own personal permaculture journey is one of endeavour, persistence and perseverance. I continue to enjoy the journey, but like many other pilgrims, the path has not always been clear and easy to follow.

Jill Finnane

Originally a science teacher, Jill Finnane is the Eco Justice Coordinator for the Edmund Rice Centre in Sydney. That includes coordinating the Pacific Calling Partnership, an initiative of the Centre, which has brought together organisations and individuals with the aim of working in partnership with Pacific Island countries in the face of climate change. She is also a founding member of the Sydney Food Fairness Alliance. Previously, Jill worked with Action for World Development as a community educator on sustainable development. She is a permaculture practitioner and teacher of permaculture which she has taught in Sri Lanka, Vietnam and Australia. She has written two books: *When You Grow Up*, which she co-authored with an Australian Indigenous woman, Constance Nungala McDonald and *Lawns into Lunch*, which is about people growing food in the city. Whenever possible, Jill cycles the 15 km to work.

Chapter 19 | Bringing Knowledge to Life

My grandfather, 'Pop', was still growing vegetables when he died at the age of 85. He spent many years as a poultry farmer and vegetable grower at Greystanes in Sydney's west when it was part of the green belt. In his retirement he supplied vegies to the family and to a delicatessen in Merrylands. Just as Sydney hasn't valued its rich heritage of market gardening so I didn't value the food growing heritage my grandparents tried to share with me during my childhood holidays at their place.

One day, when I was about ten, my grandmother, 'Ma', told me that newspaper was a useful ingredient in compost. Unwilling to show my ignorance I didn't ask her what 'compost' meant. On top of the sink in the kitchen they kept a bowl for the chook scraps. I knew about that because Mum and Dad kept chooks. Underneath the sink there was another bucket. They called it the 'compost bucket'. Ma gradually filled it with the less tasty scraps and peelings and then mysteriously, Pop would take it to the 'compost'. I never saw this happen, I didn't ask where he took it and I never ever wondered what 'compost' was or what it did. I did carefully watch him at six in the morning down a raw egg straight from the shell before heading off to the garden and I did quiz him about that, but never about 'compost'.

Something of Ma and Pop's closeness to the earth must have stayed with me though, because I became a science teacher and took great joy in leading my students to investigate and understand the world around them. Keeping chooks stayed with me too. My husband Michael and I both grew up with chooks so we bought some as soon as we moved into our own home with its large yard. We wanted eggs with bright yellow yolks, less lawn to mow, natural entertainment for our children and somewhere to put food scraps. Neighbours encouraged us by insisting on buying our excess eggs. However, we didn't appreciate the other products of our chooks and yard. We carted manure, along with dried leaves and grass clippings to the tip!

As our children arrived, I found myself spending more and more time outdoors and my mind turned to the possibility of growing vegetables. I got out the spade, opened Yates Gardening Guide, bought some artificial fertiliser and walked right up to the very back of the yard where a previous owner had laid out two rectangular beds. I began digging, planting and spraying. Sadly, the tiny plants produced tiny vegetables that were not worth cooking.

So I started using organic methods and found out what Pop had been doing with the compost bucket. As he must have done, I started to mix those non-chook-food kitchen scraps with the dried leaves, grass clippings and chook manure. At last I began to understand Ma's advice; I included some newspaper. The composted gardens produced vegetables we could eat but it was exhausting work and I hurt my back digging into our hard packed clay.

In the meantime, my life had been on another journey. I grew up in a Catholic family that gave to the missions 'over there' and to the St Vincent de Paul Society closer to home. As a child, I wondered how it came about that I always had a roof over my head, a family and a full stomach while other children in other places clearly didn't. My parents and the nuns encouraged us to put pennies in the mission box but it just didn't seem enough. I could see from the photos on the box that the children didn't have any hair – mine was long – I had plenty. And so my parents discovered they had an activist in the family when I cut off my hair and wrapped it in little newspaper parcels to send to the missions. As I grew up I felt a mixture of sadness and incomprehension whenever I saw people suffering from hunger, war, rejection and deprivation. It was as though I was incomplete as a person while ever these things continued.

When I left school, the Campions, a Catholic prayer, study and action society of 'intellectuals', nurtured that passion for justice and fostered deeper inquiry. In 1972, the Campion Society helped set up Action for World Development (AWD) to encourage Christians to work together ecumenically to address the causes of hunger, poverty and injustice. I had stopped teaching science and we were beginning our family. AWD began its life as an Australia-wide study program around two slogans; 'Take your charity and stick it up your jumper' and 'Live simply so all may simply live'. In Sydney it became an organisation whose members formed into local or cause-related groups. What we investigated at AWD confirmed that poverty was not a result of ignorance and bad management but that many of its causes lay with the history that had created the richer industrialised nations. As my involvement in AWD increased and I became immersed in its Paulo Freire inspired action-reflection-evaluation and structural analysis processes, it became clear that the reason for the persistence of 'poverty' often lay with structures and processes operating in the 'rich' world. Thus the focus of AWD's community education and advocacy work was to find out about the causes of injustice operating in Australia and to explain them to other Australians.

The Tea Group was a small AWD group that looked into the relationship between colonisation, economic theory, unfair trading practices and poverty. It prepared a Fair Trade education campaign based around selling, for the first time in Australia, tea that was processed and packaged in its country of origin – value added tea. It gradually gained so much support that it set up the World Development Tea Co-op and later Trade Winds Tea and Coffee Pty Ltd to continue pioneering Fair Trade in Australia. For a few years, I coordinated the education work of the Tea Co-op where I delved into some of the complexities of trade in agricultural commodities and its effects on small growers and food security.

Bankstown AWD was a small local group that met for about 20 years, usually in our home, educating ourselves and campaigning in our local community about issues like Aboriginal land rights, unemployment, nuclear disarmament, apartheid, and black deaths in custody. A member of that group, Sue, lent me a copy of Esther Deans' *No Dig Gardening* book because she was dismayed at the damage digging did to my back.

I had no sooner started to build my first no-dig garden than I heard Bill Mollison talking in visionary terms on ABC radio about how permaculture could change city landscapes and reduce the amount of work in growing food. As I stood in the kitchen listening to his voice filtering through the din of the children playing, my imagination started to run wild with images of a future that went beyond more tar and cement. I was captured by the seemingly utopian but utterly commonsense possibilities he talked about and found myself agreeing with him. Why shouldn't we grow food using some of the land, sunlight and water that cities and suburbs abound in? I began to see how easy it is to limit our thinking on issues to what we know and what we have seen, and how easily we discard simple solutions.

The magic was I could start with the quarter acre that we lived on. A few weeks later, I was talking to the Mayor of Bankstown suggesting to him that one way he could address the high youth unemployment in our area was by getting them to plant, cultivate and harvest fruit trees on the nature strips. Sadly, he laughed at me.

Over the years I've dabbled in many things and 'dabble' is just what I did when I first tried permaculture. I had only a rudimentary idea about it but each new bit I tried, first of all made sense, and second of all, worked. Fruit trees and herbs, perennial vegetables and vines gradually found their place in the yard and their produce on the table. Contradicting the advice of gardening books that fruit trees only grow in well managed orchards I found that it was possible to grow productive, pest-free fruit trees in a suburban yard, that compost and mixed cropping did help prevent pests and disease, and that mulching did keep roots moist. I tried every possible way of constructing 'no-dig' gardens. Different combinations made up each plot. Layers of damp newspapers, worn carpets and underfelt, old clothes, weeds, straw, hay, prunings, manure, leaves, grass clippings, lime and compost created vegetable gardens that didn't need digging.

The children seemed to find it all amazing, if a little weird. They complained about the smell in our van the time we carted home bags of seaweed after a weekend at Ettalong Beach and objected when banana passionfruit appeared one way or another in every dessert. We developed a system that rotated four chook runs with vegetable beds so that the chooks became an efficient tractoring unit. Spending a month or two in a run, they dug up the grass, weeds and spent crops, dropped their manure and scratched around in the straw. When given a rest from the chooks, each run took its turn to produce corn and beans or pumpkin and potatoes.

It all brought an earthy joy into my life. I delighted in the vigorous exercise while I helped nature convert smelly food scraps and messy garden waste into rich dark compost with an aroma reminiscent of a forest floor. I discovered the leisurely pleasure of watching organic matter ever so slowly turn hard clay soil into a place teeming with life. I still get a thrill like that of any six-year-old as I listen to the frogs on a moonlit summer's evening knowing that as well as mating they are dining on garden pests. And I think I will always get a flush of joy when I see that first egg in the nest when the chooks come back after being off the lay.

Breeding too provided its share of earthy joys. One time the bantams kept flying over the chook yard fence, taking the whole yard as theirs. For weeks no one could find their eggs. Then, early on Easter Sunday morning, we heard tiny 'cheep cheep cheeps' coming from underneath the prostrate grevillea just outside our bedroom window. What excitement as we rushed to wake the children and tell them of their Easter present from mother hen! And it cost us, and the earth, nothing.

Permaculture gave me a bond with nature and with people, starting with our children. The children grew up around my efforts at growing food. As I got dirt under my fingernails they played football, cricket and dress-ups, added their guinea pigs to our chooks, and occasionally helped to make compost. They watched in puzzlement as the no-dig gardens gradually joined together around the outside of their play area. They negotiated with me for the right to include the little hill we had built for the avocado tree in their daredevil BMX obstacle course. They rode their bikes to school alongside me when they were young (and hoped people didn't notice me cycling as they grew older.)

At the same time I was learning through AWD that the introduction of cash crops had pushed communities away from providing their own basic needs and into a dependence on market prices dictated by industrialised countries. My food growing experiences showed me the contribution home gardens could make to nutrition and gave substance to what I was reading in development literature. I mused that any World Bank official who had grown their own food organically would not have given such destructive economic text book advice to people in poor countries. I knew from stories I learnt at my mother's knee that you couldn't depend on the market. Australia experienced food scarcities during and after the Second World War and yet Pop could get more money at the markets for flowers than for vegetables. Those who could afford to pay more for flowers had skewed the market. On a global scale, poorer countries are coerced into producing what richer countries want, rather than what their local people need.

My knowledge about permaculture had all come from a small Sydney newsletter published in the '80s and later from the *Permaculture International Journal* and books like *Permaculture: A Designers' Manual* by Bill Mollison. All that reading was helpful but clearly there was more. I found just what the 'more' might involve when in 1987 I took my first course with Rosemary Morrow. She took us step by step through the ethics and design principles and their practical application in a home garden. It was so exciting as I discovered that the theory actually worked. The beans and silver beet were quicker to pick and really did get picked more often when they were close by – every time I went out to hang washing on the line I pulled out a few weeds. At last I understood that permaculture design is about how you choose to arrange the parts of a system based on permaculture principles and that what I needed to do was establish as many beneficial connections as I could think of. Even so, it took me a long while to fully grasp all the implications of permaculture design and ecosystems thinking on a broader scale.

The science teacher in me didn't disappear. I observed and experimented and looked more comprehensively at practices that built up the life of the soil, the biodiversity of the system and the management of water. *In situ* science, I found, was no less rigorous than laboratory science – the outcome after all meant the presence or absence of food on the plate. Like Bill Mollison, I discovered that ecology was in operation all around me and I could choose to work within its principles and have nature working for me or I could ignore its principles and continually fight against it. My knowledge of science, especially physics and biology, came to life in the systems I was setting up. Gravity was not just 9.8m/sec². Creatively employed and understood in the systems I set up, it provided energy to distribute tank water, move heavy loads, and help drainage in heavy clay soil. Ancient strategies like terraces, swales and bunds around rice paddies had always harnessed it.

In the late '80s I began to work for AWD mainly with the Agriculture and Food Group. I studied Systems Agriculture at the UNSW and finally did the full Permaculture Design Course. I admired the worldwide system that had evolved for the teaching of permaculture and the formal and informal networks that enlarged and extended that body of knowledge and experience. I began to remember the questions that Ivan Illich in *Deschooling Society*, had raised years before in 1971 about the kind of education that was foisted on people. He questioned development goals that saw education as good no matter how irrelevant it was to the local life style and traditions.

By 1993 it had become increasingly obvious to the AWD team that sustainability was being sacrificed in development directions favoured by big companies and funded by bodies like the World Bank and the IMF. Permaculture, my colleagues reminded me, offered not only a critique of their approach but came from a worldview that valued the wisdom of past generations while taking into account the rights of future ones. Permaculture demands an intelligent connection with the earth where the whole is seen as greater than a sum of the parts. It develops the mind. Applying permaculture is not about learning rules or a set of techniques – it's about learning principles and how to apply them in different ways and different situations. Students of permaculture learn skills in observation so that wherever they are they can get in touch with all things that are happening in a system before they plan to make changes. They learn to take time to reflect on those observations. They do exercises to develop skills in analytical thinking and synthesis so that they can work out what they can learn from their observations and reflections.

Though I valued all these things for myself and liked the idea of teaching permaculture, I was at first unconvinced that permaculture courses were a valid way of getting Australians generally to take a constructive interest in the situation of developing countries. Mostly, the AWD courses were for city people. With their links with the wider community it was important for me to listen to what they wanted and one thing they wanted was to save work. The easier food growing is, the more likely city people are to keep at it. Often too, people were looking for an easy recipe for success; yet permaculture is not about the unthinking transfer of a successful technique from one situation to a different one. In many ways it is intellectually demanding and thinking-rich. The courses showed them that growing food did not have to be mindless, degrading or servile work. At times students would ask, "What is the permaculture way to approach this problem?" I'd make it clear: "There is no one permaculture way. Think through the principles, see what approach will work for you in this situation, and that is permaculture." Many city students discovered (I hope) during the courses, that a garden is there to serve them and their needs and isn't answerable to the aficionados of home and garden magazines. From what I could see, the ethics-based approach resonated with all students, who gladly accepted a duty of care to the little patch of earth that was entrusted to them and to the people with whom they were connected.

The courses and workshops met the desire of students to learn how to grow food and AWD's keenness to give them an awareness of the broader issues of sustainability in their local community, their bio-region, and the global community. So I set up a permaculture program which continued, with contributions from many people, for the next ten years. We made use of quality resources such as a video directed by Joss Brooks about permaculture in South India called *Growing Together*, cameos of Bill Mollison from *The Global Gardener*, or slides Tony Jansen gave me about his experiences in the Solomon Islands. Badri Dahal, who had set up permaculture systems in Nepal, teamed up with me to present a couple of Permaculture Design Courses and courses to prepare people who wanted to work in developing countries. It's true that many people were so delighted to learn introductory permaculture ideas and found that implementing them brought such a change to their lives that they didn't see the need for going deeper. Nevertheless discussion and evaluations showed that even they appreciated learning about the links between local and global issues and the opportunities there were for them to 'think and act locally AND globally'.

The next section describes some of the permaculture projects I have been involved in or learnt from, around the world.

Long Bay Gaol

Not all courses were conducted at the AWD Centre. Long Bay Gaol (in eastern Sydney) provided an eight-week lifestyle program in healthy living for prisoners with HIV, AIDS and hepatitis C. Six to eight men would live under unit management doing their own cooking and planning. They had a small, fully concrete exercise yard surrounded by high walls topped with razor wire. From 1997 to 1999 I taught introductory permaculture courses there for a couple of hours a week.

I usually prefer flexible designs when beginning a new garden but in a gaol garden edges can't be constructed from loose bricks; the bricks must be cemented firmly together. The first course conducted a site analysis and drew up some possible designs. When I arrived the next week I found that the officers had organised them to build a garden with bricked edges right on top of an open drain with a down pipe emptying into the middle of the garden. It took a long time before I could get the officers to realise that a few adjustments were needed to that garden.

Eight weeks was an ideal length of time for the Lifestyles program but frustrating for garden involvement and ownership. Though most of the prisoners were keen enough to set up their own vegie patch or worm farm they weren't fired with enthusiasm to tend those set up by the previous group. The men would be just beginning to see some results for their labour when it was time to go back to a section of the gaol where they could apply none of what they had learnt. In the last week of the course they became somewhat detached and restless. It may have been because they were leaving behind half-grown vegies but more likely they were preparing to fit back into the aggressive mainstream of gaol life.

Most of them had reading and writing difficulties so it was important to make the learning activities simple, concrete and to include exercises that built up their skills in reading such as matching or sorting cards. This kind of activity also helped them to see how much they had learnt. For most of them, school had been a time of failure and distress, so demonstrating their achievements was wonderfully affirming. I was surprised how involved many of them became in talking about ethics and principles and how many of them had happy memories of food gardening with their grandparents.

It was hard to keep the momentum going with a once a week class. Very few had the motivation or confidence to do things on their own between classes and it wasn't the officers' job to encourage them. There were some surprising exceptions though. One fellow who was nearly rejected from the program found it hard to relate to the other inmates but spent hours tending the worm farm. With all that extra aeration the worms multiplied rapidly and he beamed with a wonderful sense of achievement. He told me that when he got out (of gaol) he planned to go back up country and set up worm farms so he could sell worms to local fishermen. Another, who I was told had been quite dangerous in his time, was so caring of his marigolds that he staked each one individually with tiny sticks and tended them every day. He told me that when he was working in the garden he forgot about everything else.

It seems to me that for permaculture to work well in a gaol setting it needs access to land that is not covered with concrete so prisoners can have the satisfaction and vigorous exercise of digging, and the opportunity of keeping chooks. Involvement needs to be optional and they need supported access to the garden on a little and often basis over a long period of time.

Sri Lanka

One great strength of the Sri Lankan food system is that most people still live in villages, still own their own land and still farm their land or at least grow their some of their own food; jakfruit, papaya, banana, spices, curry leaf and so on. Michael and I have a long time connection with Sri Lanka because of our involvement in the Trade Winds Fair Trade campaign. Consequently, I have been asked to run permaculture courses and workshops there for organisations like Gami Seva Sevana Organic Farm (GSS) and Satyodaya, an interfaith organisation that works with poor villagers and tea plantation workers.

Farmers and NGO workers attending a course at GSS in 2001 talked about how they felt pressured to grow mono-crops and use heavy applications of chemical pesticides and fertilisers. They described how chemical companies had given them advice and assistance. They also spoke about the widespread problems with deforestation and drought. Historically, drought has been a part of the cycle of life in Sri Lanka and there were traditional systems such as tanks in the Dry Zones and conservation practices to handle those bad times.

What Ranjith de Silva, coordinator of GSS hoped, was that the students would carry into all parts of the country the message that there is an alternative to the current notion of growing mono-crops, dependent on chemicals. He told me that farmers wanted to hand to their children farms that were in good condition but their more immediate concern was 'can we make money?' He explained that for people in developing countries they want to be able to make money from their land as well as grow food to eat. "Unless permaculture helps them to make money they are not interested in it," he said.

So it was important to show how they could do that and for me to learn to think in those terms. Lowering costs was one way. Permaculture reduces costs by designing useful interconnections that save energy, waste and water, and systems that work with the natural environment to provide pest control. So does establishing local markets. Another long-term economic benefit comes from designing for resilience against disaster. One way to do this is to learn from traditional approaches that preserved the natural balance of micro-organisms, insects, animals and plants. Similarly I tried to show them that by employing strategies that treat soil as a precious and living resource, they would be building up not only current assets but future ones too. The Kandy forest gardens are a world-recognised example of how perennials not only protect soils from the force of tropical rain but also provide a more easily harvested, year round and varied range of products. A simulation game that showed up the dilemmas of choosing between cash crops and subsistence crops led the students into discussions about the fluctuation of prices for cash crops on the world market.

My Sri Lankan experience taught me the power of short courses and community learning. In 2003, one of the courses I ran was a week-long Tamil/English course at GSS mainly for NGO workers, many of whom were university graduates. A coordinator of Penn Wimochana Gnanadayam (PWG) 'The Dawn of Wisdom Comes with Women's Liberation', had attended previous courses in 1997 and 2001. This time, PWG sent a couple of animators from the tea plantations. Plantation workers live in isolation from villages and are among the poorest of the poor and most malnourished in the country. Until recently they were not allowed to grow any food on the land around their line huts (this rule continues in many estates). Till then, their only experience of agriculture was picking tea leaves and carrying them down to the weighing station. Their only education was from a range of short courses and other informal learning processes organised by PWG. I worried that they might be out of their depth and wondered how they felt about spending a week learning alongside academically educated people.

I needn't have worried. In 2007, I again visited PWG and took part in a joint meeting of animators, field workers and coordinators. I gave them a set of posters explaining permaculture principles that the Organic Garden Project in Ha Noi in Vietnam had just produced. Siva Packiam, one of the animators who had attended the 2003 course, had now become a field worker. She held up the posters enthusiastically and started explaining them to everyone, stressing as she did that she found it especially important to let estate women know the role that perennial plants play in food security and the protection of soil. She said that part of her work involved teaching short courses in food growing to women on five estates and that the most effective way to motivate them was to show them by doing it herself. Her garden had become so productive that she was selling vegetables and plantain suckers and making and selling curry powder made from her own herbs and spices.

Bangladesh

Nayakrishi Andalon is a Bangladeshi farming movement set up by UBINIG (UBINIG is the abbreviation of its Bengali name 'Unnayan Bikalper Nitinirdharoni Gobeshona'. In English it means Policy Research for Development Alternatives) under the leadership of Farhad Mahzar, who had been a student of Bill Mollison's. When I visited Bangladesh in 1999, I met with Nayakrishi farmers and some school students who were learning about the Nayakrishi approach to farming in the school holidays. Ruma, a sixteen-year-old student, told me that during the years she had been growing up, the farmers in her region had been pressured to change from using traditional growing systems to chemical farming methods. "Farmers are growing more mono-crops to sell to the rich world and it has simply made them poorer," she said.

Vietnam

Though it started later, the promotion of chemical farming methods in Vietnam has been just as destructive. In 2007 I spent 3 months at the Vietnam Friendship Village near Ha Noi as a mentor and advisor with the Organic Garden Project. The Village provides rehabilitation to victims of Agent Orange and sees organic agriculture as an important part of the program. Three years earlier they stopped using chemicals and began to build up the soil with compost and manure, planted orchards and vegetable gardens, and introduced a small amount of mixed cropping. Permaculture design skills would have made a huge difference at that earlier stage but there was still much that permaculture had to offer. I hoped to find other organic and permaculture projects around Ha Noi and use them, and the people running them, as a reference point and teaching resource. To that end, John, the project founder and part-time coordinator and I made investigations through various NGO networks. All of them told us the same thing: "There is nothing." John visited a couple of initiatives further away from Ha Noi but he found they were too narrow in their focus. It was disappointing but made us realise just how substantial the Organic Garden's contribution was.

At first, I worked mainly with Huyen, a woman from a farming family with little formal education, who had been the garden manager for two years. Huyen was keen to build up her knowledge, but was planning to move on soon after I left, and help set up another organic project at Tyin Quang. Thus in the latter part of the project I worked mainly with Xiem who though only a few years out of school was preparing to take over as manager.

One of my tasks was to help them devise educational materials so that they could continue to educate others after I left. Community education type materials seemed the most useful. After careful analysis they decided they wanted a series of illustrated resources with descriptions in Vietnamese and English. Vigorous discussions at team meetings eventually came up with the most significant messages they wanted to give to garden visitors and workers. A local printer then stuck the words onto eighteen wooden signs and five local and international volunteer artists painted pictures on them. Huyen suggested we design posters depicting the permaculture principles that had been of most use in the garden. More work for volunteer artists!

Our little cottage looked more like an artists' den than a permaculture haven. Less exciting, but equally important resources, were the simple tasks sheets we designed, giving step by step instructions, the photograph albums with detailed explanations and, for the those who were really serious, in-depth materials translated into Vietnamese such as Rosemary Morrow's *Earth User's Guide to Permaculture* and the 'how to' boxes from my book *Lawns into Lunch*. A key person in all of this was Hien the enthusiastic young interpreter, whose increasing understanding and active involvement led her to tell me she would love to be able to play a role in getting permaculture more widely practised throughout Vietnam.

'Quan sat' means 'observe' and 'quan sat' almost became our mantra as that fundamental permaculture skill opened the students' eyes to the ecological connections happening or being thwarted in the garden. As their permaculture thinking developed they became very good systems thinkers. I could see that they understood the interlinking of systems, hierarchies of systems, and that very few systems, including the ones they were working with, are closed systems. Huyen designed and named a 'biodiversity model' (a variation of the banana circle) that replaced a section of a poorly performing tangerine orchard. She also planned for and organised increased planting density and biodiversity in the buffer zone between the garden and the surrounding wetlands. Xiem drew up a diagram showing not only the smaller systems within the garden and the village but also its connections with the larger systems beyond that played a role in the viability of the garden. Inspired by this she then created a design to improve the demonstration potential of a mango garden that adjoined an area where international visitors (and possible donors) often walk. She increased the biodiversity by inter-planting the fruit trees with nitrogen fixing trees and flowering plants and agonised over which of the new interpretive signs she wanted to display in that garden.

When I think of the many forces operating to modernise agriculture in developing countries along industrial lines I have two conflicting responses. On the one hand, I feel disheartened and fear for the future of food security, especially for the poor. On the other hand, I feel a sense of the huge importance of a small initiative like the Organic Garden Project, as a beacon to the world of just how agriculture can be. I am further heartened when highly respected organisations like the Food and Agriculture Organisation (FAO) put out a report, such as *Organic Agriculture and Food Security* by Nadia El-Hage Scialabba in September 2007, which gives authoritative endorsement of the Organic Garden Project's relevance and cost saving.

To quote from the report:

> *Working with natural processes increases cost effectiveness of food production... These low-cost farming practices reduce cash needs and thus, credit dependence... Organic agriculture breaks the vicious circle of indebtedness… which causes an alarming number of farmer suicides.*

The FAO Report concludes that a broad-scale shift to organic agriculture can produce enough food on a global per capita basis to feed the world's population over the next 50 years.

The FAO document also points to something I believe will become increasingly relevant: "…organic agriculture attracts new entrepreneurial entrants into farming who have more optimism about their future due to the value of their jobs in the local economies." And it quotes research that shows that carbon dioxide emissions in organic agriculture systems are 48% to 60% less than in chemical-based agriculture.

By 2003 I began to feel frustrated at the limitations imposed by working for a small cash-strapped organisation. I felt that I was getting stale and needed to step back and examine where I was headed. Writing the book *Lawns into Lunch* gave me the opportunity to visit former students, look at what they had been doing, and interview them about what permaculture had done to their lives and gardens. I got quite a buzz from planning, working and agonising over the shape of the book with a young photographer and recent permaculture graduate, Huni Bolliger. On the strength of the book I have since been making presentations to lots of ordinary people who are beginning to think about growing food. The most exciting of those has been playing a small role in the Green Home program organised by the Australian Conservation Foundation for local councils and businesses.

One of the myths that *Lawns into Lunch* addressed was that you needed to own your land in order to grow food. Ten of the people interviewed for the book did not own the land they were gardening, though they did have varying degrees of control over it. This control over the land you are working with seems to be an important issue generally, if people are to be motivated and able to grow food.

Since 2004 I have been working for the Edmund Rice Centre (ERC), an NGO with a redoubtable reputation for championing human rights. Its dynamic, challenging and totally professional team take seriously the need to 'speak truth to power' and to do it effectively. The ERC signed the Earth Charter in 2003 so my task has been to gradually introduce a program that links eco-justice with human rights. The Earth Charter offers a vision and sets forward ethics and principles for sustainable living. It encourages a systems approach and draws clear connections between efforts to achieve sustainability, peace, human rights, and social justice. It invites people to look at the future we are creating and asks them to care about the kind of world we are bequeathing to our children, our grand children and our great grand children. Several of the projects I have set up challenge the community to listen to the science about climate change and take responsibility for it. Even within the centre that was difficult at first. The need to address other urgent social justice issues made it hard to take the time to introduce ecological changes. One of the difficult messages that climate change is throwing up, is that we in the industrialised world have taken far more than our share of the earth's resources. People find it hard to see the connections between ecological devastation, loss of livelihood and the need for us to make big changes as a society.

Working with a team multiplies the effects of the good things I can do and makes up for my inadequacies. Organisations get similar benefits from working together. Cooperative relationships and strategic alliances have been a key focus of the Eco-justice Program. I have played a role in setting up coalitions like the Sydney Food Fairness Alliance, a growing and vigorous association of disparate practitioners who want to make Sydney's food system more equitable and more sustainable. The Pacific Calling Partnership grew out of a permaculture-inspired project conducted in Kiribati by an ERC intern. It brings together organisations that want to respond to calls from our low-lying Pacific Island neighbours. Building relationships with Pacific and Torres Strait Island communities is not only a fundamental strategy of the partnership it also creates opportunities for cultural exchange and demythologising the materialistic values and attractions of the western lifestyle.

Though the ERC is happy to work with people of all faiths and none, its ethical base draws on Catholic social justice teaching. Social justice endeavours bring many frustrations and disappointments so I have found support in my personal Catholic faith and worship and I gain strength from being able to hand over of my life and my inadequacies to God. People who take a positive approach, value fun and who are able to talk through ethical dilemmas, get me over the humps. Big people like Nelson Mandela inspire me. Local people like Doug and Betty Bailey keep me balanced. Permaculture gurus like Rosemary Morrow keep me questioning and thinking. My husband believes in me and people from developing countries remind me how crucial the struggle is.

The Earth Charter states, "As the world becomes increasingly interdependent and fragile, the future at once holds great peril and great promise." We can't predict the future but we can guess and we can contribute to shaping it. If even the most mild climate change predictions are right then the future will require radical changes in the way people live and obtain the necessities of life. We need to be people of hope and vision.

Permaculture thinking will help us to be more creative, resourceful, practical, and to use more commonsense. It can provide a pathway to living fulfilled lives that are happy, adaptable and sustainable because it changes people by expanding their thinking skills so they become ecosystem thinkers. A sustainable future will need courageous governments and enlightened people who value fresh air, fresh local food, the rights of all people, and ensure that science is at the service of the Earth and its people. It will need businesses that recognise and operate within ecological boundaries and ensure that global trade is fair. And it will need encouragement and assistance for developing countries so they can move towards self-reliance and interdependence, and away from dependence or exploitation.

Children will delight in seeing how plants grow and change, producing food that's fresh and tasty. They will enjoy sneaking out at night to catch a glimpse of frogs hunting garden pests. Adults and children can have earthy fun together counting the different decomposers in the compost – allowing their souls to become embroiled in the smells and invertebrate wonders of an evolving ecosystem. They will enjoy the flowers scattered through the food gardens to provide nectar, a key component of the integrated habitat that 'good bugs' need. Children will have no trouble finding new and more challenging places for games of hide and seek. They can hide beneath the passion vine that is sheltering the chook shed, behind comfrey that is growing near the compost bins, among the climbing beans that are nurturing the fruit trees or in the branches of a native tree that forms part of a wind break.

Grandson Noah learning how to harvest green manure crops and use them as mulch around citrus trees.

Ian Lillington

Ian Lillington is a community activist and teacher of permaculture, based in Castlemaine, Victoria and author of *The Holistic Life* – a colourful introduction to permaculture published in 2007. The book features the sustainable house that Ian designed for an eco-village in South Australia, where he lived with his family until 2006.

He first heard about permaculture in 1982 when Bill Mollison did a tour of the UK, where Ian was managing a city farm and community garden. It wasn't until 1989, when Ian visited the Holmgren-Dennett property in Central Victoria that he saw living proof that permaculture provided sustainable food and shelter on a small block in a cool climate.

Ian did a PDC the next year and has been teaching permaculture since 1992. He also worked closely with David Holmgren, co-teaching courses and editing some of David's books including *Permaculture: Principles and Pathways Beyond Sustainability*.

In 2002, Ian began work with a small volunteer team to develop what has now become the nationally-recognised Accredited Permaculture Training, offering Permaculture 101 through to Diploma level permaculture in the VET sector.

Ian is busy raising three teenage sons – helping them with soccer training, film-making and moving house. He works with an informal network of sustainability activists, especially through community sustainability and water groups based in Castlemaine and continues to teach Permaculture Design Courses and Accredited Permaculture Courses.

Chapter 20

From Urban England to Rural Australia via Permaculture

Permaculture and my life have been interwoven since about 1983. That's more than 25 years! Permaculture for me is much more than gardening. The choice about where I buy my food is as significant as whether or not it is growing in my garden. Choosing to buy produce grown or made by locals is at least as significant as growing or making it myself.

My parents grew food in a small back garden in England, so I was exposed to a culture of local food from my earliest years. They were children in the 1930s depression and lived through the Second World War, when 25% of eggs came from backyard birds. But in the 1970s, when I was a teenager, something different started to happen. The consumer generation was beginning, and while television, plastic toys, listening to records and going to discos diverted most of my peers, I continued to be interested in the garden.

My parents were pleasantly surprised, I think, by the new turn of events. My dad's job was in landscaping and sports turf, and when he got home, he didn't want to have to grow vegies as well. My mum was impressed with all the new packet food in supermarkets, a god-send for a working mother with four small children.

So I can't explain exactly why I continued to grow food when almost everyone around me was heading to the shops. Even at university in a 19th century terraced house in Liverpool, I found one square metre of soil, planted strawberries and went out at weekends to pick blackberries from wild bushes.

I was a student of Geography in Liverpool in the 1980s when I began to do some community work on a new city farm/community garden. We were part of a National Federation of City Farms and they hosted a workshop by Bill Mollison in Bristol in 1982. One of my colleagues went to it and came back with notes about tyre ponds and herb spirals, but it didn't grab me; I was focussed on completing final exams and writing funding submissions for the project. I took a bit more notice in 1985 when the Federation asked me to go on a weekend introductory permaculture course, taught by some of the early-adopters of permaculture in the UK. From then I began to understand that permaculture was a design system and could be applied anywhere.

It was a slow start; I was a community worker with an interest in the environment and didn't see myself as a designer or a teacher. But, looking back, I realise that I was searching for a structure to allow me to make sense of what we'd now call sustainability. In the 1970s and '80s there were a lot of people concerned about environmental issues, but there was no coherent approach.

What permaculture offered me then, as now, was a way to link big picture ethics and principles (which I knew were needed), with day-to-day life choices. Permaculture didn't pretend to have the answers to all my questions, but at least it was a way to start, that combined my interest in landscape, environment and community.

It's no accident that community gardens and farms in the city were my way into permaculture. Through the 1990s and up to the present, these projects provided a way to combine hands-on work with environmental and community politics. It was the city-farm link that took me to work at CERES (the Centre for Energy Research and Environmental Sustainability) in Melbourne in 1986, and again in 1989. In '86, I went armed with a list of permaculture contacts including one David Holmgren in Carlton, but as I hadn't even seen the book *Permaculture One*, that name was meaningless to me. But back at CERES in 1989, David and I worked together for a short time and became friends, without me even knowing that he had co-originated the permaculture concept.

My visit to David and Su's place in Central Victoria in 1989 was a big leap forward in my understanding of permaculture. They convincingly showed that it could be done on a small scale and in a cool climate, and I returned to England in 1990 determined to find out more.

The Itinerant Teacher

I studied for a Permaculture Design Certificate (PDC) in Coldstream, Scotland in 1990. As my course ended, Kuwait was being invaded for its oil – the beginning of the oil wars that continue to this day. Soon after finishing the PDC, I began to teach introductory courses in north-west England with another graduate. I coordinated a PDC at Manchester University taught by two more experienced teachers from Devon, as an optional extra for architecture and landscape architecture students. This only happened because of the enthusiasm of one of the students and my readiness to work, unpaid, to put together the pieces.

By now I was 31, working for Friends of the Earth in London, and with a pregnant partner, Jo. We began to talk about raising children somewhere other than inner-city Liverpool and listed Wales, Ireland and Australia as possibilities. We tried Australia first, as we were not sure if we could get residency. By late 1991, we knew that we were welcome, thanks to Jo's qualification as a social worker, and we migrated the following April. From 1990 to 2000, teaching permaculture became one of my main activities. It was once I settled into Hepburn and began to work closely with David Holmgren that I became a fully-fledged permaculture teacher, learning an enormous amount about Australian landscapes and techniques, as well as teaching and meeting many wonderful people.

Sometimes teaching managed to earn me some money, such as when I taught with the Brookmans at the Food Forest in South Australia. But more often teaching was a hobby, an interest. I wasn't able to find or create enough permaculture work to make a living. So through most of the '90s, I was working as a part-time landscaper, part-time community development worker as well as an itinerant permaculture teacher.

During this time I enjoyed the image of myself travelling around Australia with the tools of my trade in a small backpack. These were a few books and notepad, whiteboard pens as well as course notes gleaned from lectures and workshops I had attended and revised as I delivered courses and discussed approaches with friends and colleagues. I also had those most vital of tools – my mind and my body. And although I enjoyed that time, just as now, I saw myself from the outside, as a bit of an odd ball!

Perhaps it is just one of those things about me, but I didn't want or expect a formal career. Jo was earning some money, I was earning some, but we were well below the official poverty line. It was a happy place to be as we lived in a passive solar house, had good food in the garden and good friends who could fix things, who knew about the internet, or who ran the local barter system.

We discussed how we could generate and sell power locally from wind farms and methane generators, and how at least part of the sales would be in local currency through a Local Economy Trading Scheme (LETS). 15 years later, Hepburn is developing a community wind farm, and academics like Shann Turnbull are proposing local currencies that are based on a kilowatt hour of power. The wind farm and the alternative to federal currency are still not fully in place, but we were on the right track back then. We also discussed how food surpluses could be shared, through local markets and subscription farming and now, a few years later, we have farmers' markets, food box schemes and permablitzes.

This was a very productive time in permaculture in Central Victoria. I was part of a network of permaculturists and we had plenty of time to test out theories and chat; Lots of talk over the washing up, over the weeding, over the phone, and over cherry brandy into the late evening. Looking back it already feels like a bit of a golden age. These were the days before laptop computers. Email was slow, a curiosity that only the brave would use. We did have more time before the digital age, I am sure of it, and the internet is both a blessing and a curse. I miss the opportunity for a quick google when I am away from the computer; it is a very powerful tool. But like everything, it has to be used in moderation.

By the late 1990s, changes were in the air. I had three young children and was building a house and garden, later featured in my book *The Holistic Life*. I had found a reasonably steady job training housing associations and housing co-ops, sometimes training them in ecological work but more often in tenancy management and conflict resolution. The itinerant teacher had begun to settle down.

My children all had their hands in the dirt, but personal stereos and DVDs meant that they were indoors more than I was as a child. None of them are (yet) enthusiastic gardeners, but if and when the time comes, they will have a basic connection and starting point. Although they don't like cabbage, they have usually eaten at least one thing from the garden every day, in fact they are most likely to eat lettuce if it is straight from the garden.

Migration for me hasn't been easy. Although I chose a country with a common language and some similar culture, being a migrant is still a defining part of my life. But I am pleased that I made the move and for my children, the opportunity to be part of living with permaculture-minded people has been a good way to grow up.

The Birth of Accredited Permaculture Training (APT)

In 2002 I went back to the UK for the first time since 1995, and saw how permaculture and accredited training were being combined in formal education through the UK equivalent of the Technical and Further Education (TAFE) system. This was an idea that had interested me for a few years and had been discussed among permaculture teachers in Australia. In the UK, progress was mainly due to the work of George Sobol, one of the unsung heroes of permaculture development. George's persistence enabled the Workers Educational Authority (WEA) to become the accrediting training body there.

I felt that accredited training was one of the ways for permaculture to move forward in Australia. I didn't see that it would replace the other training, but it would offer something to those who need to be part of 'the system'. My motivation was to show that permaculture could fill a gap in mainstream training. The gap was, and still is, a comprehensive holistic approach to sustainability, rather than simply labelling other courses as 'sustainable'.

I contacted people back in Australia, convinced that the time was right for permaculture to get accredited there too. In late 2002, with other permaculturists, I met Guy Rischmueller, who was the right man in the right place. Guy was perfect for this job. His company, a Registered Training Organisation (RTO), specialised in accrediting courses while still allowing the developer to keep control of course content and of who was allowed to deliver that content. This was the magic key to accredited training. While Guy lived near me in South Australia, the others in the group were in other states, so we 'met' through a daily stream of emails grinding away on dial-up connections. With Guy as mentor and guide and with zero budget, we formed the group which would become the Permaculture industry body, Accredited Permaculture Training (APT). This body was able to pursue the design and accreditation of more than 50 units of permaculture training, from Permaculture 101 through to Diploma level.

Ian with graduates of the first Diploma of Permaculture at Eltham College.

I really liked the coming together of new technology with good solid permaculture design. With email on my second-hand, medium-sized computer (1998 vintage) I did not have to print anything, while I processed the equivalent of thousands of sheets of A4 paper. APT proved to be a huge administrative task, but it was also exciting to be able to get to a point where courses could be recognised and funded, to allow a new sector to be added to permaculture training.

Though some permaculturists expressed fears, APT has not diminished the role of the permaculture design certificate course (PDC). The PDC has a global recognition that an Australian-based accreditation cannot have, but at the same time APT has recognition in the domestic mainstream educational system that the PDC has not been able to obtain. It is typical of the weird world of permaculture that we have two diplomas. The 'Mollison' diploma that has been available (unaccredited) for the last 25 years, and the APT diploma, recognised by the government and the mainstream adult education system. Both can be obtained through Recognition of Prior Learning (RPL). I'm one of a handful of people who has both, though I couldn't tell you where the bits of paper are – maybe mulched, maybe eaten by silverfish!

The challenge, in my opinion, is not to argue about accredited training versus unaccredited, but to achieve wider recognition of the fact that permaculture offers some of the best training in sustainability that can be found. Permaculture is a school of thought that is backed by some of our leading researchers on peak oil and climate change, and it offers practical and immediate steps for anyone to take.

Castlemaine Dreaming

Through 2004, the APT team were essentially unpaid education bureaucrats as the accredited training gradually began to get underway. But by 2006, a new Registered Training Organisation (RTO) was needed and my role with APT ended. It was a time of great change in many aspects of my life. When I came to Australia in 1980s, my first home had been Central Victoria. Increasingly I felt a pull to move back there, and the town of Castlemaine began to reappear in my thoughts and dreams.

This attraction, I think, is partly something that is my destiny, but partly because this is an area that is facing very immediate change through lack of rain. We have hit peak water before we hit peak oil. We are having to work out how to respond, as a community, when one of our basics of life runs out. It is similar to my choice to move to Liverpool in 1979. That was a city undergoing enormous change as it adapted to the post-industrial age.

I remember back in 1986, not long after my arrival in Australia, I bought a map of Victoria. I clearly recall looking at the map, and really wanting to go to Castlemaine, although I knew nothing about it. I got my wish back then, but only as a passenger in a car as we drove straight though the town. I didn't get to live in Castlemaine until the winter of 1992 when Jo started a job as a social worker there. I kept coming back, even when living in South Australia.

In 1992 I began to visit and explore the 19th century streets and laneways of the old gold mining town, and experienced the quick transition from August cold to November heat, with not much spring in between. By early summer Jo and I had lived in Australia for six months and we had a one-year-old son, Rowan. My mum was visiting from England. The Castlemaine State Festival was on. We went along, and watched a performance of Wind in the Willows in the Botanical Gardens, an incredible arboretum, with Mr Rat and Mr Mole rowing across Lake Joanna.

I spent the first minutes of the New Year of 1993 in Castlemaine, meditating with a group of friends, and soon after, taught an introduction to permaculture course at Mischa Grupp's stony hillside block at Barkers Creek where he was building a big passive solar house. I visited Fryerstown and walked a piece of land that was later to become the Fryers Forest Ecovillage designed by David Holmgren. Castlemaine was the home of *Green Connections* magazine that had pioneered bringing together strands of permaculture and Landcare, and laid the foundations for the transition and sustainability groups that today are so strong in this area. We might have ended up living there then, but Jo and I had already decided to move to South Australia.

By 2005 Jo and I had three children, and had built a rammed earth house in an ecovillage in Willunga south of Adelaide. We had also reached the decision that we wanted to live separately. These were difficult times, and it was no surprise that I found myself drawn back to Castlemaine. I was so certain that I needed to move back to Victoria that I was prepared to go alone, but Jo decided she would come too. We got back together, gave up our jobs, enrolled the kids in new schools and found a house to rent for six weeks. We put our possessions in a small truck, waved goodbye to the removalists and drove ourselves over to Castlemaine. I remember feeling very light, and was tempted to drive to Queensland, to just let go of all our possessions, but I wasn't brave enough. Anyway, a good friend was already waiting in the new house, with log fire burning to welcome us on a frosty evening.

Castlemaine meant a fresh start in many ways. I soon joined the community choir for men (The Acafellas) and Jo joined the women's equivalent (The Chat Warblers). I started to study shiatsu. Only few months earlier these things seemed impossible, when viewed from the desert lands of South

Australia only 800km to the west. We rented a Victorian mansion with a huge garden but no water, as the reservoirs were almost empty. Increasingly, I felt like I had lived in that house in a previous life, when it had first been built in the 1860s, with a dark-haired wife who was Irish. Perhaps this explains why I feel so at home here?

Later we bought a house on one acre in Campbells Creek, 3km from the centre of Castlemaine. I liked it because it had soil (a rare thing around this gold mining country) and because it sloped gently to the north. Maybe the gentle wooded hills reminded me of north Somerset where I grew up? I found work with the local Sustainability Group and revelled in the biennial Castlemaine Festival, mixing 'green' events in with the traditional orchestras and opera. My long awaited book, *The Holistic Life*, finally rolled off the press.

We had achieved all of these things, seen massive floods – the second time above the 100 year flood level in 3 years, we had avoided terrorist attacks, but we were not happy travellers. After a year of being back together, we were ready to live apart again. I still really liked Campbells Creek, but Jo moved closer to town. She didn't want to be part of another possible eco-development. Our ideal was to have two small separate houses as part of a community where the children could grow up among fruit trees and friends. That ideal is still strong and we still live with food in the garden, low power bills and a cycle ride to school and work.

The land around Castlemaine feels like it will support me, and although the frosts can be harsh, we have a longer growing season than many places. At night, I walk up to the forest a kilometre to the east, and revel in the starlight and the kangaroos. I have good friends here, more friends within 10km than I had in the whole of South Australia. This is a place where people come to build passive solar houses, to run organic restaurants, to make good cheeses, to grow great wine grapes and to run sustainability projects. This town of only 8,000 people had 250 participants in last year's Ride to Work Day. It is a community that wins awards, that runs big festivals, and has a sense of humour. But we recognise that like every other Australian town, it is also living way beyond its means and we are trying to address that too. There is a project called Maines Power, working with the big industries in town to help them manage their energy and water needs. There is a Transition Towns group, a food co-op and a LETS Scheme. There are great ideas for dealing with the fact that our major reservoirs are empty. But there is a long way to go and many more changes to be experienced and endured before we are anything like sustainable. I have entered my 50th year. What will the next 50 bring?

Jane Scott

Born in England, Jane Scott now lives in the Dandenong Ranges outside Melbourne. She completed an Environmental Science degree in 1979 and discovered permaculture while she was mowing the lawn 12 years later. Her partner came to the back door of their home saying, "There's a guy on TV who reckons you don't have to mow lawns anymore!" Life was changed forever as permaculture became her framework for the future. Facilitating numerous Permaculture Design Courses and short courses brought many happy and satisfying years.

At the same time a passive solar, water efficient home with a chemical free orchard and vegie garden took shape. Today Jane reaps the benefit of good initial design and can occasionally lay back on the banana lounge and eat mandarins from the tree (the ones the possums and birds don't get!!) In reflecting on nearly 20 years of involvement in permaculture, Jane considers the many 'cultural creatives' who have shaped her life and is profoundly grateful for their influence. She included permaculture education in her Master of Environmental Science thesis in 2004 as one of the great examples of transformational education in Australia and continues to practice permaculture as the fundamental philosophy of her family and community life.

Chapter 21 | To My Great Grandchildren with Love and Hope

Dear Great Grandchildren,

On Origins and Happiness

I am writing to you as I contemplate the world you will inherit and I want to connect with you to share a passion that has motivated and shaped my life and reaches out to you across the years. If I have fulfilled my life's goals reasonably well, you may recognise the word 'Permaculture', attached to your Great Grandmother, Jane. Just as my Great Grandmother was a herbalist and healer in the small Irish village of Avoca, perhaps you could imagine me as a 'permie', a flawed but dedicated advocate for the future, from the Dandenong Ranges outside Melbourne in the late 20th Century.

I was born in Bristol, in the UK on 'May Day', May 1st 1958. My mother said that the nurses decorated my crib with spring daisy chains. I have always loved English daisies, making my father mow around the little patches in our lawn once we came to Australia. I wonder if it is such simple introductions as a child that tunes us in to nature for a lifetime. My first seven years were spent in a classic Victorian home with an ideal children's garden. Looking back I realise that the garden was a beautifully contained ecosystem, with an orchard over the rear wall where I played with the boy from next door. The orchard was a place where we could be away from adults, a place of independence but also safety, knowing that just across the wall was my own garden with all its delights. It contained a goldfish pond, a swing, a massive array of brightly coloured flowers, a rabbit, (and a fox!), a tortoise, a dog, a cat and children; great elements of a thriving ecosystem.

My mother had tuberculosis at 18 and as a result was not supposed to have children, but she had three. I spent a lot of time with her in the garden learning such things as breaking open pansy seed pods to spread the seed for the following year. We used to regularly walk to my grandfather's past a mysterious dry stone wall that I was too short to see over. I constantly wondered what was on the other side. My mother must have lifted me up to see but I could never remember. Going back to Bristol, 30 years later, I did look over the wall and found a hillside of allotments, people's miniature permaculture ecosystems. It filled me with joy and answered many questions about the origins of my passion for nature and permaculture.

Another strong influence on my childhood was the nature 'rambles' with my aunt and uncle. Unlike my parents, they owned a car and used to come on Sunday afternoons to take my brother and me to explore caves, beaches, woodlands, rivers and classical music. These were magical times imprinted on my memory until this day.

Moving to Australia

It is 2009 as I write and I live on a north facing, gently sloping acre of land with predominantly indigenous vegetation that would have been thriving when the Wurundjeri people hunted here. Our 'Nobelius Hill' as it is now called, would have been cleared for farming in the mid 1950s and then again for building houses in the 1970s. Fortunately the couple who built this house knew about orienting it to the North so that the sun would heat and cool it naturally. This is one of the main reasons your Great Grandfather, Tom, and I chose to buy it in 1994.

Previously we were living on a south facing quarter-acre in a tiny seven square cottage. It was cold and dark throughout the winter and had 17 deciduous trees and a lawn which kept us mowing and raking all through the spring and summer. My permaculture journey began on a spring day as I was mowing. Tom came to the back door and called, "Hey there's a guy on the TV and he says that you don't have to mow lawns anymore!" How is it that a small statement like this can precipitate such a life-changing journey? I stopped the mower and went inside to watch the rest of Bill Mollison's *Global Gardener*. I wonder if you can still dig up a copy in your library today. Australians Bill Mollison and David Holmgren co-originated a brilliant design strategy for sustainable human settlement, beginning at the most practical grass roots level – replacing lawn with food!

I was trained in Environmental Science in the 1970s. One good thing that came out of my studies was that I fell in love with ecology during a field trip to the Little Desert when we intensively explored why the desert grass *Triodia irritans* grows in rings. I became so immersed in the ecology of those grass circles, that I believe the English girl became an Australian creature at that point.

But overall the environmental degree really knocked me about a bit. I had learnt all the problems and the things that were wrong with the planet, but did not know where to go with it. At the end of the four years most of us said we didn't really care as much about the environment as when we started, and we knew we had all this unresolved passion and energy locked up inside us. So I left my course consumed by a picture of doom and gloom. There were very few environmental jobs and I was not sure of what contribution I could make toward a sustainable future.

Ten years later, permaculture was an amazing gift that gave me a connecting vision. Suddenly I belonged to a world of people who wanted to challenge the dominant paradigm of consumerism and to find ways to provide local food, share resources and build community: to act not only in the present but sustainably for the future – for their descendants – for you.

The permaculture principles provided a framework for the jigsaw pieces of my education and life experience. I had many questions in my heart and mind. How do I make a difference in my daily life? What influence can I have as an individual? What kind of house and home do I create? Who are my allies? How do I deal with the grief and loss of our natural environment? What responsibilities do I have to create a better future? Permaculture teachers began to answer these important questions for me as I began a journey of personal transformation.

The complex and yet ultimately simple definition of sustainability evolved in the late 1980s – to provide for the needs of today without compromising the needs of the future.

Through permaculture I had a clearer way forward, a way to make good decisions and to evaluate them against principles of integrity, a way to act not only in the present but meaningfully for the future as well. My previous knowledge about the plight of the world, the destruction of our environment and people's ignorance came together hopefully and positively as I began to create a new reality. Permaculture focussed on solutions, not problems. It matched the experiences of my life and gave me a purpose and direction.

Within a year we had formed a small dedicated group of people, the Mountain District Permaculture group, to explore permaculture and to begin a community garden project. We met monthly to learn more about growing food, saving seeds, acting as a community and connecting to the future through our plans and visions. Permaculture wasn't something that I just did – it was who I continued to become – a very exciting time for me. We chose a home that provided an opportunity to tread more lightly on the planet and we began to educate ourselves in the way of nature.

There are a couple of stories worth the telling from this early time because they impacted on me significantly. The first is when my mother and I went to hear Bill Mollison speak at Hawthorn in Melbourne in the early '90s. Our Mountain District Permaculture group was struggling with the question of growing in very shady gardens. It seemed an almost impossible task and yet we wanted to be responsible for providing our own food and caring for our land. I publicly asked this question of Bill, "There are a group of people who want to practice permaculture in the Dandenong Ranges but we all struggle with the issue of shade from the large dominant trees. Do we have a chance of success?" People who were at that meeting tell me that Bill made me quite famous with his answer which was typically brutal: "You bloody hippies, you settle in the most beautiful areas, build your houses next to the waterfalls and then proceed to destroy what you went there to enjoy in the first place!"

Silence everywhere, I sat down, mortified, while he continued. Ultimately though, the point of his response was what permanently changed my thinking. Bill was asking us to consider the land before we considered our own needs. It was degraded land that needed permaculture, not ecosystems that were relatively pristine, healthy and productive. From that point on I began to develop a land connection sensitivity and 'humility' that otherwise may not have emerged. I hope that this sensitivity has somehow travelled down to you through your parents. In later years I spent considerable time at Peppermint Ridge Farm in Gippsland learning more about Indigenous Permaculture which was relevant to the beautiful piece of bushland that I care take here in Emerald. But I digress.

The second important story is about your Great Grandfather. His journey always focussed on the practical and he built sheds, gathered resources, redesigned and refitted the house and eventually was the catalyst and prime mover in creating the new home we found on the sunny side of the hill. One summer morning I woke up to find the bed beside me empty and a clattering outside in the garden. It was very early, pre-dawn and I was a bit confused by the activity below. I went out to the balcony where I saw Tom, dressed only in a T-shirt and his underpants. Unusual! He was busily constructing tiered worm farms from milk crates which he must have somehow visualised in his sleep. I had rarely seen him look so delighted. At that moment I realised that having a vision for the future like the one we had created though permaculture, could make us happy. It continues to do so.

Finding Inspiration

Where do we find inspiration to continue the difficult journeys in life? Your GG and I are separated now but we continue to be friends and raise our children, your grandparents, on strong foundations. Our daughter Aurora studies Environmental Science and Biology and our son Jensen, as a result of the fantastic gardening program at his primary school, considers becoming the new Peter Cundall (who has just retired as the long standing 'Father' of ABC Television's *Gardening Australia*). They are the ones who inspire me.

Years ago listening to a talk by Bill Mollison, I heard him say, "You don't have to be part of a structure; you are itinerant teachers. Go and share your knowledge and skills wherever you are." I was absolutely captivated and inspired by these words in which I visualised lifelong community learning. I will always be a student, sometimes a teacher but most satisfying of all, a co-creator of the universe in our place.

Thanks to Government grants, my house on Sunnyside Terrace has solar hot water and electricity. I cannot express the level of satisfaction I felt at finally being able to generate my own power. I am inspired by positive changes in our society. The orchard planted over 10 years ago continues to flourish despite the fact that I rarely tend it these days. I see the benefits of good initial (not perfect) permaculture design, and nature works with and for me to provide fruit (mainly for the birds!!). The patterns of nature inspire me as I see how I belong in nature through them. One needs to be open to the messages of nature – this is part of what walking humbly on the land means.

The permaculture principles are inherent in my life now and I feel prepared for whatever is coming. I have changed along the way, embedded the principles as preparation for the future. I am also grounded in the present and hope you, my great grandchildren, will know me for what I did in my garden and community – that my claim to fame will be tomatoes! This means of course that you also inspire me – to be as loving and as proactive as I can be in a world with so many fears. I cannot hold you in my arms but I can work for you now so that you may inherit something coherent and healthy both in your external lives – a garden and less CO_2 emissions, and in your inner lives – the resilience to deal confidently with whatever has been handed to you. As futurist and mentor Robert Theobald said, "Things are getting better and better and worse and worse at the same time."

Patterns in Nature and Limitations

In the summer I grow tomatoes really well but not much else, despite having 22 thousand litres of water at my disposal. I am aware of the 'energy in versus energy out' principle of permaculture and so I continually evaluate where my input of energy best lies on this property. One year I grew a single head of broccoli from at least ten plants and over a thousand litres of water. I was really proud of that one head but it made no sense. I have learned now to do my internal energy audits, to become more efficient like nature. Nature inspires me with her practical commonsense, her beauty that is ever changing and her gifts of colour, texture, fragrance and nurture.

Our limitations in life are about what we do and do not observe. Someone wise once said, "We are only limited by what we do not already know." For a while I have been trying to figure out why my summer vegies don't grow particularly well given the great soil and the worms I have cultivated and the abundance of water available. Quite independently I have also been considering removing the massive Irish strawberry tree that is beginning to dominate the eastern side of my vegetable garden. It was always a good screening plant but now it is growing so fast and so tall and lush … then an 'Aha' moment! Perhaps those twenty thousand litres of water I gave my vegetables last summer…!

I am still trapped in linear thinking and thinking in webs is sometimes difficult – no surprise given the world I inhabit. A wonderful permaculture friend and mentor, Terry White used to inspire us with his biological metaphor of 'observation and interaction', the first principle of permaculture. When we observe carefully we gain new insights and knowledge which in turn gives us the ability to get our 'antenna' up so that we can begin to see what was not evident to us before. Once our antenna is up we are sensitive to and inhabit a new world of understanding and connections and we begin to see the web of life more clearly. Part of this new revelation may be confusion – go with it – let 'blessed' confusion be a stage toward new conceptions of the world. As Einstein once said, we must think differently, outside the box in which we created our problems in the first place.

Despite the wonderful holistic training in permaculture I sometimes slip back into linear thinking and am oblivious to the incredible web of life that surrounds and nurtures me. I would consider this one of my greatest limitations in making sense of the world, not just in a permaculture context either! My great grandchildren – get your antenna up and allow your mind to break free of preconceptions! Look beyond the obvious and see what is really happening in the world around you. Connect with nature every day – it is the greatest teacher.

Go With the Flow and Learn to Let Go

Until permaculture, 'going with the flow' was to me a hippy concept from the '60s and conjured up an image of a laid back, almost self-indulgent way of being. Two of permaculture's most prominent teachers David Holmgren and Graham Bell showed me a new definition. There's more to be achieved working *with* like-minded people than *against* those who are not. Bamboo is strong because it bends with the wind and does not break…

Nothing during my time as a permaculture teacher and facilitator was an effort. The first time I taught an Introduction to Permaculture course I was nervous; I baked choc chip cookies to placate the participants if things didn't go well! But it just flowed, I belonged, I was in the right place. I had found my tribe. Over the subsequent ten years, brilliantly skilled and knowledgeable participants and tutors emerged and willingly offered so much of themselves. There was a constant diversity of people in my life bringing richness I had not imagined possible. Our catch cry was 'we are not alone and we are not mad!' as we recognised and challenged the dominant paradigm together. My confidence as a facilitator and teacher grew as many diverse and yet like-minded people became my community, interwoven in my tapestry. I've always believed in diversity. It is one way of acknowledging the Flow. My great grandchildren, find your tribe as I have. Find the ones who help you shine and fulfil your goals and dreams.

In the early nineties at a permaculture conference I was transformed by the work of a man called Alan Savory. He authored a concept called Holistic Resource Management. He recited a wonderful story about elephants and people coexisting in Africa. In my memory it goes like this. A small village was located on a river where the people would come to wash clothes, collect water, and commune and sing their songs as they worked. Inhabiting the same area was a herd of elephants that came to the river to bathe and drink in the evenings after the people had left. Local land managers, aware of the ever growing threats to the elephant population, decided that the area should become a reserve. The villagers were excluded and had to move to another location on the river, giving the elephants free range. This was considered the 'right' decision within the context of elephant conservation. Within months the river bank began to erode and degrade. Without the villagers and their singing the elephants were constantly at the river until it was no longer a viable habitat. Until the intervention to 'save the elephants' the villagers and the elephants had lived in ecological harmony.

The point of this story led to a principle of Holistic Resource Management that has stayed with me. Make a decision on the best information and evidence available at the time and then assume it is wrong. This means that monitoring and feedback are continuing, essential components of the process. No longer can we say this is the right decision, we have fixed a problem, and walk away. Instead we need to continue assessing the situation and readjusting as new information comes along. This is how nature works to maintain her dynamic equilibrium. This ensures that we are in the Flow.

Another gift permaculture gave me was to practice letting go. Permaculture defined me for so long that, in fact much to my discomfort, people in my local community began to label me 'Permaculture Jane'. There came a time after the birth of my second child where it seemed important to let go of the formal groups and processes. Fewer night time meetings and weekend working bees were what my family needed. This letting go was extremely difficult but profoundly liberating when it happened. It was another seminal time of integrating permaculture into who I was – a time of change, in part dominated by confusion. Internally I was breaking down old ways of thinking. During this 'letting go' period I ran one Advanced Course in pattern language. There was a moment in the course on the second day when we became intensely confused about where our explorations were taking us. As the 'teacher' I tried to fix it for people but I couldn't. There was obvious discomfort as people grappled with the new ideas.

In our debriefing later we decided that the group had in fact reached an important place of transformation that I didn't need to fix. We were moving from linear thinking and starting to think in a web and this was confusing. As a teacher I'd never been there before and instead of letting go, to allow a new reality to emerge I had tried to keep it safe for the participants and have answers for their confusion. I know now that 'Blessed confusion' is a sign of transformation. Pattern language based on Christopher Alexander's book, *A Pattern Language* provides new structures for people's emergence as creatures of the earth. Here's a language that we can use to 're-design what's been done to/for us', taking into account the natural patterns that create and sustain us.

The Present

In May 2009 I was diagnosed with breast cancer. I am still in treatment as I begin this letter to you. My immersion in permaculture all of these past 17 years has been important preparation for dealing with this life threatening (and affirming) challenge. I am observant and I interact with my health carers, asking questions and setting my own pathways of sustainability. I am aware of my energy inputs and outputs, meditating on the time of rest, the fallow period, I must have in my usually full and active life. A blue tongue lizard came in a meditation to teach me about hibernation and energy conservation. I have become even more reliant on the healthy organic foods now available to me.

I have asked how I will live more fully as a result of this experience. What is the yield at the end of seven months treatment? The yield is rich and I continue to harvest. And what an important time it has been to apply self regulation and to accept feedback – to regulate my sleeping patterns, my emotions and my thinking. I have watched each of the friends, old and new, who have chosen to journey with me (some have not) and from their feedback

I have created a new world, both inner and outer. I believe that it is a more sustainable inner world as I tune into what is truly important in life and let peripheral worries slip away. I have a new deep respect for the people who design, particularly those responsible for the nurturing, comfortable recliner chairs we rely on for the long periods in our lives when we must rest.

The patterns of my days and nights during chemotherapy were my focus; I had to respond to the rhythms of my body as it assimilated the chemicals necessary to fight the cells that were out of equilibrium. I learnt to recognise the pattern of the symptoms in my own body. If I concentrated too much on the detail of possible symptoms or the often torrid testimonies of individuals on the internet, I became distracted with unnecessary fear. The pattern of my own response and the bigger picture were where my focus needed to be. This way I could manage the most difficult times.

The other profound response to my cancer is reflected in the 'integrate rather than segregate' principle. From the beginning I accepted my journey – the language of fighting it did not work for me. I wanted to understand 'nature's machine' and to go with the flow in the process of healing. This is an individual journey that we all must take in our own ways but the principle of 'integrating' a serious illness into your life journey is sound. For me it was a natural process and I am thankful for the years of learning, through permaculture, to recognise that I am part of nature. This understanding has allowed me to travel gently through this illness.

It is still too early to fully comprehend the changes in my life. I finished treatment three days ago. I know however that I am undiminished by the experience of the last months, surrounded by the love of my community, connected to my garden as spring wove her vibrant magic in the orchard, garden beds and native vegetation. My creativity is begging release like a newly sown summer garden after a time of fallow and a good rain! I am filled with love and gratitude for life. I hope that this sharing will be an inspiration to you during the tough times. I love you.

The Future

My dream of the Earth includes a mighty reinvigoration of connectivity between ourselves and the Earth. In this we leave behind whatever religious, cultural and political mandates have turned us into 'Masters of the Earth' and we return to the original blessing of the Garden of Eden where we are a part of all the wonder that is. My darling children, if the Biblical stories still influence your lives, know that we have never fallen from grace; we are part of grace, of an abundant universe where each element, living and non living is inherently sacred and valuable.

Every atom that coalesced to form planet earth at the beginning is constantly re-formed into new, ever more complex and wondrous nature. When my body returns into this process I will be part of the eternal goodness which some call God. Permaculture is a process, a lifestyle; a way of being that has allowed me to tap into the Flow of Life. It connects me with you my Great Grandchildren in a never-ending cycle of birth and death and rebirth. If the wildfires have increased in your world because of global warming, look beyond them as a threat and see that they have been part of a natural cycle in Australia's evolution since the beginning of time. Take responsibility if you still live amongst the forest, knowing that you are a creature of the forest. Use knowledge and wisdom to adapt to your home and above all see the blessing of renewal that a fire brings.

There are so many things I despair about in this world and my fears for you are real. What are we leaving you? How will you live with the effects of global warming, where will the nuclear wastes be held in relation to your home, will a civil community persist around you, will you have food security? At times my fears for you have become oppressive and my hope weak. I am sorry for the things I haven't done for you.

In all of this despair however, there is a light. I hold in my heart an ancient proverb which says "…and the candle said to the darkness, I beg to differ." Permaculture is a light. I see this image of holding light in the darkness with each new seed I plant, each relationship I build within a like-minded community of people and within each fear for the future we express and try to conquer. We are thinking of you and we send our love through time. We can only hope that our actions light your way through difficult times. I know that if you follow some of the example of your Great Grandparents in planting a seed to grow nutritious food, or getting to know the blessings provided by an earthworm, you will experience happiness.

My offspring, I am handing you a light – it is called permaculture. It can guide you through the mystery and challenge of our evolution. If you haven't heard of it already – Heaven forbid – please discover it now.

With Love,

Jane Elizabeth Scott, 2009

Jane Scott facilitating a Transition Town workshop.

Josh Byrne

Josh Byrne is an environmental scientist with a passion for sustainable gardening, appropriate technology and innovative environmental design. Best known for his role as the WA presenter and writer for ABC's *Gardening Australia* TV program and magazine, Josh is also engaged in research, teaching and consultancy work in the areas of sustainable landscape design and urban water management. He is an Ambassador for a number of community and government sustainability initiatives including the national Smart Approved Water Mark program and the WA State Government's Living Smart Household Sustainability Program and Nature Play initiative. He is also a Patron for the Conservation Council of WA, the Organic Association of WA and Sustainable Gardening Australia. Alongside gardening, his hobbies include cooking and travel. He lives in Fremantle with his partner Kellie and son Oliver.

Josh shows students at the Swan Valley Community School how to train trellised fruit trees.
Josh designed the garden and helped build it step by step on ABC TV's Gardening Australia program.
Photo courtesy of Acorn Photo Agency.

Chapter 22 | The Magic of Gardening

I discovered the magic of gardening at an early age, looking after Dad's herbs at home. As a keen cook, he kept quite a collection. There must have been thirty different varieties with each one growing in an individual terracotta pot, neatly lined up in rows on a tiered planter shelf. It was my job to keep them in top shape. Hand watering was a daily task in Perth's hot, dry summer. Regular feeding to keep the growth soft and palatable, and occasional re-potting, division and replanting – all simple tasks that made a big impact on me. These were the first things that I nurtured and the result of these efforts ended up in our dinner most evenings.

My next garden project was a bit more ambitious. At fourteen I set up a four-square-metre 'no-dig' vegie patch on top of the lawn in our backyard. I was fascinated by how a few bales of straw, a load of compost and a handful of seeds could be converted into bundles of fresh vegetables within a matter of months.

I often tell people that growing food is highly infectious. Once you make a start the sense of satisfaction and reward is overwhelming. I can still vividly recall the moment when it really hit me. It was late one warm afternoon and I had been working in my little patch, which by this stage had expanded to include a number of large tubs filled with tomatoes and climbing beans trailing along the fence. I sat on the lawn, leaning back against the warm red brick wall of our shed and basked in the contentment and satisfaction with what I had created. From that point on I've considered myself a gardener.

Dad picked up on my new fascination with gardening. As a keen book collector he was thrilled to see me fossicking in second hand book shops as I slowly developed a collection of my own. One evening he gave me two books that he had bought years before and tucked away in a forgotten section of his library. They were first edition copies of *Permaculture One* and *Permaculture*

Two. They made immediate sense to me. The ideas were logical, the principles were practical and the ethical basis underpinning it all was fair and just.

The more I read, the more I started to question things, especially life in a comfortable suburb, living in high energy, wasteful homes, surrounded by high maintenance unproductive landscapes. It was all wrong. Clearly there must be a better way to do it.

I realised that I needed to learn more about the natural world and the way that it worked. This realisation influenced my subject selection in high school and eventually led me to study environmental science at university, but it was always the practical aspects of permaculture that kept me inspired.

Suburban Permaculture

My interest in permaculture got a major boost at the age of nineteen when I stumbled across a neglected suburban block with potential. It was a classic run down rental property that still retained the features typical to the original 1950s homes in the area. There were old fruit trees, a hen shed, a dilapidated glass house and a workshop. There were remnant native species that, given their age, would have pre-dated the development of the area. It was like a time capsule. All around it, time had moved on. Dalkeith was now a prestigious riverside address noted for its leafy gardens and proximity to private schools. Meanwhile number 63 Circe Circle was still like something out of the BBC classic *The Good Life*.

My brother was renting the place with his mates. The first time I visited I thought to myself, "I've got to get this place!" Timing was on my side and it wasn't long before he moved out and I moved in. And so began my first serious garden project.

Being my first rental property, I wasn't exactly sure of what the rules were when it came to working in the garden. The place was a complete mess so I assumed that any work tidying up was bound to be appreciated. I started off tentatively. I was busy studying at university and was also running a part-time local gardening round, so my early efforts were mainly limited to exploring the various nooks and crannies of the block. Each session was like an archaeological dig. The more things I'd discovered including old tools and forgotten pathways, the more intrigued I became with the garden. It wasn't long before I decided the best thing to do was to drop down to part-time study and take this project more seriously. Whether it was my growing passion for gardening or an avoidance strategy for chemistry and statistics is beside the point – it was the best thing I ever did.

Over the next six years, my housemates and I rejuvenated the property and created a remarkable garden that combined friendship, learning, fun with a real feeling of independence and self-reliance. We pulled the lawn up, repaired the glasshouse, repaired the sheds, installed a rainwater tank and set up trellises for growing fruit trees and vines. We built new pens for ducks and aviaries for pigeons and quail. We experimented with solar ovens and composting toilets, brewed beer and baked bread. It was brilliant.

Being poor students we learnt how to be resourceful with salvaged materials and created a space with rambling pathways, extended poultry runs, homemade garden seats, worm farms and compost bays. We basically overhauled the place and it wasn't long until I actually felt as if the property was ours.

What we did buy, we bought locally. There was a small group of shops around the corner called the 'Dalkeith Village' which had been there (in more humble form) since the suburb began in the 1950s. There was a bakery, a little supermarket, a bottle shop, a nursery and even a small father-son owned hardware store where you could still buy individual nails. Of course they charged like a wounded bull, earning the proprietor the nickname 'Old Man Fleece' amongst my housemates, but that didn't really matter. It was more important for us to know that we were supporting local businesses, which provided the convenience, comfort and security that we had what we needed near by.

In time we struck up great relationships with our elderly neighbours. Most had been there for some forty plus years and remembered a time when everyone had fruit trees and chooks in their backyard. I'm sure that they thought we were odd to start with, but when they realised we were really having a go, their interest grew. It wasn't long before I was passing eggs and homegrown produce over the fence. In return they would make jams, give tips on the best local plant varieties and dear old Mrs James up the back would throw her scraps over for our chooks. There was a real sense of community.

My friends were also fascinated. When people visited they'd head straight for the chooks. At night I'd give tours by torchlight to explore the jungle like garden and spot frogs. Being born in the mid 1970s, we were part of a generation that didn't automatically connect with gardening and self-reliance. Perhaps that was part of the buzz for me, in that we were doing something different – something inherently practical and something that just made sense.

Leaving the Backyard

In 1996, the 6th International Permaculture Conference (IPC6) was to be held in Perth, and with it a special Permaculture Design Course (PDC) was to be run involving a number of respected WA permaculturists as guest teachers. I'd been planning to do my PDC for a while, but as a student, time and cost were always an issue. Attending IPC6 was an opportunity too good to miss. I deferred from university for a semester, cranked up the gardening round to save up some money and registered for the design course and conference.

By this stage, I was confident with many of the practical aspects of permaculture from the hands-on skills that I had developed in my backyard and on my gardening round, but I was eager to explore new concepts in design and relate them to my academic studies. The PDC, conference and site tours that followed certainly provided this and made me realise that it was time to leave the comfort of my backyard and neighbourhood and expand my skill set by working on other projects.

As always an opportunity soon presented itself. One of the site tours we took during the PDC was to the Environmental Technology Centre at Murdoch University (the university where I was studying). It was a fascinating project which had been set up as a research and development facility to explore the applications of small-scale environmental technologies in the areas of water supply, sanitation, energy, shelter, and food production. The research group's main focus at the time was the application of appropriate technologies for improved environmental health in remote indigenous communities, however they had also fostered an active local community group to help with the development of the 1.7-hectare site.

It was just what I was looking for – a fresh, practical learning environment at an exciting scale to develop new skills. The infrastructure was impressive including solar powered bores, a hybrid (wind and solar) remote area power system, composting toilets and a range of wastewater reuse systems. There were demonstration buildings made from strawbale and mud brick, and the beginnings of a productive landscape including aquaculture ponds, food forest and vegetable gardens.

I signed up as a volunteer and was soon spending most of my time there. My home garden was established and thriving and I was enjoying putting my energies into something different. Before long I was employed as site coordinator and spent the next five years managing the ongoing design and development of the site whilst completing my undergraduate degree.

Like many good projects we were under-funded financially but were able to achieve a great deal due to the commitment of staff, students and community volunteers. There was a strong shared vision for the importance of demonstrating low impact ways of living and increasing self-reliance. We ran Work for the Dole programs, hosted Corrective Services participants and kept the doors open for anyone who wanted to participate or learn about permaculture.

Alongside my site work, I was extremely fortunate to be mentored through a broad range of academic experiences. It was a unique situation where I was expected to teach, present at conferences and undertake field work around WA, all before graduating. This was possible thanks to a progressive group of academics who saw the value in practical experience, recognised passion and provided me with the opportunities to grow.

Time to Talk About It

By 1999, there was considerable interest in my home permaculture garden. It started with an article in our local newspaper which picked up the novelty of a group of Uni students with a garden packed with food, in a suburb better known for immaculate lawns and prize winning roses. Next was a double page spread in *The West Australian* newspaper and shortly after that, a spot on Channel Seven's infamous *Today Tonight* television program.

Around this time I was approached by the Australian Open Garden Scheme to include our garden in their program. Back then, the Open Garden Scheme was very traditional and I saw it as an opportunity to demonstrate practical and accessible ideas on local food production and sustainable living through gardening. The weekend was a big success, attracting over a thousand visitors, which was a result that surprised everyone. Clearly people were eager to see working examples of permaculture at scale they could relate to.

The real excitement came when the ABC contacted me to do a story on our garden. We did a two part story over two episodes and the response from the public was incredible. To my astonishment, the ABC contacted me and asked if I would be interested in doing a regular gig. This was a total surprise. I had never had aspirations to get into media. I was simply doing something that I loved and obviously it was at a time when there was an appetite for it. Unfortunately the timing wasn't right for me. Having finally finished my undergraduate studies, I had planned a trip to Africa and the Middle East. My ticket was paid for and my backpack all but packed. I decided to turn it down with the suggestion that I'd be keen for the opportunity when I return in a year's time.

For the next eight months I travelled through southern and eastern Africa, including a three month stint in Malawi volunteering on a village farming demonstration project. From there I travelled through Egypt into Israel to work on a kibbutz and finally via England and Thailand home. Any thoughts of a media career were well and truly gone, instead I was re-inspired to continue my research and development work, commence postgraduate study and most importantly get back to my garden.

Within a week of returning to Perth, the ABC called again. I was both flattered and tempted but again decided the timing wasn't right. Motivated by what I had seen during my travels, I was determined to fully commit to my Honours degree which would then make it possible to undertake a PhD. I was still convinced that practical research and development work, backed up with academic credentials was the greatest way for me to effect change.

In a meek voice I asked if they'd be kind enough to contact me in a year's time. The response wasn't encouraging and I thought that was it. This was probably the most responsible decision I had ever made. I went on to complete my Honours and gained entry to undertake a PhD. Remarkably, eighteen months later the ABC called again. The previous WA presenter had finished up and *Gardening Australia* needed a new WA presenter. The timing was finally right.

I must admit that I got off to a nervous start with filming. Contrary to popular belief, there's no intensive training or coaching involved – they simply throw you into it. It's also important to flag that there was no predetermined plan by the ABC to push Permaculture as such, but rather to present a variety of stories across a whole range of gardening topics. True enough – the interest in sustainable gardening was on the rise within the community, but the fact that my stories have a strong green bent is more to do with my personal beliefs than a conscious programming decision.

In my first half a dozen shoots, I did my best to present practical ways people could garden with less impact on the environment, while secretly trying to decipher jargon like 'piece to camera', 'action in vision', and 'cut away' which were dotted throughout my shooting script!

The feedback was mixed – *Gardening Australia* is an institution with a loyal audience who naturally take a while to warm to new presenters. My confidence was already shaky and when I received my first letter of complaint I was devastated. I had done a story on reusing tyres in the garden (a long debated and contentious issue in the organic gardening/permaculture community) and someone had taken issue with it, decided to write to me telling me that I was clearly an idiot with no understanding about organic gardening because tyres are poisonous and if they're used in the garden then you'll get cancer and die! Of course I was aware of the issue of cadmium and lead being used in the tyre manufacturing process, and had also discussed it in depth with my former environmental toxicology lecturer who agreed that provided the tyres were in good condition that there was minimal risk and that he was unaware of any published material that says otherwise. None the less my confidence was shaken.

I was mortified to discover that the letter had also been sent to Peter Cundall, the well-known presenter of *Gardening Australia*, and I immediately thought that I was at risk of getting the sack! Having not actually met Peter at that stage, I wrote him a long email offering my sincere apologies for bringing the program into disrepute. His response was brilliant – telling me not to worry and to stay on the path of doing what I was passionate about, which as far as he was concerned was the right path.

At the end of my first season on *Gardening Australia* in 2003, the presenters were asked to submit their story ideas for the following year. I put forward a proposal to cover the transformation of my next home garden project. Tragically, my utopian student garden had come to an untimely end when the owners eventually decided to redevelop it (a fate that was inevitable given its prime location and land value). The up side was this provided an opportunity to start something from scratch and to cover it on national television. I had come across another rental property, in the suburb of Hilton near Fremantle. It belonged to one of my former university lecturers (the same one who advised me on the tyres!) who kindly agreed to give me free reign on the designing and implementing the garden. At 900m^2, it was about the same size as the old one in Dalkeith.

I pitched the concept to cover the transformation in a step-by-step series: story one – clean up and design; story two – setting up vegie patch; story three – chooks; story four – home orchard and so on. I devised a program detailing how one would go about doing it, and that is how the permaculture series started. They scheduled the first story which we completed, and we were planning to do one a month, over the course of the year. After the first one went to air the feedback was incredible. Interest in the project was so strong that the Executive Producer decided to immediately double the production schedule.

From the beginning of the project my intention was to try to demystify the term 'permaculture' to help bring it into the mainstream by presenting the concepts in a way that looked good. I'd spent enough time working in snazzy gardens with my gardening round over the years to realise that people have expectations about how they want their landscape to look, and that these expectations and the presentation of the average permaculture garden are poles apart. This is a fact of life, and is unlikely to change. It doesn't mean that form has to dominate function, it just requires a little time, thought and effort to bring the two together.

The response from the permaculture community was mixed, ranging from extremely positive to quite critical. By this stage I had toughened up enough to realise that this is part and parcel of being in the media but it's still challenging to hear negative comments from peers in a movement which is meant to be about making change.

On a more positive note, the word was getting out and the ideas were being talked about in the broader community. Feedback indicated that enrolment in permaculture courses was on the increase and interest was on the rise. Audience response was fantastic so I was confident that I was on the right track.

At the end of the series and with the completion of the garden, we opened the garden as part of Australia's Open Garden Scheme and received around 3500 visitors over three days. The response was amazing with people eager to learn about organic food and how they could tackle critical environmental issues starting in their own gardens.

With the completion of the project and my lease coming to an end, I was once again on the lookout for a new project. Still a long way off from being able to afford my own place, it would have to be another rental. Having done two large gardens, I decided to challenge myself with a small inner suburban garden and thought it would be a great opportunity to demonstrate what could be done in a small area. Time and luck was once again on my side, with a friend's house becoming available – an old 1905 limestone semi-detached South Fremantle, with a small and run down backyard.

Again I pitched the idea to my Executive Producer to cover the project on the program with the with the angle that limited space is a challenge facing lots of gardeners around the country, and here's a chance to demonstrate that it is still possible to harvest water, recycle waste and grow food, and just as importantly it can look fantastic!

Over the next two years we created the garden on camera, including underground rainwater tanks and greywater system, trellised fruit trees and vines, feature vegie beds and herbs pots, mobile chook pen, worm farm and composting bins. The elements of the garden are carefully considered and integrated, plantings are diverse, organically managed and provide habitat, most of the materials are locally sourced and salvaged, but I never used the word permaculture. On completion of the project we opened the garden as part of the Australian Open Garden Scheme. In anticipation of the crowds, we blocked off the street, set up stalls, invited guest speakers and buskers and devised a way to move people in one side of the garden and out the other. We had over three thousand visitors over two days and again the response was terrific.

The encouragement that I received from people visiting my gardens over the years, as well as the experience gained from working on other projects along the way gave me the confidence to set up my own landscape and environmental design and consulting practice with the aim of applying the ideas and skills that I had been developing to more diverse projects and at a larger scale.

It wasn't long before opportunities arose to design school and community gardens, as well as contribute to the design process of housing developments, mining camps and town sites. The larger commercial projects are far from what would typically be described as permaculture, but they are real opportunities to apply strategic and systems based thinking for better design outcomes.

This is where I feel the permaculture movement has a lot to contribute and indeed has a moral responsibility to get involved to provide creative direction to mainstream projects. This can be challenging and requires pragmatism to engage in a process where only incremental improvements may be achievable. The reality is that development is constantly happening around us and we need passionate people to get involved and come up with ways to do it better. This is how change happens, through practical demonstration and gradually raising the bar in a way that people are comfortable with and in a way that they can understand.

Personal Reflections of Permaculture in a Changing World

We seem to be at a point in time where two things could happen. We could see a radical shift in political will, corporate responsibility and individual behavioural change at the scale and pace required to avert a looming environmental catastrophe. Or, if this doesn't happen, then the fallout combined with the impact of peak oil will force humanity into a corner where we have to change. Either way, massive changes *are* on the way and there will be a new operating model for society as we know it.

I consider myself a very optimistic person, but I have growing doubts that the momentum of modern capitalism can be slowed down without a major shock, and am increasingly thinking that it is likely to be the latter of the two scenarios described above that will play out. This is, of course, a stupid and tragic situation because we know that a large amount of loss and suffering will be incurred.

The way I have personally come to terms with this reality is by accepting that 'sustainability' is a journey, not a destination. It's a process of continually trying to do things better and more efficiently by learning from our experiences. The types of design strategies, innovative technologies and social models that are needed to avert a global environmental catastrophe, are also the tools that are needed to enable us to build resilient and self-reliant communities that will stand up in the wake of significant change. By applying the permaculture approach, we are doing what we can on an individual level to make a difference. Yet we are also empowering ourselves to be able to respond positively in the face of an uncertain future.

Tony Jansen

Tony is a dedicated and highly motivated sustainable agriculture and food security specialist. He is a strong communicator who has worked with sustainable agriculture systems in the Pacific, South America, and Australia, particularly in Solomon Islands where he helped initiate the successful NGO the Kastom Gaden Association. Tony's expertise is in program and project planning, management and participatory consultation. He works in extension services based on farmer to farmer approaches, and on improving agriculture through incorporating and building on local knowledge and worldviews of different groups.

He has lived in Solomon Islands for many years and has developed the ability to be a bridge in mutual understanding, communication and developing partnerships between groups from different cultural, social and economic backgrounds.

Chapter 23 | Learning from Other Cultures for Sustainable Development

Early Years

My family had a small vegetable plot when I was growing up and I can remember when I was very small the fascination of going up to this plot, not at our house, but at the community allotment garden. So I guess I've been interested in gardening from a very young age, I liked doing that sort of thing – growing plants, starting with culinary and medicinal herbs, working outdoors, bush walking, the ocean. A lot of my friends have gone in similar directions, so now I have a lot of friends who work in bush regeneration, landscaping, home gardening, development work, wilderness-based activities or in other ways on sustainability and creative 'edges'.

Also because I moved a lot as a child I got a taste for travel and experiencing different cultures and ways of living; how people grow and use food was a connecting thread. In fact, since my late teens I've almost never lived without a food garden growing at least some of my food, nurturing my soul and giving me recreation and exercise.

Education

At university I studied anthropology and geography. I was always very interested in international development issues – global inequality and problems with the environment. I did some really excellent courses, for example Ted Trainer's lectures on the links between the environmental and the poverty crisis, and the need for both rich and poor to move towards a more sustainable way of living. Ted has done a lot of thinking about these issues, and I found his lectures really inspiring. One of his quotes that has kind of been a mantra for me was 'we must live more simply so the rest can simply live'. The 'we', being the affluent, developed countries and the rest the majority of the world's people. But then I'd go to the next course which was about business-as-usual, where they did not even seem to be aware of what seemed to me to be a pressing crisis. It seemed to me that too few people were pulling these disparate ideas together.

I had first heard about permaculture at university and also on TV when Bill Mollison was featured on the *Visionaries* program. A friend called me and told me to turn on the TV. It clicked immediately. I guess what attracted me to permaculture was that it seemed to be an integrative philosophy that allows you to do very practical hands-on things like growing your own food and improving human settlements in a thoughtful way; making a contribution to bigger picture world issues through practical local action. I continued to study for a while and then I deferred because I was frustrated. I never did get back to university to finish my degree; I've just been too busy ever since.

The Start of Overseas Development Work

I went travelling around India and other parts of Asia, and while there I did some voluntary work. For a while I was a volunteer English teacher in a Tibetan refugee settlement in the Himalayan region of Ladakh. While I was there I went to the amazing Ladakh Ecological Development Centre in Leh, initiated by the social and environmental activist Helena Norberg-Hodge. Up on the Tibetan plateau, reading about sustainability in their fantastic little library, I connected in more depth to permaculture thinking. I realised this was the direction that integrated problems of food production with wider issues of inequality in society. In India I also connected with Buddhism through the Dalai Lama and the Tibetan movement. The Buddhist principles of connectedness and helping other beings have been a very strong influence in my life ever since.

Back from overseas I made an attempt at going back to university but I never finished – permaculture was partly responsible for that! A friend rang at short notice and said he was going up to Byron Bay in northern NSW to do a PDC with Bill Mollison and Jude and Michel Fanton from Seed Savers. I booked a spot and jumped in the car! Their course was very inspiring, with an amazing, diverse and active group of people. I'd already got a taste for volunteering in developing countries on my travels, and had become convinced that good work was about helping others and contributing to environmental and social problems through a focus on food production. I wanted to do more of that kind of work.

Back Overseas – Learning from Other Cultures

Permaculture gave me the opportunity to gain skills that I could offer. During the PDC I met people who were looking for volunteers to join projects, and got the chance to work as a volunteer for a year in Ecuador. This was at the Rainforest Information Centre's partner CIBT (Centro de Investigación de los Bosques Tropicales). In my hand I had a reference from the Permaculture Institute and a one-page letter saying, "Come over and we will give you a chance, but no promises and you will live on a local wage"! So I bought a round-the-world ticket and went to Ecuador.

For me this was the completion of my tertiary education. I spent a year in the field in a remote and low income community, totally immersed in all the complexities of permaculture on the ground in a developing country. There were some successes and lots of amazing and inspiring people, but I think I learned most from the contradictions, the failures and the difficulties.

I soon realised that we have to challenge our assumptions when we go into another culture and another place. We think we have something to offer, because in permaculture there are these great ideas which would seem to be fairly universal and can be applied everywhere. But there is the danger of coming in with a missionary zeal – that this is the 'New Way' – this is what people should be doing instead of what they are doing now. The organisation I worked with was based on western volunteers coming in from the outside and was environmentally focussed. Rainforest conservation was their aim. As I understood it, working with very disadvantaged communities living on the edge of, or in, the forest was not their core purpose. But they soon realised they had to work with the local communities in order to conserve the rainforest, so sustainable agriculture became important. In my opinion there were many contradictions. It seemed to me that PDCs were being run for the locals without really acknowledging what these people already knew, and without really thinking through the change process involved in moving to sustainability within the local worldview.

I saw a lot of different approaches in the various projects I was involved with or visited. I worked at a very impressive farm in its early stages, which had been designed on permaculture principles. It took the whole year to progress just a few modest and frustrating steps as a foreigner trying to establish a 'model farm' in a foreign culture. It was a valuable experience but it was, I decided, the first and last time I would invest my energy in 'models' for other people, based on the ideas of outsiders.

Interestingly, there was another CIBT project in the Andes that had a completely different approach. The person working there had completely immersed himself in the community, and would not do anything without their full understanding, involvement and participation in the decisions. Whereas, as I saw it, we were a bunch of 'gringos' down at the other site rushing ahead and hiring people and doing all this tree planting trying to establish things on the ground, he spent one year laying the foundations for work to begin. In that one year they had a lot of problems with grazing animals, maybe goats, that were getting into the area and destroying the trees. The community had agreed this would be an experiment with permaculture ideas, involving the local school. They already had a fence around the garden that the community had built, and all they needed was a gate. It became a joke that it took a year for the community to get themselves organised enough, and have enough commitment and understanding, to put a gate on the area which meant that planting could proceed for the nurseries that the community was establishing for itself.

But in the end the project was much more successful because it had community ownership and involvement. Time had been invested in the social structures and community ownership rather than rushing into the physical planting. That was a good learning experience for me. So that year, even though I probably wasn't very useful, was an important learning time for me. I realised the principles I had learned in anthropology about recognising my own culture-tinted eyes and trying to see and understand things through the culture and worldview of others was as important as the technical methods of permaculture.

When I came back, reluctantly, to Australia, I knew what I was looking for, and went as a volunteer to the Solomon Islands with the small Australian organisation APACE (Appropriate Technology for Community and Environment). I was looking to spend at least five years in my next place as I had realised a year was needed to just to begin to understand the local context. If I wanted to be really useful I was going to have to stay for a long time. I had done some voluntary work with APACE before going to Ecuador and knew a little, but not much, about this area of the world known as Melanesia.

In my work in the Solomons we approached village food production as part of the bigger scheme of things. I helped start and have nurtured and seen mature, a local non-government organisational called Kastom Gaden Association (KGA). It's now one of the larger and more successful non-profit organisations in the country. I have been involved in the production of numerous KGA publications and manuals. There, a permaculture approach means you don't separate agriculture from the forest or the water catchment or the livestock, whereas a lot of development work tends to do just that. It focuses on one narrow area and does not look at the whole system. Subsistence farming and traditional agriculture are systems approaches. They are really complicated systems and people are aware of and involved in lots of different activities as part of their livelihood. They manage many different kinds of resources and maintain complex social systems that provide the basis for healthy communities and strong cultures. I see a permaculture approach as looking at the whole system and understanding each element's impact on the others, trying to create links and synergies.

There are different buzzwords in development circles, like 'sustainable livelihoods' that, while similar to permaculture, are perhaps not quite as practical and do not necessarily have a central focus on food production. I think the latter is a very important feature of permaculture. In the Solomons we have traditional agriculture based around shifting cultivation, which involves walking out into the forest, often quite a long way, clearing the forest with small hand tools, planting food – mostly root crops and other plants – in a mixed poly-culture garden, then moving on. Unfortunately this approach ends up mining the soil fertility until the soil is depleted.

Eventually the farmer comes back to that site twenty or more years later when the tropical rain forest has re-grown and the soil fertility returned. It's a system that we westerners tend to rush into and think, "That's an incredibly destructive way of farming – people are ruining rainforests and planting root crops, this can't go on!" But when you really start to understand the system you realise it's actually an incredibly sustainable system that maintains extraordinary biodiversity and provides a very high return for the labour involved. However, it comes under a lot of pressure when the population grows beyond a certain point and needs to grow more food, so that the long fallow periods can no longer be maintained. This is now happening all over Melanesia. I guess what I am saying is that things rarely as simple as they seem. Sustainability is evolving. It is time and context based. What was a very good practice in the past may no longer be so good today. In these changing times, having the tools to observe, understand, plan and organise for change can be as important as having the technical tools.

At the end of one year in Ecuador I felt like I was just starting to understand things a little bit. When I went to the Solomons I always had it in mind that I wanted to spend five years there in order to see some of the fruits of my work and to get to the point where I felt I could make the best contribution through understanding the local situation. In many cases with people who work in developing countries, you're just settling in and then you go. Five years is probably what you need, to get to know a culture and be useful in some way. I've been in the Solomons a lot longer than that – over 14 years.

There's an English guy called Charles Rogers who I respect and know quite well. He works with a farmers' organisation in Vanuatu. He's much older than me and pretty much spends his whole life there doing practical agriculture work with local people. I don't think he would ever call it permaculture, but a lot of the things he does are quite similar. He said to me once that when people come to do development work and have been in the community for a year they think they can write a book about their experiences. When you've spent five years you realise that maybe you can write a paper on what you know, and when you've spent 25 years you feel like you might be able to write a paragraph but you're not even sure about that anymore! For me it's been a very similar experience. With experience you become more confident because you know and understand more. At the same time you become less confident in all the ideas that you were so certain of earlier on, and in your own understanding of local realities because you start to understand the way other people see things, the way they view the world. You realise how often you really don't understand or know the full picture yet and maybe never will. That's been one of the key learnings for me. It makes you more humble over time.

People in those countries have a lot of difficulties in their lives. A lot of the things that we in the West take for granted are just not available, or are difficult. Like getting water or energy such as electricity or having enough money to put kids in school, or making a phone call. In our society we have extreme wealth in the other direction. But in over-developed countries, there are not the same notions of happiness or satisfaction that I think are common in more traditional and lower-consumption societies. I guess that's one of the things that I struggle with; sometimes the simpler things are, the richer they are. That's something we need to learn from these cultures and societies. There needs to be much more of a meeting ground in the middle. These places could do with a little bit more to make the lives of their people more comfortable and less difficult, but at the same time if we could simplify our lives and be more connected to our land, our local trading and food production in our own communities, we'd find that we would have much richer lives and much less wasteful resource consumption.

As I spend more time in development work I think increasingly about scaling up. How do you get the small local successes to become a wider success? We know that small, incremental measures work better than big ones in most development interventions. But how you get lots of small things to grow into big and positive change, is a challenge. Now more and more, I am trying to network and link people, groups and places, so they can learn from each other.

An obvious area for scaling up is to work with schools and educate the next generation. But because the complicated curriculum system and teaching needs are outside of our area of expertise, Kastom Gaden has avoided working with the formal school system, although we have had groups, schools and government departments approach us. Instead we've decided to stay focussed on the core area of training for people in the villages, and have had some involvement in curriculum in the Rural Training Centres which are meant to provide village skills-focussed schools for rural people. In general, informal knowledge is very practical and in our efforts to add to that knowledge we need to continue to be practical. If our approach is more curriculum based and formal, there's a danger of ignoring the really successful village-based approaches and models. For example, often the best instructors for farmers or young people are those with no formal education that couldn't be 'teachers' in a curriculum-based model.

That's one of the biggest dilemmas in the Solomons – that there's this incredible focus on access to education but it's a transplanted education system that leads people to aspire to leave the village. Until that's sorted out – until there's a vision in education of a future *within* the village, we have a major problem developing. The more educated people have been taught to want to do 'modern agriculture', and engage in agriculture theory at a 'higher' level to that of the grannies in the village; to ignore the fact that they are living in one of the richest and longest lasting agricultural societies in the world. In the village you have these real 'professors of agriculture' – the wise and experienced farmers who have been doing it their whole lives and learning from their ancestors. They are completely unacknowledged by the formal education model. It's difficult – we have tried to bring bush foods and traditional knowledge about bush tucker into the schools but in the end we fell back to working through informal village and farmer networks. All this is complicated even more by the fact that you have 100 languages in the Solomons!

Adapting Permaculture, Seeing it Differently

One of the things you learn about from this development work at the grass roots level is the sense and processes of change. Learning about change, observation and experimentation is sometimes more important than the change itself. It's not so much about re-designing whole farming systems, but about helping people to see the bigger picture regarding issues like pest management, soil fertility, biodiversity, nutrition, health and other things they might not have understood or connected together before. Innovation and adaptation are key in this rapidly changing world. Permaculture often starts with visionary ideals but it doesn't always focus enough on how those ideas are adopted and adapted in a change process as it spreads through a society. In my work the process of facilitating farmers in learning from each other, often becomes the building block to improved agriculture. The urgency is still there but I'm much more aware of what it means for change to occur in fairly conservative traditional societies. This makes me see things differently, because whereas I would have been quite critical before, I now see that small steps are important in enabling people to move from where they are now to where they are headed.

Sometimes permaculture can be quite disengaging from traditional agriculture and has this very Western design template idea to it. It's more and more about networks and how people share their ideas and information and opportunities. My early overseas experiences made me cautious about using the word 'permaculture' too freely. For me it has been much more about listening to people and using a permaculture approach to understand their needs through an integrated framework. This helps them to look at food production systems and how they can improve them, as against the kind of zeal that says 'this is the magic answer'. For me, using a permaculture approach means looking at whole systems, including local knowledge and worldviews and reflecting on how you can understand and improve the connections. It means sharing ideas that would not have been shared without an outsider's input. I think there has been a history of bringing the latest 'solution' into traditional societies and developing countries in general, rather than looking at and valuing what they already have. The latest new idea from the West, even if it is a good one, runs the risk of undermining and eroding what is already there. The poorest and those on the margins can't take these kinds of risks. Empowerment for change is about building on people's own knowledge and experience. It's exploring all these questions and issues that has been my work since those early days – how small, traditional farmers learn and change, how to try and improve these processes and what organisations and networks are needed to support them.

In Melanesia you can't say, "Why don't you just adopt a permaculture design approach and have a permanent garden around your house with your livestock and your forest?" The way permaculture zoning works doesn't necessarily fit. Instead working with permaculture zone 1 there might mean looking at how people can have a little bit of food closer to their house. Having a few vegetables and beans and maybe a few more fruit and nut trees really close to the house would really help nutrition. At the same time it would provide an opportunity to learn how to build soil fertility and see soil as something you can improve, rather than as just a wild system where you go out and use the more fertile soil. Gradually these methods can be adapted the bush gardens. But in Melanesia, the bulk of the food is going to continue to come from shifting gardens in the bush because that's where the productive land is, and that is a system that still works reasonably well for most people.

And where would livestock be in a Melanesian permaculture system? There, livestock are usually running around like crazy in zone 1 – around houses and daily living areas, and fencing material is very expensive so this is unlikely to change. We have had some success in traditional kitchen gardens but there's also a lot of resistance. You learn as you go. Forest gardens are a problem because it uses a lot of the women's time, carrying things back and forth and being away from small infants who are left in the village. At the same time, there are important reasons for people being out walking on the land, such as the connection to the forest, the bush. It's a time when women can be away from the village, out together out working in the garden, that kind of thing. Here a simplistic, inflexible idea of zones 1, 2, 3 radiating out from the house as is used elsewhere, doesn't really work.

The permaculture approach, of putting things in the right place in terms of distance, the energy involved, how often you need to go there, the connections between them, is good. But when you look at a village situation it's so much more complicated. Some of the permaculture zone ideas are based on a Western nuclear family or homestead kind of idea. How do you fit a village into that? How do you allow for social relationships and every day interactions of a kind that are not so common in western society? While permaculture sounds good, it also creates all kinds of challenges, and we've had plenty of those.

Permaculture and the Mainstream

I'm not sure why permaculture has so little visibility in the West. It varies from place to place. In some places permaculture is much more talked about, more central and closer to the mainstream. Last year in India, we were staying with my friends Narsanna and Padma Koppulla who do amazing work in Andra Pradesh with low caste landless women in difficult semi-arid environments. They call what they're doing 'permaculture', so we became aware that the word permaculture was used quite a lot. Narsanna has adapted the approach and has been amazingly successful. I've known him for over a decade, and from being a quite radical activist on the fringe, he's now a consultant to the Andra Pradesh government. A lot of the ideas such as watershed approaches and empowering people through land restoration have become mainstream and they use the word 'permaculture' quite a bit.

Every once in a while someone from Malaita in the Solomons will ask, "Well, what about permaculture?" because they've heard about it. And I think, well actually, that's what we've been doing all these years. There's now a Japanese organisation working in a few parts of the Solomons that's openly calling what they do 'permaculture'. My tendency when working in traditional cultures is to not to use outside labels too much, but the idea is there and that's the important thing.

Back to Australia

Being based in both Australian and the Solomons, I travel back and forth quite a lot. I've worked with Kastom Gaden for so long, and I now want them to have some distance and find their own way, but I'll be supporting them. Maybe I'll do more of this kind of work in other countries, but I'm sure I'll stay involved in Melanesia.

Being back in Australia I have found an amazing change in the social climate with the change of government from the Howard era. My previous short visits to Australia were sometimes quite depressing. There was a strong alternative movement but there were also a lot of negative trends, a less tolerant society. It's amazing how that has shifted. I would have thought that would have taken a long time, but it's quite an exciting time now.

I really enjoy the mood in northern New South Wales where I live. We have friends here who are avocado farmers and we grow our vegetables at their place. It's quite strange to move away from zone 1, but the place we rent doesn't have any sun, so we drive to our vegetable garden, but it's very nice, it's a beautiful place. It's been very interesting to work a little with Australian agriculture. I'm so used to working in agriculture systems where labour is abundant and where labour constraints are not an issue. Here, herbicide use is a labour substitute. It's amazing to see the size of these farms that one person is responsible for, and the challenges that this creates. In Australia, the difficulty of scaling up in permaculture for conventional farmers would be in understanding the process of step-by-step change. Thinking about how people use the land. I've been working with my avocado farmer friends over the years and they now have a stall at a farmers' market, selling their produce into the local economy. They are not certified organic but they are close to it in their farming methods. That's another reason to be really hopeful. Who could have imagined the local trading and flourishing of local markets that's going on across Australia? Clearly these things can happen quite quickly, without big changes in policies from the top down.

The Future

My vision for the future would be a world with more equality, where the majority world has access to enough, a reasonable level of comfort in terms of resources, communication and infrastructure. I'd like to see the rich countries consuming a lot less and it would be nice to see economies become much more locally based and much more connected to their bioregions. I'd like to see people in towns and cities much more connected to their rural hinterland and regenerating sustainable landscapes. In that world resource consumption and food production would be locally based. We would have a really dynamic global system of communicating, networking and sharing ideas that would allow people to be much more involved with each other. A general move in that direction would be good! But with technology it's so hard to predict the way things will go. I go through waves of being more or less optimistic. But I think there's enough to be optimistic about, though a lot of the problems we face are overwhelming.

I guess my philosophy is that we have to work for positive change. That's our role on this earth, and doing that is enough. Whether it will lead to a changed world in twenty years or not is beside the point. This kind of work is important and needed and satisfying in itself. The more I think about it the more I wonder how the Solomons will be when they'll have twice as many people in twenty years. They'll definitely have a big urban shanty town regardless of where things are going.

My experience with ecosystems is that they can be more resilient than we sometimes imagine. I've seen so many amazing examples in different parts of the world showing how things can be restored. With the right thinking and the right opportunity and support, nature does come back. So while I do think gloom and doom scenarios have an important role to play, I tend not to go into them too much. I do feel that with a changed attitude, things can be restored quickly and nature might have more adaptive abilities than we give her credit for. That's my hope anyway. I think I see a change in mindset beginning to happen. Having worked so much in Melanesia, I see a lot of changes there. People there have a challenge trying to reconcile two different worlds – their traditional world colliding with the modern world. But I think there's a real strength in people as they increasingly realise they need to maintain their cultural self-reliance as well as embracing some of the news things in the world around them. They need to keep some kind of balance.

In the West it's amazing the way climate change, from being this incredibly frustrating issue of stone-walling for more than a decade, has suddenly rocketed up the agenda and there's a lot of positive things happening. Maybe there could be more. But there's definitely big change happening. There are so many overwhelming problems, but I will continue in this work in the hope that enough positive seeds have been sown that will bear fruit at some point in the future. I guess that's part of my philosophy. There's a lot to be hopeful about.

Morag Gamble

Morag Gamble lives and works from her owner-built sustainable home at Crystal Waters Permaculture Village – the United Nations world habitat award-winning settlement in the Sunshine Coast Hinterland – where she also leads the kids' community garden project. Through her organisation, SEED International, Morag teaches permaculture design, and designs edible landscapes and sustainability centres in urban and regional contexts. Since the mid 1990s this work has taken her to 20 countries.

Morag is also a lecturer in food politics at Griffith University, is the founding President of Sustain QLD, and a Director of the Ethos Foundation. She is the co-founder of the Australian City Farms and Community Gardens Network and the Northey Street City Farm. She has also produced/directed the documentary, *Think Global: Eat Local – A Diet for a Sustainable Society*, and has written a thesis for her Master of Sustainability Education titled *Characteristics of Quality Permaculture Education*. For the Brisbane City Council, Morag has produced a comprehensive *Urban Agriculture Report* and the *Community Garden: Getting Started Guide*.

Chapter 24 | Think Global: Eat Local

Growing Up

I grew up in an outer suburb of Melbourne where most people had a quarter-acre block and two kids. Over the years, I have heard a lot of criticism about suburbs, and I am as critical of sprawling car-dependent developments, especially those with huge houses on tiny blocks. But I had a great childhood and I think my suburban upbringing has strongly influenced me and why I have embraced permaculture.

My suburb, my street, was my village, my community. It was a great place to grow up. I knew everyone in the street. All my friends lived there. We left our doors and cars unlocked. There were few fences between the blocks. We played and rode and skated a lot on the street. It was also our canvas for chalk drawings and games. The 'tribe' of children roamed from house to house until we were each called home for dinner. We collected tadpoles in the local creek, made cubbies under the gum tree in the spare block and hid our treasures in the hydrants.

Our house was designed and built by my family and was solar passive. I didn't know this at the time, hadn't heard the terminology, but I did know that our house faced north, not the street, and that my father had sought our block because of its northerly aspect. Mum showed us how to actively work the windows and doors for cross-ventilation in the summer.

My parents planted out the block mostly with local natives to encourage native birds and wildlife. One of my earliest memories is talking to a kookaburra on our verandah. I also remember seeing a koala in our backyard, and I enjoyed befriending the possums who came to sit on my shoulder in the evening.

In the kitchen courtyard, we grew pots of herbs. It was usually my role to tend and harvest these as Mum's helper. We had an area for a vegetable garden too, as well as some fruit trees and chooks. The addition of chickens to our place started as a school project. My brother and I raised them from day-old chicks in our bedrooms, and we always took them on holidays with us to East Gippsland. Our holiday place there also had fruit trees.

Almost all my school holidays were spent there on an island or a boat, in the Gippsland Lakes. The island was home then to a small fleet of tiny wooden fishing boats. The fishermen headed off late in the night to fish the waters and bring them back to sell at the Fisherman's wharf, the smell of which I'll never forget. Our place was next to the lake, and I often heard them bringing in their nets in the shallows in the early hours. I feel a strong family connection to that place as my Dad had spent his teenage years on the Island and built a small cabin there for his family when he was just 13. His father was sick with malaria after the war and died in the local hospital.

After my father finished high school, he moved to Melbourne to live with his uncle, Judge Freddie Gamble, who I understand to have instilled a high level of ethics in my father, which he has passed to me. The ethics of permaculture – earth care, people care, fair share – resonated well with the ideas I had grown up with.

Encountering Permaculture

It was my father, a land surveyor, who introduced me to Permaculture. He bought a copy of *Permaculture One* when it was first released. He raved about permaculture, as he did about the work and writings of people like E.F. Schumacher, Fritjof Capra, Ian McHarg and Masanobu Fukuoka. I held them in high esteem as reference points of good thinking and action. I think this is what led me to study Environmental Planning and Landscape Architecture at the University of Melbourne, and later to be so strongly drawn to Schumacher College in Devon, England.

My mother, a primary school teacher, raised us as vegetarians which I understood to be for health and animal welfare reasons. She was very interested in natural health. In early shopping trips with my mother, I remember going to the bulk foods stores and reading labels. In our house we had no sugar, salt, caffeine or meat – the food was simple and natural. I was the kid with 'black bread' with homemade nut butter in my lunchbox.

I began my activism in the peace movement at the age of 14 and later in my teens shifted my activism to the environment movement. After some time, I became tired of being against things, and began to realise that people's eyes glazed over when I started to speak about the issues that concerned me. I was keen to find a way to be 'for' something and to find a practical way to bring all aspects of my life and thinking together.

While I was studying Environmental Planning and Landscape Architecture I began to explore permaculture properly for the first time. Rather than designing for people, I was interested in helping people and communities design for themselves. Permaculture seemed to provide a useful framework for this. One lecturer opened my mind with his statement, "Design is not just an end in itself; it is means to something greater." I was interested in being involved in creating positive change rather than designing lovely gardens for the affluent. Permaculture seemed to provide a good basis for this, but it wasn't until a gap year journey around the world exploring sustainability projects that I really discovered permaculture and embraced it as a way of life and work.

Schumacher College, Ladakh and Fritjof Capra

While in my postgraduate year at University I was handed a book by a friend that really started me connecting ideas – *Re-enchantment of the World* by Morris Berman. This and other books I was reading at the time kept referencing ecological scientist Fritjof Capra, a name I had heard often from my father. He was the author of *Tao of Physics*, *The Turning Point* and *The Web of Life*. At last I read his books, back to back, while I was taking a break by myself at our island place. My life and my thinking changed from that point. Like the title of Fritjof's book, *The Turning Point*, so this retreat period was for me. It was catalytic, extraordinary. I felt like I was existing in another dimension. I felt the fire in my belly burning so strongly. I saw unexplained lights hovering around me one evening. And during my evening walks I was often accompanied by owls. During this retreat in reflection, I felt I had discovered my sense of purpose, clarified my thoughts, and began to see the world differently. Capra's works just made so much sense to me. His writings introduced me to the concept of paradigms, in particular the emerging ecological paradigm, to systems thinking and interconnectedness. As I read I thought, "This is where I am coming from." His books articulated what I was feeling and presented me with a scientific and philosophical framework. Until then I think I had felt somewhat out of place.

Soon after I discovered that Capra was teaching a five week residential course at Schumacher College in England and I was accepted into the program. Schumacher College is an independent ecological college which has been at the forefront of ecological thinking and sustainability education since 1991. Its visiting scholars include Capra, Vandana Shiva, Amory Lovins and many other well-known ecological thinkers and activists. I have stayed in touch with Fritjof Capra and visited him many times in the USA. Together we have explored the links between the principles of ecological systems he developed with international scientists, and the principles of ecological systems design we present in permaculture. In his more recent book, *The Hidden Connections*, which connects ecoliteracy and eco-design, Fritjof makes direct reference to permaculture.

What began as a ten week trip to the United Kingdom – the five week course with Fritjof and five weeks travelling to eco-projects (Findhorn, Centre for Alternative Technology and a variety of community gardens) ended up being a 10 month journey. I stayed longer at Schumacher College as a resident-helper and experienced community living. I met and studied with others who were inspiring scholars, mentors and leaders such as eco-activist Vandana Shiva, deep ecologist Arne Naess, eco-spiritual leader Satish Kumar, eco-architect Christopher Day, relocalisation advocate, Helena Norberg-Hodge.

I left Schumacher College with an overwhelming sense of hope, positivity and clarity. I had met people from all around the world who were embracing the ecological paradigm and creating livelihoods, lifestyles and projects based on this. I had a sense that even though I was one person doing small local projects, that I was part of a much larger movement of change globally – a positive change – and I felt honoured to be acting as a catalyst for change in my local community, and ever since I have always tried to help link our local projects with others around the world.

In June 2003, I had the honour of returning again to Schumacher College, but this time as a scholar-in-residence to co-teach their first permaculture course – the annual Ecodesign Course. During the course, the participants redesigned the college landscape and infrastructure. Their ideas were taken into account during the retrofitting of the college. The course was part of the Master of Holistic Science offered at the College. The students of this program continued the practical permaculture projects started during this course and handed them on to the incoming students. I co-facilitated the second part of this course with Janine Benyus, author of *Biomimicry*. Again, I had the opportunity to explore the connections between permaculture and other theories and principle sets. I think it is critical as permaculture teachers and practitioners that we keep ourselves open to new and other ways of thinking and to cross-pollinate ideas and strategies.

Being at Schumacher College had set off a positive ripple effect in my life and gave me the courage to follow my heart in my life and work, and evolved in me a commitment to live with integrity. This experience helped me to clarify my thinking and values while also opening me to the joy of living in community and being connected to others. I also learnt much about myself. I had been so taken by Helena Norberg-Hodge's work that I volunteered to spend a summer working with her and her organisation ISEC (International

Society for Ecology and Culture which took me to Ladakh on the far side of the Himalayas working in sustainable community development and sustainability education. In Ladakh I began to understand what living sustainably REALLY meant, and began to understand more deeply the value of being connected to place and community.

From Ladakh, I came home to Australia with a sense that my thoughts of wanting to live more sustainably were not a utopian dream but were actually possible. I had experienced it and I had seen an old culture living that way. I began to realise that it was not against human nature to live more lightly and respectfully, but that it was human culture, the concepts, ideas, policies, and economic systems, that prevented us. Therefore, I thought, if the crises we are facing as a society are created by learnt – not innate – behaviour, it is possible that things can change. When I came home from Ladakh I wanted to find a way to express those values and practices in the context of my culture and community. This gave me much hope.

As soon as I arrived back in Australia I enrolled in a Permaculture Design Course at Crystal Waters Permaculture Village. Permaculture provided me with a practical platform to live the thinking that Schumacher College filled me with, and to experiment with the low-impact ways of living that I experienced in Ladakh.

I met my husband, Evan, at that permaculture course and have been an active permaculture practitioner, teacher, designer, community gardener and international advocate ever since. Within a few months of meeting Evan I had moved from Melbourne to Brisbane in Queensland to be with him. Immediately I met up with local activists and became a co-founder of Northey Street City Farm.

Northey Street City Farm

Northey Street City Farm is situated on the banks of Brisbane's Breakfast Creek in the inner suburb of Windsor. More than 1500 exotic and native fruit trees, bush tucker (native food) plants, shrubs and ground covers have been planted on the four-hectare farm site since its inception in April 1994. The aim of City Farm was to create a working model of a community-based urban permaculture farm that demonstrates, promotes, educates and advocates for environmental and economic sustainability in a healthy, diverse and supportive community. As well as gardening, building and doing community education, I was the City Farm coordinator for a couple of years, secretary for another, newsletter editor, events coordinator... I lived up the road and was active in just about every aspect of the city farm's development and management from 1993 through to 1998.

At Northey Farm, mostly as a volunteer, I led community education programs in sustainability, launched the solstice and equinox events, organised local food feasts and music events, facilitated study circles, established an organic food collective, organised the first permaculture workshops at the city farm and did a lot of outreach work through radio, TV, newspaper, seminars and talks. I organised the Green Fair at the city farm in 1995 that was attended by 3000-4000 people. As a permaculture teacher I actively promote community food systems as a core strategy to address many of the problems we are facing globally. Community food systems help us to bring the food economy home, strengthen local communities and economies, reduce our ecological footprint and improve the nutritional value of the food.

At this time, I also started Permaculture Brisbane with a couple of friends and organised an Australian tour for English eco-activist, John Papworth. He was E.F. Schumacher's friend and with Schumacher's support and encouragement, founded the fabulous *Resurgence* magazine.

I was also helping to launch other community gardens in the region. I travelled extensively around Australia visiting community gardens and city farms and was involved in the formation of the Australian City Farms and Community Gardens Network, as well as its Queensland branch. This organisation has grown from strength to strength and now has regular national gatherings. With the aim of sharing our collective experiences, helping each other and offering support, I organised the first regional network gathering in 1995, then the first Queensland gathering in 2002 which led to organising the first national gathering with a couple of friends in 2004. The national community garden/city farm event in Melbourne in 2006 attracted Vandana Shiva and Helena Norberg-Hodge as keynote speakers. (It's amazing how things all seem to connect.)

To make ends meet, I worked part-time in a food co-op, offered an urban permaculture design service, took a few flute students, and worked one morning a week in a local cafe. At one point I worked in State Government as an Environmental Planning Officer, but left once I'd saved enough money to go back to Ladakh. I was offered a position on my return, but could not bring myself to take it. At the time I felt I could make a better contribution through permaculture, and I wanted to give myself the chance to create a livelihood as a permaculture educator and designer.

In 1998 I moved to Crystal Waters Permaculture Village near the cooperative town of Maleny in south east Queensland where Evan and I now live together with our children.

Cultivating a Permaculture Career

I recognise and value the role of volunteerism in developing my permaculture experience and understanding. Through volunteering, creating new projects, following my heart and intuition, I have been able to cultivate a permaculture career. I was able to work this way for so many years by living simply, sharing and exchanging resources and skills within our community and generating a small income from community workshops. One of the most surprising things was the offering of tithings to support Evan and me. A local woman directly tithed 10% of her income to support our work and help us continue. Two others followed her lead. This type of support gave me much more confidence in my work, helped me value what I was doing, and understand that our actions mattered. Volunteering is a very free way to work. Being tithed, gave an added sense of responsibility and accountability which I think was an important step for me at the time.

The fact that permaculture, as an informal community education system, has global recognition and value is remarkable. It is one of the features of permaculture that attracts me professionally to remain de-institutionalised and autonomous, yet within a strong international network and framework of common understanding.

Permaculture Education

Permaculture is education, design and action for ecological sustainability and sustainable community development. It is an open and accessible form of design that embraces systems thinking, ecological values, and indigenous knowledge. It provides a practical framework within which to place my values, ideas, and experience and enables me to offer something of value to the communities within which I work, both in the majority world and the 'West'. I see that permaculture is also about training trainers and inspiring and enabling community leaders and activism, to foster a positive ripple effect. It's about building resilience and sharing and spreading knowledge and ideas, rather than holding onto information for power and personal gain. Through my work with permaculture around the world, I have been encouraged to see how it can be relevant to both people in a rural village of Indonesia, and apartment dwellers in the urban centres of developed nations. I have seen how it can make such a difference to people's lives and community wellbeing in so many different contexts, and support local and bioregional self-determination and decision-making.

Coming from a landscape architecture background, I value the whole systems approach permaculture offers to ecological planning and design, how it cultivates ecoliteracy, and how it facilitates connection.

I like too, that permaculture is not a recipe, but an approach and a process; a way of thinking guided by the ethics; a way of seeing the world and ourselves. The permaculture design course helps to teach people how to think systemically and to design sustainable systems using core principles of ecology and ecological design. Unfortunately I think the popular perception of permaculture is still that it is a collection of gardening techniques and that you are doing permaculture if you have a compost heap, worm farm, herb spiral, mandala garden and a couple of fruit trees in your back yard that looks a bit messy with a lot of mulch around.

I see permaculture in a much broader context than this and am particularly interested in teaching permaculture from a holistic perspective. In 2006, I completed a Master's thesis in Environmental Education titled, *Characteristics of Quality Permaculture Education*. The following is a summary of the elements that I concluded were important to offering quality permaculture education, derived from participants of permaculture design courses and leading thinking in sustainability education.

In summary, my research concluded that quality permaculture education:

1. Embodies the principles of ecology.
2. Is informed by the threefold permaculture ethic (earth care, people care, fair share).
3. Is guided by the principles of permaculture design.
4. Is guided by the principles of sustainability education.
5. Supports the development of ecoliteracy (understanding of natural systems).
6. Is respectful, inclusive and open to the ecological wisdom of indigenous perspectives and other ways of knowing and experiencing.
7. Encourages critical and systemic (ecological paradigm) thinking and problem solving.
8. Is holistic and cross-disciplinary – is based on whole systems approach – addresses local and global issues from local and global perspectives, both in and between majority and minority countries. It addresses the whole person. It connects and draws from a wide range of disciplines.
9. Is practical and action-oriented – supports development of knowledge, skills, values and attitudes required to work toward positive, integrated solutions for ecological sustainability.
10. Is transformative – encourages a shift to ecological paradigm thinking

and action focussed for social transformation toward ecological sustainability – inspires, motivates, empowers, gives hope and confidence.

11 Is values-oriented – questioning values, attitudes, assumptions, consumption patterns, lifestyles and economic growth – addresses root causes.

12 Is future-oriented – has a long-term perspective, recognises intergenerational equity and encourages visioning a sustainable future. Demonstrates possible alternatives.

13 Is locally-relevant and accessible – supports community initiatives and community self-determination – is embedded in and relevant to community, culture and place, encouraging development of locally-specific knowledge, approaches and solutions. Presents accessible approaches to solution-finding that can be addressed at a local level.

14 Is inclusive, participatory and collaborative – engages participants in participatory decision-making, planning and development, and promotes cooperation. Facilitates clear and collaborative goal-setting.

15 Is learner-centred, creative, dynamic and flexible – builds on knowledge and experience, is human-scale – addressing the whole person. Is concerned with process and the building of positive and mutually beneficial relationships.

16 Facilitates connection and reconnection between individuals, nature and community, between urban, rural and natural environments, and with oneself.

17 Demonstrates ecological integrity at all levels.

International Work

Over the past 15 years, Evan and I have been fortunate to have had the opportunity to work around the world on independent community-based projects. We have been invited to lead programs, present and consult in 20 countries. I have gained enormous inspiration, invaluable experience and deep insights from our global exploration of ecovillages, permaculture projects and urban agriculture projects, and the insights gained while teaching permaculture programs in many of these.

In many places, people are seeing permaculture and community food systems as a non-violent political statement against the current 'oil wars'. I felt this particularly strongly when I was in South Korea in early April 2003. People were afraid that the war against terror might bring war to their peninsula again. Our lifestyle is oil dependent and our food drips in oil. For every calorie of food we consume from conventional agriculture, it takes, on average, 10 calories of fossil fuels to produce it – in transportation, machinery, petrochemicals, storage, processing and packaging. Permaculture offers solutions to reduce our dependence on fossil fuels and live well. With our food and lifestyle choices we vote.

Permaculture for Peace was a central theme woven through our talks, workshops and courses around the world in 2003 – particularly in the USA where there was a lot of fear. People needed positive news and a constructive direction for action. I found in that political climate, the motivation to change patterns of living and contribute to sustainability was high. Here are some of the highlights of our international work in permaculture.

Cuba

Permaculture helped to feed people in Cuba when the economy collapsed with the dissolution of the Soviet Union. The 'Green Team' of Australian permaculturists who travelled to Cuba in the '90s to assist in the 'special period' have left a positive impact. Pocket permaculture gardens can be found tucked away throughout the capital of Havana. Also, many school grounds are filled with edible and medicinal gardens. The young students learn much from the old farmers who tend these gardens.

Although the economy is now recovering, permaculture is still important as it is helping to provide direction for the new wave of development, and the Government has initiated a Sustainable Planning Unit for Havana. We were hosted by this Unit and invited in February 2003 to present a theoretical and practical workshop about permaculture at their centre. At the end of the workshop, we had collaboratively designed a community permaculture demonstration garden on the land next to the centre and had started work to develop it. We also discussed the water and waste-water management systems, patterns of development, the need to protect urban agriculture zones, urban ecology and corridors of green.

Our understanding was that while food security is still an issue in Cuba, the crisis situation had diminished. Permaculture had moved from being primarily a food security strategy to a public health and urban re-greening strategy, and also a way to build and maintain the strong culture of urban agriculture. Urban agriculture and permaculture had become part of urban design and urban planning, thinking and practice. This, I think, is critical. Most Australians (and the global population) live in cities. We are starting to see urban designers think about local food production opportunities but just in small ways. There is a need for integrated design thinking about food in our cities and suburbs.

Turkey

On the west coast of Turkey the first community garden was developed in June 2003 using a participatory permaculture design approach. We were invited by the Izmir community and municipal office to facilitate a design charette – a participatory design event. Over 40 local residents gathered in the cultural centre to envision and design the half-hectare public land that was set aside by the Mayor. The enthusiasm was so high, that within 10 days of the beginning of the whole design process, the gardens had begun to be planted, a compost system set up, a pond dug, a community garden association formed and a waiting list of gardeners formed. It was such a successful project that the Mayor immediately announced the garden would be expanded.

We met first with the mayor and local government team, and listened to the key project drivers from the community. From this we developed an initial presentation of projects from around the world to show some possibilities and provide starting points for discussion. This was followed by 2 weekend workshops with a group of locals interested in being part of the garden. The first weekend included walking the land, assessing site and community needs, clarifying the vision, collectively creating a wish list and setting the parameters. Then, in small teams, they created designs for the site. There were so many similarities that overnight we were able to distil these and establish the design framework that the group accepted.

During the week, we took the design to the mayor for his approval, and then drew the design on the land with lime. This meant that at the start of the second weekend workshop the participants could walk the design, get a feel for it and make a few final changes. This second weekend was primarily a hands-on workshop to set up the gardens – a permablitz – but also a time to explore ways of effectively managing the gardens and setting group agreements, communication strategies and so on.

It was a very productive 10 days and we regularly receive photo updates of the gardens which show just how much a group of people can transform a derelict piece of vacant land.

Bahamas

In February 2003, we led the first permaculture design certificate course in the Bahamas. During this 3 week course at the Island School, the inaugural Sustainability Summit was held. This event drew the Prime Minister, National Ministers of Environment, Health, Agriculture, Education, Local MPs, government officers, representatives of the environmental organisations and universities. Also in attendance were leading ecological thinkers, educators and designers Amory Lovins, John and Nancy Todd and David Orr.

The course participants were involved in many projects during the 3 weeks. They prepared a comprehensive document of sustainable development strategies for the Bahamas, but more specifically for the South Cape of Eleuthera. This document was presented at the Sustainability Summit and well-received. The participants also worked with the local community to help establish a permaculture garden in the middle school, and worked on permaculture design projects for the Island School campus and grounds of the new Cape Eleuthera Institute, the country's leading sustainability research institute. Their designs were presented at the end of the course and received feedback from Amory Lovins and the Todds.

Many of the research projects conducted by the Institute will also be permaculture focussed with a participatory and community-centred approach. The Island School, established a decade ago, now has dedicated staff positions focussed on Permaculture education and site development, and each semester at least one intern works with them.

The Prime Minister of Bahamas was inspired and encouraged by the Permaculture demonstrations at the Island School on Eleuthera Island. He said that he saw great value in Permaculture for the sustainable future of the Bahamas, which currently imports 99% of its food and is too dependent on the tourism dollar.

South Korea

Permaculture is starting to take off in South Korea. Many people want to reconnect with their land and with their traditional culture and recreate a sense of community. We were told many times that South Korea developed too fast from a rural village-based culture to a high-rise, hi-tech one. While there are many benefits from these changes, many express a deep sense of loss – that in the rapid change something very important was left behind. There are many parallels between permaculture and the traditional culture of South Korea so it resonates well with those who discover it. It is not about going back to the old ways, but a desire to move forward in a direction which embraces the past while contributing to a safe, positive and healthy future.

We led Permaculture Design Certificate courses in South Korea for 3 years in a row each August and now local teachers lead regular programs. The courses attract farmers, teachers, architects, landscape architects, planners, engineers, earth scientists, writers, IT specialists, students and retirees. In the final year we also led an urban permaculture course and ecovillage design workshop at Seoul National University. The Korean Permaculture Institute has developed permaculture resources written in Korean.

Permaculture is now also being integrated into a number of schools in Korea, both primary and secondary. One university has established an Eco-Community Unit after a Professor was inspired by Permaculture. Seoul National University hosted this year's Ecological Design Summer School which was focussed entirely on Permaculture Design.

National and educational TV channels as well as national newspapers have covered these courses and generated wide interest and awareness of permaculture. Some of the first permaculture students have formed a company which consults around the country as well as in China. Their company operates like a cooperative. They receive invitations from government, universities, schools, and NGOs to assist in design of everything from ecological villages to ecological curricula. More than half of the 50 staff are PDC graduates.

One participant from the course we led there ventured to Laos where they helped to start an ecological farming school on 4500ha of land donated by the government. This land had been severely damaged during the war and was littered with land mines. Another moved to Cambodia to work with local communities to help them develop greater levels of self-reliance and sustainability.

Barcelona

Barcelona City Council has embraced Local Agenda 21 (LA21) and sees permaculture as fitting very much with their vision for a sustainable city. The Council sponsors an LA21 Eco-centre, a fabulous resource and sustainability education centre, which offers regular programs for schools, helps to facilitate the design and development of school-based environmental programs – including food gardens, composting and worm-farming. They also produce free educational literature and multi-media resources on how to live more sustainably in Barcelona.

As part of Barcelona's Year of Design events, we were invited to give a public lecture at the Barcelona City Hall to introduce permaculture design. Over a hundred people attended and many signed up for the permaculture workshops which followed. The Barcelona City Council supported a 4 day Urban Permaculture workshop in the Labyrinth, one of city's oldest public parks. Together with the more than 40 participants we created a demonstration permaculture garden which will be shown to the thousands of children and other visitors who come to these gardens on excursions. Many who attended the permaculture workshop were gardeners, including municipal gardeners. They are now exploring ways to integrate some of the basic water-saving, soil-improving, low-input and seed-saving techniques into the landscapes they manage in and around Barcelona.

Barcelona City Council is innovative in many ways. They have even planted citrus trees along the city streets which local residents and pedestrians pick and the Council harvests the rest to make preserves for Christmas gifts.

The UNESCO Master of Sustainability program at a university just outside of Barcelona is interested in incorporating permaculture into its program, both in a practical and theoretical sense. We met with the leading Professor of the program and a number of the lecturers to talk about the possibilities. We also met with the students and shared information on how permaculture can be applied in both urban and rural contexts. The postgraduate students who come from Spanish speaking countries around the world, showed great interest in Permaculture and its relevance to their home communities. A national permaculture network is being led by the group in Barcelona, and new projects are emerging throughout the country, including communities, ecovillages, eco-retreats, community gardens, ecological magazines and local currencies.

Italy

Like Spain, permaculture is gaining momentum in Italy. We were invited to give a week of short introductory talks and workshops about permaculture across the north of Italy, from the Ligurian Coast to Venice. Back then, in 2000, the first response when people heard the word was "Perma-what?" In just a few years the situation has changed dramatically, and the interest in permaculture is growing throughout the country. The Italian Permaculture Academy has been launched, based in the restored medieval village of Torri Superiore. Now, local teachers are emerging and advanced courses are being held. Courses are always fully booked and we have since been invited back a couple of years later to lead an Ecovillage (Social Permaculture) course.

Locals suggest that globalisation, EU control of agriculture and the impact of climate change is encouraging farmers and landowners to rethink. Also the food safety scares in Europe over the past few years have led consumers to think more deeply about the source and production processes of their food and want to reconnect with it. The local food movement is growing rapidly throughout Europe. Community food systems – farmers' markets, co-ops, Community Supported Agriculture (CSA), box schemes and community gardens are going from strength to strength.

Indonesia

I first travelled to Indonesia to lead a 3 week Permaculture Design Course for women. The participants travelled from each end of the country, from Aceh to West Papua. I have since returned many times to continue supporting the work of the Indonesian Development of Education for Permaculture (IDEP). A primary project was the development of a cross-disciplinary curriculum, inspired and guided by permaculture. This is being developed for Indonesian Primary Schools and is still in process, as it was put on hold while IDEP focussed its energy on supporting the victims of the Bali bombing and the tsunami. IDEP is now experienced in supporting communities in crisis and does so from a permaculture foundation.

Prior to the tsunami and the bombing, I was involved in a sustainable community development project in the west of Bali which was working to link villages around the base of Bali's only National Park. The project aimed to create a bioregional strategy for nature conservation, ecological restoration, low-impact eco-tourism, sustainable livelihood generation, food security, micro-enterprise development, water conservation, catchment management and environmental education. The program is led by the locals to meet their needs and stands to be a valuable model for other communities.

The 10 core lessons I have learnt through engaging in permaculture work internationally are:

1. Fit projects to local context, climate, culture, environment, needs rather than impose them.
2. Reduce the ecological footprint – water, waste, energy, resources and maintain healthy soils.
3. Strengthen and build local resources/abundances; recirculate, plug the leaks.
4. Become involved in food security and support sustainable agriculture and community food systems.
5. Community is security.
6. Join hands with others. There's more you can do together than as individuals. Community led projects lead to greater commitment, ownership and meaning.
7. Engage in counter-development work and become counter-experts. Rethink notions of development and empower people to solve their own problems locally to be more self determining and self-directed, without dependence on so called 'experts'.
8. Our efforts contribute to community and environment, and a sustainable future; therefore spend energy wisely to make a difference – work where it counts.

9   Work with community and with nature in a cooperative and equitable way, be humble, lose the ego (earth care, people care, fair share – the permaculture principles).

10 Spread hope, inspiration and good news. Network, put local action into a global context, create positive ripples.

Our film Think Global: Eat Local

In 2008, we made a short film about local food systems, bringing together 15 years of footage from 15 countries in 15 minutes. *Think Global: Eat Local – A Diet for a Sustainable Society* is a celebration of local food systems in communities around the world – farmers markets, food box systems, food co-ops, community farms, community gardens and school gardens. The film touches on many of the issues caused by and impacting our current unsustainable food system and points to the relocalisation of food systems as a key strategy for working toward sustainability, social justice and wellbeing.

We pulled together footage of local food projects that we have filmed and photographed both locally and internationally since 1992. The locations include Cuba, Ladakh, Indonesia, Turkey, South Korea, Spain, The Bahamas, USA, Scotland, Denmark, Slovenia, Bulgaria, Germany, Hong Kong, and in Australia, Northey Street City Farm, Maleny and Crystal Waters. The film features interviews with Fritjof Capra, and well-known Brisbane-based food activist and academic, Kristen Lyons, Northey Street City Farm's Organic Market coordinator Anaheke Metua, CSA farmer Les Nichols, a local food chef, a naturopath, and input from Evan and myself. The film was produced by our organisation, SEED International, with the help of a small grant from the Maleny Film Commission through Festivals Australia. It has been shown around Australia as well as being screened internationally.

Reflections

All this work has given me an incredibly hopeful view of the world and humanity, because our life has revolved around such positive people and projects. In every corner of the world we meet people who are working to create positive change in their local area. The collective impact of this is enormous, and I keep this perspective when I am working on my own small local projects. These local projects do make a difference to people's lives where they live, and it is the cumulative effect of all this positive action that can tip the balance.

In our experience with permaculture internationally, especially over the past three years, we have noticed there is a trend away from predominantly individual interest (landowners) to community interest (sustainable community development projects – social, economic, environmental and agricultural). In urban areas, permaculture activity is being directed more towards the development of community gardens and community food systems. In rural areas it is toward revitalising rural villages and towns, and the formation of eco-communities. Permaculture continues to spread around the world and has great value in many different climates, cultures and economic contexts. These are very positive steps, which indicate to me a maturation of permaculture itself. It is also becoming increasingly noticed and important here in Australia. The growing awareness of and concern about peak oil and climate change combined with the economic crisis is highlighting the value and relevance of permaculture in our communities. I think the community approach is important, which is why I have embraced the Transition Town initiative and have helped to launch Transition Town Maleny, and a similar eco-action group at Crystal Waters Permaculture Village where I live.

Morag's husband Evan and daughter Maia at Northey Street City Farm in Brisbane.

Through permaculture I have been able to create a sustainable and meaningful livelihood for myself and my family, and been able to share this approach locally and internationally.

In essence, my learnings have been:

Live simply. Follow your passion. Live with integrity. Do it now. Lead by experience. Work where it counts. Regularly question your actions and direction – reflect, rethink.

- Avoid fundamentalism, dogmatism, and 'isms' in general.

- Do not go into communities and impose ideas – wait for the invitation. Facilitate the revaluing and revealing of local knowledge and experience. Humility rules. Gently, gently.

Stuart B. Hill

Professor Stuart B. Hill is Foundation Chair of Social Ecology at the University of Western Sydney, where he taught qualitative research methodology, social ecology research, transformative learning, and sustainability, leadership and change. Stuart retired in 2009 and is now an Adjunct Professor.

Stuart's PhD was one of the first whole ecosystem studies that examined community and energy relationships (1969); and he received awards for Best PhD Thesis and Best PhD Student. In 1977 he received a Queen's Silver Jubilee Medal for his community and social transformation work.

Prior to 1996 Stuart was at McGill University, in Montreal, where way back in 1974 he established Ecological Agriculture Projects, Canada's leading resource centre for sustainable agriculture.

While in Canada Stuart was a member of over 30 regional, national and international boards and committees. Stuart is Founding Co-editor of the Journal of Organic Systems and is currently on the editorial board of five international refereed journals. Until 2004 he represented professional environmental educators on the NSW Council on Environmental Education. Stuart has co-authored four books and has published over 350 papers and reports.

Stuart has worked in agricultural and development projects in over a dozen countries. His work in the Seychelles to make a whole coralline island self sufficient in food and energy is particularly significant. Stuart's background in chemical engineering, ecology, soil biology, entomology, agriculture, psychotherapy, education, policy development and international development, and his experience of working with transformative change, has enabled him to be an effective facilitator in complex situations that demand collaboration to achieve situation improvement. These skills were used extensively in his recent role as Provocateur for the Victorian Government (2004-2005).

Afterword

Four Key Features of Permaculture (applicable to 'everything'); and an Opportunity for the Future (also applicable to 'everything')

If you have ever said to yourself, as you watched what was going on around you, or reflected on your own situation and actions, "This is not right," or, "There must be a better way," I suspect that as you read this important book you will have been filled with hope for a better future; and may even have found some answers specific to your particular concerns.

Most of us, most of the time, follow these 'thoughts of discontent' by just continuing to 'plod along'. Perhaps this is because the change needed seems to be too enormous, or we feel that whatever we might do wouldn't be enough to make a difference, or perhaps deep down we feel too afraid to change.

On some occasions, however, most of us have chosen to do something to address the 'wrongs' and to 'find a better way'. I think of this as 'choosing to act on deep caring and love rather than on fear' – a choice that we actually have every day, moment-to-moment.

Throughout history we can recognise significant moments when individuals and groups have done this in ways that have the potential to benefit everyone. In Kerry Dawborn's reflective chapter – 'The New Frontier: Embracing the Inner Landscape'– she described how in 1955 the acts of Rosa Parks, and in the 1980s those of Muhammad Yunus, led to such transformative, significant and meaningful change. You probably have your own special examples of such inspiration and hope for better futures; not just involving famous people, but perhaps someone in your own family, as well as some of your own past courageous acts.

It was such hopeful and defiant acts in the 1970s that gave birth to permaculture; brought to life by two unlikely collaborating 'midwives': the in-your-face experienced university lecturer, field ecologist and activist, Bill Mollison, and the more introverted landscape design student, David Holmgren (Mollison & Holmgren 1978).

Since then permaculture has become a global movement for the improved design and management of food and energy systems (and much more). Caroline Smith has provided some of the background to the movement in her helpful 'Introduction'.

I became familiar with permaculture in the 1970s while I was a Professor of Zoology at McGill University in Montreal; when I was also associated with 'The New Alchemy Institute' in Cape Cod, Massachusetts (Quinney 1981, 1984). I, like John Jeavons (1974), had already written about a need for a 'permanent' and more sustainable agriculture (Hill, 1976, Hill & Ramsay 1977). I attended one of Bill's first workshops in North America at Samuel Kayman's Stonyfield Farm in New Hampshire; and both of us subsequently spoke at several events in Canada. It was not until I immigrated to Australia in the mid-1990s that I had the pleasure of meeting David; I was able to record part of his story in our book on the history of ecological ideas in Australia (Mulligan & Hill 2001), and was privileged to be asked by him to write the Foreword for his excellent book on the key design principles of permaculture (Holmgren 2002). I have also attended numerous permaculture gatherings and workshops; and have been fortunate to have been able to learn from the authors in this collection of inspiring stories.

Because of this exposure, I now feel as if there is a bit of permaculture in every cell in my body. So reading these stories has for me been particularly rewarding.

To help my students to focus I often ask them, "What are the three most important things for you about... (whatever it is that we are discussing)?" I have listed below the three things that make permaculture really important to me. In some ways they summarise the foundational elements of permaculture that I recognised in each of the stories in this book. I encourage you to reflect on their possible relevance to you, particularly in the context of 'finding a better way' to think about and act on this perspective, and to do this in relation to your particular circumstances.

1. Firstly, as you will have found clear in these chapters, permaculture has its roots in a passion for nature, social justice and wellbeing for all. It is **grounded in cutting-edge ecological science and effective collective and personal action,** all underpinned by **a clear set of ethical 'testing questions'**; to help us to relate our actions to our shared values (see also my expanded list of such questions in Fig. 3 below). These aspects of permaculture are discussed in more detail in the many inspiring stories in this book, and in the writings of Mollison and Holmgren.

2   Secondly, permaculture is an evolving framework for the development of sound theories and practices in relation to the **design and management** of not only our food and energy systems, but also our individual and collective lives, including all our institutional structures and processes. As such, it can help us address, and more importantly avoid, the many crises that currently face our species (from all aspects of sustainability, to the specifics of climate change, water management, pest prevention, biodiversity conservation, and nourishment for all). This focus on designing systems that can work sustainably is in contrast to our society's overemphasis on 'deceptively simple' endless 'problem-solving' interventions within mal-designed and mismanaged unsustainable systems.

3   Thirdly, permaculture is **a collective endeavour**; a movement that all of us can become a part of. As such, it offers training programs, supported by a rich literature of books, magazines and articles, inspiring websites with downloadable talks, YouTube videos, and other materials, outreach projects in less developed parts of the world, local working bees to establish permaculture gardens within our neighbourhoods, and a full range of political and social activism initiatives. Again, as you read this amazing collection of stories, I'm sure you would have been inspired by the ways in which the authors' lives have been transformed and given greater meaning and purpose through their association with permaculture.

Usually, after my students have done the above exercise, I ask them the following additional question: "If you could add one more important thing about whatever we are discussing (in this case 'permaculture'), what would it be?" This is because I have found that this fourth feature is usually the most important one; it just takes offering this 'extra opportunity' for it to surface.

4   So, for me, the fourth key feature of permaculture is its **ongoing co-evolutionary change (change that benefits all involved) and improvement**; as it continues to journey forwards, along an 'upward spiral path'. This involves a progressive process of experimenting with small, meaningful, cutting-edge initiatives, and then learning from these by paying attention to all of the outcomes; this is in contrast to the more common linear approach to change, with limited attention being paid to unintended outcomes.

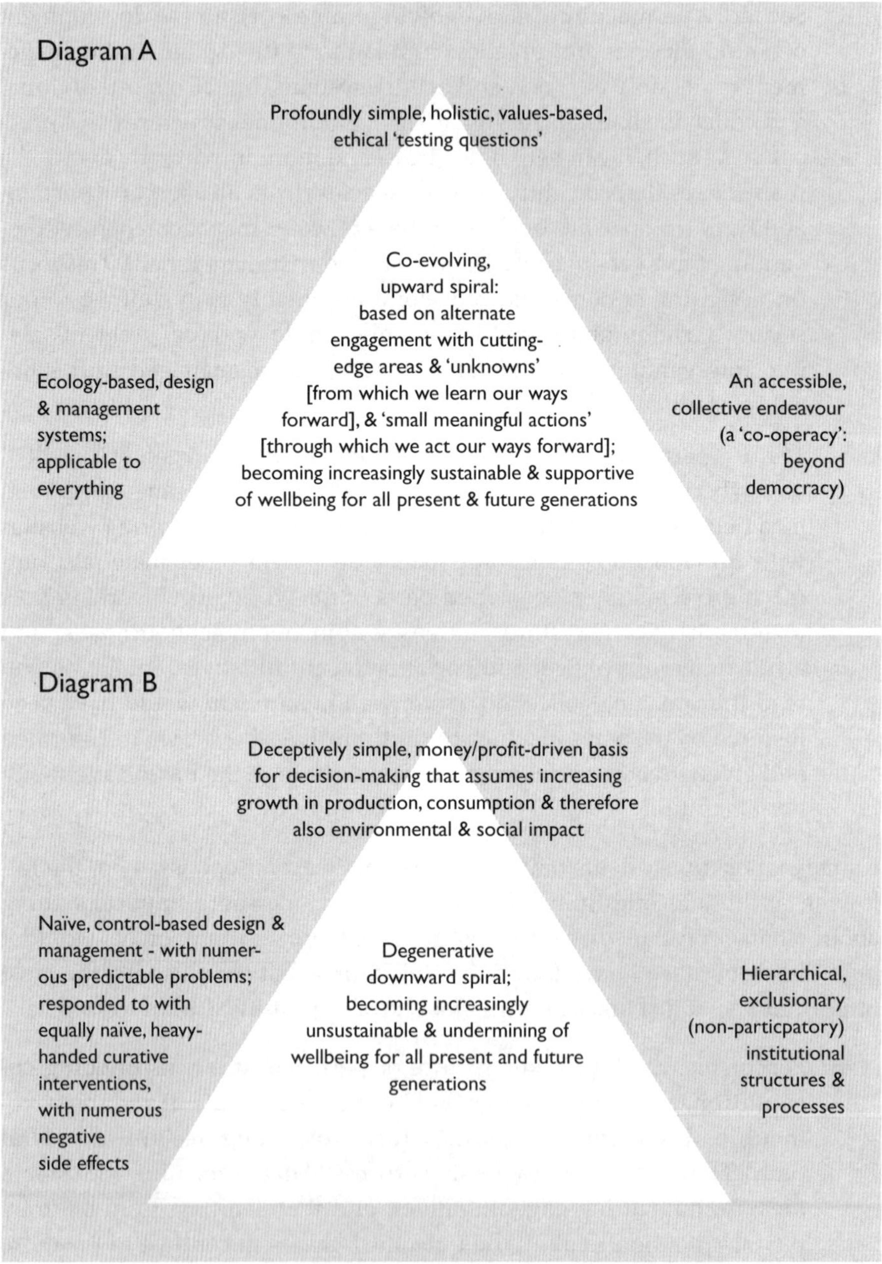

Diagram A

Profoundly simple, holistic, values-based, ethical 'testing questions'

Ecology-based, design & management systems; applicable to everything

Co-evolving, upward spiral: based on alternate engagement with cutting-edge areas & 'unknowns' [from which we learn our ways forward], & 'small meaningful actions' [through which we act our ways forward]; becoming increasingly sustainable & supportive of wellbeing for all present & future generations

An accessible, collective endeavour (a 'co-operacy': beyond democracy)

Diagram B

Deceptively simple, money/profit-driven basis for decision-making that assumes increasing growth in production, consumption & therefore also environmental & social impact

Naïve, control-based design & management - with numerous predictable problems; responded to with equally naïve, heavy-handed curative interventions, with numerous negative side effects

Degenerative downward spiral; becoming increasingly unsustainable & undermining of wellbeing for all present and future generations

Hierarchical, exclusionary (non-particpatory) institutional structures & processes

Fig. 1. Key features of permaculture (diagram A) contrasted with those of most systems in current industrialised societies (diagram B).

I believe that if the above four key characteristics of permaculture were also common to our systems of government, business, education, health (etc), all of these areas would benefit enormously and be in a much better position to contribute to the wellbeing for all.

The challenge in every area facing all cultures is how to best enable change from systems that don't work to ones that can work better. In Figure 1 I have contrasted these four qualities of permaculture with the dominant characteristics of most current industrialised societies. What I am advocating are actions to enable progressive change from the situation characterised in diagram B to that in diagram A.

The areas where I believe permaculture – and most other hopeful initiatives – need to develop further relates to our 'psychological healing and development'. In the case of permaculture, I refer to this as working on '**permaculture of the inner landscape**'. I believe that the quality, clarity, relevance and effectiveness of our 'outer landscape work' is limited or enabled by the quality of our 'inner landscape', and that this is the area where the greatest development is both needed and will contribute towards genuine progress.

I first started thinking about this in the mid-1980s, when I ran a workshop on this subject at a permaculture conference in the USA. Since then I have been monitoring developments in this area. As part of my research for writing this 'Afterword' I searched the web and was encouraged to find that there are now a dozen or so permaculturists who have subsequently developed workshops on 'inner permaculture' or related topics (I encourage you to search the web using this term to find the latest developments in this exciting area of development).

Since the 1970s I have maintained my interest in this inner-outer connection, and have continued to develop it through my research, teaching, writing and action. See, for example, my 'Foreword' to David Holmgren's 2002 book; my analysis of some of the psychological aspects of why PA Yeomans' (1958) Keyline system has not been more widely adopted, despite its many valuable features, including its extraordinary capacity to capture carbon in the soil and ameliorate climate change (Hill 2006); and my framework for how we might better develop our relationships with nature and place (Hill 2003). Continuing to improve our competence in this latter area is one that I consider to be foundational for the ongoing development of permaculture.

Many of the stories in this historically important collection include much about 'inner permaculture' and the 'emotional' aspects of being a permaculturist, such as the many stories of how discovering permaculture enabled the authors to renew their hope in the future, and find greater meaning in the present.

PAST EXPERIENCE	HUMAN CONDITION / BEHAVIOUR	VALUES	FOOD SYSTEMS
High physcal & emotional stress & Unhealed hurts	Internalised distresses Maladaptive patterned behaviours (irrational attempts to get rational past needs met in present: needy) Unaware Disconnected, isolated, lonely Unbalanced, erratic Confused Indecisive or deciding irresponsibly Powerless, attracted to pseudopower Defensive, secretive, distrustful Disempowered, trapped, fearful Dependent, competitive Depressed or hyperactive Bargaining with death Oppressive, oppressed (hierarchical) Preoccupied with control issues Uncommitted, postponing Inflexible	Confusion, inconsistency re values Compensatory Me for me, selfish, individual focus All is fair in competition End justifies means Wants = needs More is better Short-term focus Live now, pay later, never Innocent until caught Ownership gives freedom, exclusivity Simplify, control Symbols of power = power It's their fault, blaming, irresponsible Enemy oriented Faith in curative solutions Importance determined by surface feelings Can't change till they change Natural limits are to be transcended	Malfunctioning Resource depleting Unjust systems Milk the system Over emphasis on economic efficiency (short-term) Labour regarded as a cost Manipulated demand dictates production Ignore, undervalue externalities, long-term Over specialisation Uniformity (monocultures) Large capital intensive operations High power machinery Magical bullet solutions (chemicals, etc.) Nuclear scenario Genetic engineering, space colonies Overkill of pests (biocides, antibiotics) Unsustainable
Low physcal & emotional stress & Healed hurts	Fully alive in the present Unique responses to unique situations Spontaneous Aware Integrated, relational Balanced Thinking clearly Deciding wisely Loving, caring Empowered Open, honest Autonomous & mutualistic Free, creative Embracing life, joyful Collaborative, supportive Committed Flexible	Clarity, consistency re values Universal, global, species concerns Committed to self, group, species, life, planet Considerate of needs of others, other species Concerned with long-term Supportive, giving, sharing catalytic role Distinguish needs from wants Sense of enough Pay as I go Self policing, high ethical standards, equitable access Comfortable with chaos, complexity Everyone is potentially powerful Responsible, just Treat causes, prevent problems, promote health Importance determined by rational thinking, gut feelings Providing leadership Actions limited by natural laws, values	Meeting real needs (nourishment, health, fulfilment, justice) Sustainable, conserving Environmentally supportive, regenerative, renewable, solar resource base Flexible, evolving, long-term plan Labour, knowledge, skills and natural resources as capital Regional self-reliance Participatory High autonomy, self-maintenance, self-regulation, functional diversity Complex, eg., multistorey polycultures Small (most) to larger scale Lower power appropriate technologies Preventative solutions (focus on causes, early indicators, bio-ecological & social approaches) Building on natural processes and cycles Energy efficient, resource conserving, skill intensive

Fig. 2. Possible influences of past experiences on food system characteristics (from Hill, 1991).

In the mid-1980s, to help understand the connection between 'inner and outer', I developed a framework for relating the state of the world – particularly the ways in which we produce food – to our values, our ways of being and doing, and to their psychological and cultural roots (Hill, 1991; see Figure 2). Such lists of characteristics can be used by each of us to (honestly and courageously) reflect on which ones might apply to oneself and to one's group, and to then address any issues that become evident.

In addition to paying more attention to the 'inner landscape', and with respect to enabling ongoing progress, I find that keeping the notion of paradox in mind can be very helpful. For example, an exercise that I use to challenge my students who are studying ways to become 'deep leaders' is to ask them to reflect on ways to bring about meaningful change anonymously (to contradict the 'deceptively simple' assumption that leaders must be visible heroes; other key features of 'deep leadership' are listed in my article on 'deep environmental leadership', Hill 2009). My students usually generate a list that includes:

- working in a large team that collaborates across difference and other boundaries,
- being ready to work over long-time frames (as against expecting immediate outcomes),
- using indirect, subtle, low-power, and often bio-ecological design and management interventions (in contrast to the currently dominant emphasis on heroic, instant, direct, heavy-handed, chemical and physical interventions).

I believe that such 'paradoxical reflection' within permaculture (and all other endeavours) might open up a diverse range of additional opportunities for further development.

Also, to enable such developments, it can help to reflect on what the tenets and practices of permaculture are 'in the service of', and then critically test them against such 'integrator indicators' as spontaneity (the 'healthy' opposite to 'distressed patterned behaviours'), wellbeing, equity, meaning, sustainability, joy, and the ongoing progressive co-evolution of all systems. Such 'testing questions' are conceptually similar to those used for landscapes by 'Holistic Resource Managers' (Savory & Butterfield 1999); and they relate to my extended version (Hill 2005) of David Holmgren's (2002) very valuable '12 permaculture Design Principles'. A more extensive set of 'testing questions', using a social ecology framework, is provided in Figure 3. I encourage you to test your ideas and initiatives against such a list. An additional list that

Personal – Does it (policies, programs, plans, regulations, decisions, initiatives, etc.) support:

1. **Building & maintaining personal capital – personal sustainability:** empowerment, awareness, creative visioning, values and worldview clarification, acquisition of essential literacies and competencies, responsibility, wellbeing and health maintenance, vitality and *spontaneity*?
2. **Home & ecosystem maintenance:** caring, loving, responsible, mutualistic, *negentropic* (capital building) *relationships* with diverse others (valuing equity & social justice), other species, place and planet? ['negentropic' is the opposite of 'entropic': breaking down]
3. **Lifelong learning:** positive total life-cycle *personal development* and 'progressive' change?

Socio-Political – Does it support:

4. **Building & maintaining social capital – cultural** [including economic] **sustainability:** trust, accessible, collaborative, responsible, creative, celebrational, *life-promoting community and political structures and processes*?
5. **Inter-cultural and interpersonal capital:** the valuing of 'functional' high *cultural diversity* and mutually *beneficial* relationships?
6. **Co-evolutionary change:** positive *cultural development* and evolutionary change that benefits all involved?

Environmental – Does it support:

7. **Building & maintaining natural capital:** effective *ecosystem functioning* and *ecological sustainability*?
8. 'Functional' high **biodiversity**, and prioritised use and conservation of resources?
9. Positive **ecosystem development** and co-evolutionary change?

General – Does it support:

10. **Proactive** (vs. reactive), **design/redesign** (vs. just efficiency & substitution) and **small meaningful collaborative initiatives** that one can guarantee to carry through to completion (vs. heroic, Olympic-scale, exclusive, high risk ones). Also **public celebration** at each stage – to facilitate their spread – thereby making wellbeing and environmental caring 'contagious'?
11. A focus on key opportunities and **windows for change** (contextually unique change 'moments' & places)?
12. **Effective evaluation and monitoring:** (broad, long-term, as well as specific & short-term) by identifying and using **integrator indicators** and **testing questions**, and by being attentive to all **feedback and outcomes** (& redesigning future actions & initiatives accordingly)?

Fig. 3. Testing questions for 'challenging' all understandings, ideas and initiatives (from Hill 2006).

can be used in the same way is provided as a power-point presentation on my website, under the heading of "wisdom."

Greater progress could be achieved in permaculture, as in all other individual and group 'improvement' endeavours, if more attention were paid to enabling the development of a healthier 'inner permaculture'. By this I mean one that is embodied, holographic (in the sense that 'anything you detect anywhere is likely to be found everywhere'), in-the-present, and characterised by spontaneous ways of being and doing. This would complement and better enable the contextually-relevant development of external design competencies and initiatives. It is a matter of 'going in (psychologically) to go out (environmentally)', and 'going out (environmentally) to go in (psychologically)' – a 'both' process rather than an 'either/or' one!

Greater engagement in 'inner permaculture' could, I believe, enable this important movement to progress from its current state (which I call 'permaculture I') to the next stage in its development ('permaculture II'), while daring to dream what permaculture might involve over the long term ('permaculture III' and beyond)! What role do you want to have in this ongoing cultural evolution, both in permaculture and in society in general? What will it take to enable such change to happen? And what might the benefits be for present and future generations?

Stuart Hill and David Holmgren at Melliodora, autumn 1999. Photo by Joy Finch.

Glossary

ANZAC – Australian and New Zealand Army Corps.
APACE – Appropriate Technology for Community and Environment. An organisation based at the University of Technology, Sydney (UTS), which works with indigenous communities to implement appropriate technologies that empower people in controlling their own futures.
APT – Accredited Permaculture Training™.
Biotecture – The use of living plants as an integral part of the design of buildings.
CBD – Central Business District.
CERES – Centre for Education and Research in Environmental Strategies in Melbourne, Australia.
Chook – A domestic fowl.
Energy descent – The post-peak oil transitional phase, when humankind goes from the ascending use of energy that has occurred since the industrial revolution to a descending use of energy.
Greenies – Slang term for environmentalists.
Hard yakka – Hard work.
Jackaroo – Australian term for a male farmhand. A female farmhand is a Jillaroo.
Kombi – Idiomatic term referring to a Volkswagon Type 2 van – also known as a microbus, which was configured as a camper or cargo van or a passenger van. Due to its popularity in the environmental and counter-culture movement of the 1960s-70s, it is also known as a 'hippie van'.
Landcare – Australian government-supported community initiative started in the 1970s, to help protect and promote natural ecosystems, flora and fauna on private and public land.
LETS – Local Employment and Trading System or Local Energy Trading System.
Muster – (verb) To gather together. (noun) An assembly of people or animals (that have been herded together).
NGO – Non-Governmental Organisation.
NSW – New South Wales (Australian State).
PAWA – Permaculture Association of Western Australia.
PDC – Permaculture Design Certificate.

Permablitz – An informal gathering of people over one or more days, with the purpose of creating a food garden according to permaculture principles – a permaculture make-over. The term comes from a combination of Permaculture and the idea of 'blitzing' or doing something quickly. Inspired by the Australian television gardening program Backyard Blitz.

Permaculture Convergence – A permaculture event which usually involves a conference, workshops, site tours and more. International Permaculture Convergences (IPCs) are held every 3 years in different locations worldwide. National and regional convergences are also held.

Permaculturist – Person who practices permaculture.

Permie – Slang term for a person who practices permaculture.

PIJ – *Permaculture International Journal*. Known early on as the *International Permaculture Journal*, *PIJ* grew out of the early Australian permaculture publication Permaculture, during the 1990s. *PIJ* was the first Australian permaculture magazine to make it to mainstream news stands, followed by *Green Connections*, which included a focus on Permaculture. *PIJ* ceased publication in 2000.

PIL – Permaculture International Limited.

RTO – Registered Training Organisation.

Surfie – Surfer.

TAFE – Technical and Further Education – Australian tertiary education framework with a focus on technical education and trades.

Transition Towns – Towns or regions that have undertaken to engage in proactive measures to assist their communities in dealing with rising energy prices, increasing energy scarcity through peak oil, and climate change. The Transition Town movement began in the United Kingdom in the early-mid 2000s and has since spread world-wide.

Ute – Pickup truck. Small truck.

VET – Vocational Education and Training.

WWOOF, WWOOFers – In Australia, WWOOF stands for Willing Workers on Organic Farms. This term is used as a verb, to WWOOF and a WWOOFer is a worker. Also stands for World-wide Opportunities on Organic Farms, elsewhere in the world. Movement started in England in the early 1970s, connecting people who want to volunteer on organic farms and projects, with farmers and others who are looking for help.

References Cited and Further Reading

The references here include those cited in the authors' chapters as well as a limited selection of further reading. Not all references cited appear as some are no longer available. Other references are partial only as full details are no longer available. Interested readers are encouraged to conduct their own search for additional writings by the authors.

ACRES USA. See http://www.acresusa.com.
Ad Hoc Panel of the Advisory Committee on Technology Innovation (1989) *Lost Crops of the Incas: Little-Known Plants of the Andes with Promise for Worldwide Cultivation*. Office of International Affairs (OIA).
Alexander, C., Ishikawa, S. & Silverstein, M. (1977) *A Pattern Language*. Oxford: Oxford University Press.
Alexandra, J., Higgins, J. & White, T. (1998) *Environmental Indicators for National State of the Environment Reporting: Local and Community Uses*. Canberra: Environment Australia.
Alexandra, J., Haffenden, S. & White, T. (1996) *Listening to the Land: a Directory of Community Environmental Monitoring Groups in Australia*. Maryborough: Australian Conservation Foundation.
Ball, C. (1985) *Sustainable Urban Renewal: Urban Permaculture in Bowden, Brompton and Ridleyton*. Social Impacts Publications in association with the Permaculture Association of South Australia, Armidale, NSW.
Bateson, G. (1972) *Steps to an Ecology of Mind: Collected Essays in Anthropology, Psychiatry, Evolution and Epistemology*. University of Chicago Press.
Bell, G. (1992) *The Permaculture Way*. London: Thorsons.
Bentley, T. (1998) *Learning Beyond the Classroom: Education for a Changing World*. London: Routledge.
Benyus, J. (1997) *Biomimicry: Innovation Inspired by Nature*. NY: William Morrow.
Berman, M. (1981) *Re-enchantment of the World*. Ithaca: Cornell University Press.
Berry, T. (1990) *The Dream of the Earth*. San Francisco: Sierra Club.
Brooks, J. (1994) *Growing Together*. Film.
Caldicott, H. (1982) *If You Love this Planet*. See http://www.helencaldicott.com/films.htm.
Capra, F. (1983) *The Turning Point: Science, Society and the Rising Culture*. NY: Bantam Books.

Capra, F. (1991) *The Tao of Physics: An Exploration of the Parallels Between Modern Physics and Eastern Mysticism.* Boston: Shambhala Publications.
Capra, F. (1997) *The Web of Life: A New Scientific Understanding of Living Systems.* NY: Anchor.
Capra, F. (2004) *The Hidden Connections.* NY: Anchor.
Clayfield, R. (1996) *You Can Have Your Permaculture and Eat It Too.* Maleny, Qld: Earthcare Education.
Clayfield, R. & Skye (1991) *The Manual for Teaching Permaculture Creatively.* Maleny, Qld: Earthcare Education.
Co-evolution Quarterly. See *http://www.wholeearth.com/index.php*.
Dale, A. (2001) *At the Edge: Sustainable Development in the 21st Century,* Vancouver, B.C.: UBC Press.
Davis, A. (1951, 1981) *Let's Have Healthy Children.* UK: Sygnet Books.
Day, C. (2004) *Places of the Soul: Architecture and Environmental Design as a Healing Art.* Oxford: Architectural Press.
Deans, E. (2001) *No Dig Gardening and Leaves of Life.* Pymble, NSW: HarperCollins Publishers.
Douglas, J. Sholto & Hart, R. A. de J. (1976) *Forest Farming.* London: Watkins.
Earth Garden **magazine.** *http://www.earthgarden.com.au.*
EarthSong journal. *http://www.earthsong.org.au.*
Ecological Agriculture Projects. See *http://eap.mcgill.ca.*
El-Hage Scialabba, N. (2007) *Organic Agriculture and Food Security.* Available at *ftp://ftp.fao.org/paia/organicag/ofs/OFS-2007-5.pdf* .
Fanton, J. & Fanton, M. (1993) *The Seed Savers' Handbook.* Byron Bay: The Seed Savers' Network.
Finnane, J. (2005) *Lawns into Lunch: Growing Food in the City.* Frenchs Forest, NSW: New Holland.
Finnane, J. *Pacific Calling Partnership.* See *http://www.erc.org.au/index.php?module=pagemaster&PAGE_user_op=view_page&PAGE_id=63&MMN_position=67:67.*
Fitter, R., Fitter, A. & Blamey, M. (1983) *Collin's Field Guide to the Wild Flowers of Northern Europe.* London: Harper Collins.
Flannery, T. (2005) *The Weather Makers.* Melbourne: Text Publishing.
Francis, R. Eco-Social Matrix (ESM). See *http://permaculture.com.au.*
Fukuoka, M. (1978) *One Straw Revolution.* Pennsylvania: Rodale Press.

Gamble, M. (2005) *Characteristics of Quality Permaculture Education.* Unpublished Masters Thesis, Faculty of Environmental Sciences, Griffiths University.

Gamble, M. & Raymond, E. (2008) *Think Global: Eat Local – A Diet for a Sustainable Society.* SEED International. See *http://web.mac.com/localfood/ Morag_and_Evan/Welcome.html*

Gardening Australia. See *http://www.abc.net.au/gardening.*

Grass Roots **magazine**, edited by Meg and David Miller. Shepparton, Victoria: Night Owl Publishers Pty. Ltd.

Green Connections **magazine**, edited by Joy Finch. Castlemaine, Victoria: Green Connections.

Hamaker, J. D. & Weaver, D. A. (1982) *The Survival of Civilization.* Burlingame, CA: Hamaker-Weaver Publishers.

Harney, P.J. (1997) *Changing the Social System of a Catholic Secondary School: An Examination of Salient Design Features Pertinent to the Change Process from a Permacultural Perspective.* Unpublished PhD Thesis, Department of Educational Foundations, Australian Catholic University.

Hawken, P. (2007) *Blessed Unrest: How the Largest Movement in the World Came into Being and Why No One Saw It Coming.* NY: Viking Press.

Hill, S.B. Stuart Hill's presentations and accompanying notes can be found at *http://www.stuartbhill.com.*

Hill, S.B. (1976) 'Conditions for a Permanent Agriculture.' In *Maine Organic Farming and Gardening* 3(5) pp 8-9. [Reprinted in *Soil Association Quarterly Review* 3(2) pp. 1-4, 1977].

Hill, S.B. (1991) *Ecological and Psychological Pre-requisites for the Establishment of Sustainable Prairie Agricultural Communities.* In Jerome Martin (ed.) *Alternative Futures for Prairie Agricultural Communities.* Faculty of Extension, University of Alberta Edmonton, AB, pp. 197-229.

Hill, S.B. (2003) *Autonomy, Mutualistic Relationships, Sense of Place, and Conscious Caring: A Hopeful View of the Present and Future.* In John I. Cameron (ed.) *Changing places: Re-imagining Australia.* pp. 180-196. Sydney: Longueville.

Hill, S.B. (2005) *Sustainable Living Through Permaculture: a Social Ecology Perspective.* Keynote address to the 8th Australasian Permaculture Convergence, Melbourne, 8-15 April. Available from the author.

Hill, S.B. (2006) *Enabling Redesign for Deep Industrial Ecology and Personal Values Transformation: A Social Ecology Perspective.* In Ken Green & Sally Randles (eds.) *Industrial Ecology and Spaces of Innovation.* Ch. 12, pp. 255-271. London: Edward Elgar.

Hill, S.B. (2009) 'Deep Environmental Leadership.' In *Eingana* 32(2) pp. 6-11.

Hill, S.B & Ramsay, J.A. (1977) *Energy and the Canadian Food System with Particular Reference to New Brunswick.* Prepared for the New Brunswick Government Agricultural Resources Study Group, NB. Ecological Agriculture Projects, Macdonald College of McGill University, QC. 181 pp.

Hill, S.B. & Mulligan, M. (2001) *Ecological Pioneers: A Social History of Australian Ecological Thought and Action.* Cambridge UP.

Holmgren, D. David Holmgren's writings and presentations can be found at *http://www.holmgren.com.au*.

Holmgren, D. (2009) *Future Scenarios.* Canada: Chelsea Green Publishing. See also *http://www.futurescenarios.org*.

Holmgren, D. (2002) *Permaculture: Principles and Pathways Beyond Sustainability.* Hepburn, Victoria: Holmgren Design Services.

Holmgren, D. (1985) *Permaculture in the Bush.* Hepburn, Victoria: Holmgren Design Services.

Holmgren, D. (1994) *Trees on the Treeless Plains.* Hepburn, Victoria: Holmgren Design Services.

Holmgren, D. & Mollison B. (1976) 'Permaculture.' In *Organic Gardener and Farmer: Journal of the Organic Gardening and Farming Society of Tasmania.* Launceston, Tasmania. Simultaneously published in *The Feral Gazette*, Vol, 5., No.2, 22 June 1976, student newspaper of the Tasmanian College of Advanced Education.

Hopkins, R. (2009) *The Transition Handbook: Creating local sustainable communities beyond oil dependency (Australian and New Zealand edition).* Lane Cove, NSW: Finch Publishing Pty Ltd.

Illich, I. (1971) *Deschooling Society.* San Francisco: Harper & Row.

Jeavons, J. (1974) *How to Grow More Vegetables Than You Ever Thought Possible on Less Land Than You Can Imagine.* Berkeley, California, USA: Ten Speed Press.

Kennedy, M. (1995) *Interest and Inflation Free Money.* Seva International.

Kennedy, D. & Kennedy, M. (1997) *Designing Ecological Settlements.* Berlin: The European Academy of the Urban Environment.

Lillington, I. (2007) *The Holistic Life: Sustainability Through Permaculture.* Adelaide: Axiom.

Lovelock, J. (2006) *The Revenge of Gaia.* London: Allen Lane.

Mars, R. (Supervising ed.) (2002) *Permaculture Volume 2: The Best of PAWA. Selected Articles from the Newsletters of the Permaculture Association of Western Australia.* Mundaring, WA: Candlelight Trust.

Mars, R. (2001) *Using the Submergent Triglochin huegelii for Domestic Greywater Treatment.* PhD thesis, Murdoch University, WA.

Mars, R. (1996) *The Basics of Permaculture Design*. Perth: Candlelight Trust. [Reprinted by Permanent Publications, UK. 2004].

Mars, R. with Mars, J. (1996) *Getting Started in Permaculture*. Revised from 1994 edition. [Reprinted by Permanent Publications, UK, 2004].

Mars, R. with Willis, R. (eds.) (1996) *The Best of PAWA – Selected Articles From the Newsletters of the Permaculture Association of Western Australia*. Mundaring, WA: Candlelight Trust.

Mars, R. *Passive Solar Design of Buildings*. Video. Perth, WA: Candlelight Trust.

Mars, R. *RAPS – Remote Area Power Supply*. Video. Perth, WA: Candlelight Trust.

McDonald, C.N. & Finnane, J. (1996) *When You Grow Up*. Broome, WA: Magabala Books.

McHarg, I. (1969) *Design with Nature*. NY: Natural History Press.

Miller, R. *A Brief Introduction to Holistic Education*. Available at http://www.infed.org/biblio/holisticeducation.htm.

Mollison, B. (1996) *Travels in Dreams*. Tyalgum, NSW: Tagari.

Mollison, B. (1991) *Global Gardener: Gardening the World Back to Life*. Melbourne: Film Australia.

Mollison, B. (1991) *Introduction to Permaculture*. Tyalgum, NSW: Tagari.

Mollison, B. (1989) *Visionaries: In Grave Danger of Falling Food*. DVD.

Mollison, B. (1988) *Permaculture: A Designers' Manual*. Tyalgum, NSW: Tagari.

Mollison, B. (1979) *Permaculture Two*. Tyalgum, NSW: Tagari.

Mollison, B. and Holmgren, D. (1978) *Permaculture One*. Tyalgum, NSW: Tagari.

Mollison, B., Slay, R. & Jeeves, A. (eds.) (1985) *Permaculture Design Course Handbook*. Stanley, Tasmania: Permaculture Institute.

Monbiot, G. (2006) *Heat: How to Stop the Planet from Burning*. Boston: South End Press.

Moore, A. (2009) *Sensitive Permaculture: Cultivating the Way of the Sacred Earth*. Castlemaine: Python Press.

Moore, A. (2001) *Stone Age Farming: Eco-Agriculture for the 21st Century*. Castlemaine: Python Press.

Moore, A. (1998) *Backyard Poultry - Naturally*. Castlemaine: Python Press.

Moore, A. (1987) *The Dowsing and Healing Manual*. Self-published but now permanently out of print. Available from the author as a correspondence course.

Morrow, R. (2006) *Earth User's Guide to Permaculture*. Australia: Simon & Schuster Ltd.

Morrow, R. (2003) *A Girl in the Outback.* In Six Journeywomen, *Decent Exposure: Life Stories and Poems.* Blackheath, NSW, Australia: Mountain Wildfire Press.

Mulligan, M. & Hill, S.B. (2001) *Ecological Pioneers: A Social History of Australian Ecological Thought and Action.* Melbourne: Cambridge University Press.

Norberg-Hodge, H. Ecological Development Centre, Ladakh. See *www.isec.org.uk.*

Nugent, J. (1999) *Permaculture Plants: Agaves and Cacti.* Nannup, WA, Australia: Sustainable Agriculture Research Institute.

Nugent, J. & Boniface, J. (1996) *Permaculture Plants: A Selection.* Nannup, WA, Australia: Sustainable Agriculture Research Institute.

Nugent, J. (forthcoming) *Permaculture Plants: Palms and Ferns.*

Nuttall, C. (1996) *A Children's Food Forest: An Outdoor Classroom.* Brisbane: Food Forests and Learnscapes in Education.

Nuttall, C. & Millington, J.L. (2008) *Outdoor Classrooms: A Handbook for School Gardens.* Palmwoods, Qld: PI Productions Photography.

Odum, H. T. (1981) *Energy Basis for Man and Nature.* NY: McGraw Hill

Orr, D. (1991) *What is Education For?* In *The Learning Revolution (IC#27)*, Winter 1991, p.52, Context Institute.

The Owner Builder **magazine.** See *http://www.theownerbuilder.com.au.*

Permaculture International Journal (PIJ). See *http://www.permacultureinternational.org/pcabout/pil-history.*

Permaculture **magazine.** See *http://www.permaculture.co.uk.*

Permaculture Activist **magazine.** See *http://www.permacultureactivist.net.*

Permaculture of the inner landscape. See *http://www.thelivingcentre.com; http://permacultureinstitute.pbworks.com/PermPsych; http://www.lostvalley.org; http://www.regenerativedesign.org; http://permaculturesanctuaries.blogspot.com*

Quinney, J. (1981) 'A Report From the Tree People: Introduction; Nitrogen-fixing Trees and Shrubs; Hedgerows and Living Fences.' In *Journal of the New Alchemists* (7), pp. 56-7; 60-61; 62-3.

Quinney, J. (1984) 'Designing the Sustainable Farm.' In *Mother Earth News* 88 pp. 54-65, Jul/Aug.

Resurgence **magazine.** See *http://www.resurgence.org.*

Sattman-Frese, W. & Hill, S.B. (2008) *Learning for Sustainable Living: Psychology of Ecological Transformation.* Lulu.

Savory, A. *Holistic Resource Management.* See *http://www.holisticmanagement.org.*

Schumacher, E. F. (1999) *Small is Beautiful: Economics as if People Mattered – 25 years later.* Dublin: Hartley & Marks Publishers.

Slaughter, R.A. (2009) 'Beyond the Threshold: Using Climate Change Literature to Support Climate Change Response.' In *Journal of Integral Theory and Practice*, vol. 4, no. 4, 2009, pp. 26-46.

Small-Wright, M. Perelandra. See *http://www.perelandra-ltd.com*.

Smith, C.J. (2000) *The Getting of Hope: Personal Empowerment through Learning Permaculture*. PhD Thesis, The University of Melbourne.

Smith, J.R. (1977) *Tree Crops: A Permanent Agriculture*. Old Greenwich, MA: Devlin-Adair.

***State of the Environment* reports.** Available at *http://www.environment.gov.au/soe*.

TerraCircle. See *http://www.terracircle.org.au*.

Transition Network. See *http://www.transitionnetwork.org*.

The Good Life. See *http://www.bbc.co.uk/comedy/goodlife/index.shtml*.

Wallace, G. (1984) *The Wallace Soil Reconditioning Unit*. In Riverina Outlook Conference proceedings. See *http://www.regional.org.au/au/roc/1984*.

Whitehouse, H. (n.d.) 'The Challenge of Valuing Ecological Sustainability in Schools.' In *Primary and Middle Years Educator,* pp. 14-16.

Whole earth catalogue. See *http://www.wholeearth.com/index.php*.

Yeomans, P.A. & Yeomans, K.B. (ed.) (1993) *Water for Every Farm – Yeoman's Keyline Plan*. Keyline Designs, Qld. Available at *http://www.keyline.com.au*.

Yunus, M. (2009) Interview: Elders with Andrew Denton 7th December. ABC Television, Australia.

Index

A

ABC Radio 32, 54, 249
ABC Television xviii, 9, 18, 118–119, 206, 277, 284, 289–293, 342
Aboriginal 21, 46–47, 192–194, 197, 200–202, 205–206, 249
abundance 11, 61, 82, 108, 278, 321
Accredited Permaculture Training (APT) xv, xvi, xvii, 40, 164, 173–175, 214, 224, 229, 239, 242, 262, 266–268, 334
ACRES USA 72, 196, 201, 336
Action for World Development (AWD) 246, 248–253
Aeration plough xi, 69, 74
Afghanistan 153–154
A Girl in the Outback (book) 147, 341
Agriculture and Food Group 252
Aid 86, 90, 100, 102–103, 107, 226
Albania 153
Alexander, Christopher 280, 336
Alexander, Jason 36
Alfalfa House Food Co-op 114
Alternative Energy Development Board 243
Alternative to Violence Program (AVP) 153
altruism / altruistic 34, 160
ancestors 47, 95, 301
Animals
    Argentine Ants 80
    Bettongs (brush-tailed) 140
    butcher/ing 94–95
    Cape Barren geese 140
animism 47, 199, 203
apartheid 150, 249
A Pattern Language (book) 280, 336
Appropriate Technology for Community and Environment (APACE) 110–111, 122–125, 298, 334
Argentine Ants 80
Arnold, David 127
Arundel, Phil 119

atomic bomb 165. *See also* nuclear bomb
August Investments 118
Australian City Farms and Community Gardens Network 110, 115, 121, 126, 306, 312
Australian Conservation Foundation 36, 259, 336
Australian Permaculture Convergence (APC). *See* Permaculture Convergence

B

Backyard Poultry - Naturally (book) 196, 340
Bahamas 317–318, 322
Bali, Indonesia 192, 321
Ball, Colin 114, 336
Bangladesh 3, 57, 61–62, 256
Barcelona, Spain 319–320
Basics of Permaculture Design, The (book) 239, 340
Bateson, Gregory xi, 32, 336
Bell, Graham 27, 108, 239, 278, 336
Bentley, Tom 229, 336
Benyus, Janine 310, 336
Berman, Morris 309, 336
Berry, Thomas xix, 336
Bettongs (brush-tailed) 140
big lie (concept) 13–14
biodiversity 36, 135, 144, 199, 230, 251, 258, 299, 301, 327, 332
biological inheritance 11
Biomimicry (book) 310, 336
bioregion x, 40, 45–46, 81, 91, 107–108, 145, 186, 189, 219, 304, 313, 321
Blazey, Clive 62
Boniface, Julia 81, 341
Books
    A Girl in the Outback 147, 341
    A Pattern Language 280, 336
    Backyard Poultry - Naturally 196, 340
    Basics of Permaculture Design, The 239, 340

Biomimicry 310, 336
Children's Food Forest, A 126, 341
Community Garden: Getting Started Guide 306
Dowsing and Healing Manual, The 193, 340
Earth User's Guide to Permaculture 258, 340
Eco-Social Matrix 46, 337
Field Guide to the Wild Flowers of Northern Europe 42, 337
Forest Farming 78, 337
Getting Started in Permaculture 238, 340
Heat: How to Stop the Planet from Burning 63, 96, 340
Hidden Connections, The 310, 337
Holistic Life, The 262, 266, 270, 339
Lawns into Lunch 246, 258–259, 337
Let's Have Healthy Children 168, 337
Listening to the Land 30, 36, 336
Lost Crops of the Incas 83, 336
Manual for Teaching Permaculture Creatively 186, 337
No Dig Gardening and Leaves of Life 249, 337
One Straw Revolution 64, 337
Outdoor Classrooms: A Handbook for School Gardens 164, 341
Permaculture: A Designers' Manual 26, 47, 79, 85, 103, 172, 209, 226, 239, 251, 340
Permaculture One 18, 23, 27, 43, 64, 78, 81, 113–114, 127, 130, 136, 138, 158–159, 168, 171, 206, 264, 285, 308, 340
Permaculture Plants: Agaves and Cacti 76, 82, 341
Permaculture Plants: A Selection 76, 341
Permaculture Plants: Palms and Ferns 83, 341
Permaculture: Principles and Pathways Beyond Sustainability 18, 20, 28, 175, 262, 339
Permaculture Two 23, 36, 78, 114, 171, 285, 340
Permaculture Way, The 108, 336
Places of the Soul 64, 337
Re-enchantment of the World 309, 336
Stone Age Farming: Eco-Agriculture for the 21st Century 201–202, 340
Tao of Physics, The 309, 337
Travels in Dreams 108, 340
Tree Crops 78, 342
Trees on the Treeless Plains 35, 339
Turning Point, The 309, 336
Water for Every Farm 33, 342
Weather Makers, The 85, 337
Web of Life, The 309, 337
When You Grow Up 246, 340
You Can Have Your Permaculture and Eat It Too 178, 187, 337
Brazil 100, 171, 212
British Society of Dowsers 192, 202
Brookman, Graham and Annemarie xvi, 132–145, 211, 265
Brooks, Joss 253, 336
Brunswick Valley Permaculture Group 180
building community 89
Bundjalung people 192
butcher/ing (animals) 94–95
Byrne, Josh xvi, xviii, 284–293
Byron Bay 181, 195, 296

C

Cairns, Jim 77–78
Caldicott, Helen 193, 336
Callahan, Professor Phil 196
Cambodia 116, 146, 149, 152–153, 319
Campbell, Fiona xvi, 15, 111–130
Campion Society 248
cancer xviii, 140, 280–281, 290
Candlelight Farm 237–238, 241–242, 245
Cape Barren geese 140
Cape Eleuthera Institute 318
Capra, Fritjof 308–310, 322, 336–337
carobs 133, 135, 143
Catholic xvii, 155, 248, 260

Celtic 46–47, 49
Central Victorian Greenhouse Alliance 30, 37
Centre for Alternative Technology 310
Centre for Energy Research and Environmental Sustainability (CERES) 125, 264, 334
Centro de Investigación de los Bosques Tropicales (CIBT) 297
Certificate I 175, 224, 229, 234
Certificate II 175
Certificate III 174
Certificate IV 174, 178, 185, 214, 224
chemicals xvi, 13, 42, 71, 80, 135, 148, 168, 170–171, 198, 234, 255, 257, 259, 272, 281, 324, 330–331
chemotherapy 281
Chernobyl nuclear disaster 168
Children's Food Forest, A (book) 126, 341
cholera 101, 107
Christian faith xvii
Clayfield, Robin xvi, xviii, 58, 138, 174, 178–189, 210, 224, 337
climate change xiv, xix, 3, 18, 37, 48, 55, 74, 141, 144–145, 176, 202, 212, 244, 246, 260, 268, 305, 320, 323, 327, 329, 335, 342
co-evolutionary change 327, 332
co-housing 117, 125
Coleman, Dave 211, 237, 240
Coleman, Naomi xvii, 11, 174, 204–223
common ground 46, 181, 197, 216
Commonground 158–159
Common Ground 125
Communism 20–21, 134, 159
Community Garden: Getting Started Guide 306
community garden(s) 116, 118, 121–122, 126, 127, 198, 262–263, 275, 306, 312, 317
Community Supported Agriculture (CSA) 75, 320, 322
Community Title 45–46
Community Water Grants 240
Companies 127, 227–228, 319
    August Investments 118
    Danone 10
    Green Harvest 115
    Green Life Soil Company 240
    Rainbow Power Company 45
    SEED International xix, 306, 322, 338
    Trade Winds 249, 255
complexity xiv, 4–5, 7–8, 10, 15, 71, 104, 330
composting toilet 241
Conservation Council of WA 284
consumer culture xiv, 226
convergence. See Permaculture Convergence
co-operacy 328
cooperative stewardship xiv
co-originator xv, xix, 19, 24
Cooroy Butter Factory 171
Countries
    Afghanistan 153–154
    Albania 153
    Australia. See Places
    Bahamas 317–318, 322
    Bangladesh 3, 57, 61–62, 256
    Brazil 100, 171, 212
    Cambodia 116, 146, 149, 152–153, 319
    Cuba 36, 49–50, 155, 185–186, 316, 322
        Havana 33, 36, 50, 316
    Ecuador 83, 90, 209, 297–299
    Ethiopia 146, 153
    Holland xv, 66, 135–136, 168–170
    India 31, 34, 43, 57, 153, 155, 253, 296, 303
        Ladakh 296, 311–312, 322
    Indonesia 313, 320–322
        Bali 192, 321
    Ireland
        Kinsale 175
    Italy 320
    Japan 18, 59–60, 84, 100, 165, 303
    Kenya 76, 84, 185–186
    Nepal 34, 253
    New Zealand 18, 55, 105, 185, 211
    Pakistan 57
    Philippines 55, 61
    Solomon Islands xviii, 110–111, 122–125, 253, 294, 298–299, 301, 303–305

South Africa viii, 60, 124, 135, 150, 185
South Korea 315, 318–319, 322
Spain 320, 322
  Barcelona 319–320
Sri Lanka 62, 246, 255–256
Tanzania 101–102
Thailand 64, 153, 185, 192, 233, 289
Turkey 60, 317, 322
United Kingdom (UK) 27, 60, 117, 149, 164, 175, 192, 239–240, 262, 264, 266, 273
  Totnes 129, 175
United States of America (USA) 18, 44–45, 50, 55, 60, 72, 142, 159, 202, 309, 316, 322, 329
Vanuatu 299
Vietnam 21, 41, 135, 145–146, 149, 152–153, 246, 256–258
  Ha Noi 256–257
Course Orientation Workshop (COW) 174, 224
Creative Adult Learning Facilitation (CALF) 174, 224
Creative Facilitation 185
Crystal Waters Permaculture Village xv–xvi, xix, 56–57, 59, 89, 115, 119, 182–184, 188–189, 197, 210, 243, 306, 311–312, 322–323
Cuba 36, 49–50, 155, 185–186, 316, 322
cultural inheritance 11
cultural sensitivity 101–102, 105
Cundall, Peter 277, 290

D

Dahal, Badri 253
Danone (company) 10
Dare, Pat 82, 198
Davis, Adele 168, 337
Dawborn, Kerry xiv, 2–15, 158, 163, 325
Day, Christopher 310, 337
Deans, Esther 43, 249, 337
death(s) 92, 95, 97, 102, 166, 169–170, 249, 282, 330
Decade of Education for Sustainable Development 49

deceptive simplicity / deceptively simple 4–8, 11–12, 15, 327, 331
Deep Ecology 181–182
de-forestation 101, 255
Dennett, Su 18, 27, 262
Denton, Andrew 9, 342
Depression, The. *See* Great Depression (1930s)
de Silva, Ranjith 255
Diploma of Permaculture 26, 86, 93, 195, 213, 224, 241, 262, 267–268
distressed patterned behaviours 331
Dixon, Chris 239
Djanbung Gardens 40, 49, 120
dominant paradigm 90, 275, 279
Douglas, J. Sholto 78, 337
Down to Earth Association in Australia 77
Down to Earth Association of WA 78
Dowsers Society of NSW 193
Dowsing and Healing Manual, The (book) 193, 340
dowsing pendulum 193, 200
Dryland Institute 214
Dynamic Groups 178, 185
Dynamic Learning 178

##

Earthbank 81
Earth care / Care of the Earth. *See* Ethics of permaculture
Earthcare Education 178, 187
Earth Charter 259–260
Earth Community xiv, 2, 10
Earth Garden magazine 32, 77, 201, 337
EarthSong journal viii, 337
Earth User's Guide to Permaculture (book) 258, 340
eco-architecture xii
ecoliteracy 310, 313–314
Ecological Agriculture Projects 324, 337, 339
ecological design xiv, 314, 319, 331
ecological footprint 12, 44, 49–50, 57, 94, 194, 312, 321
ecological paradigm (thinking) 309–310, 314

ecological sustainability xii, 15, 75, 313–315, 332, 342
Eco-Social Matrix (ESM) 46, 337
ecosystem(s) x, xix, 4, 12, 43, 137, 140, 181, 244–245, 251, 261, 273–274, 276, 305, 324, 332, 334
ecovillage / ecovillage movement xv, 56–59, 106, 118–119, 139, 141, 155, 162, 182, 262, 269, 315, 318, 320. *See also* Crystal Waters Permaculture Village; *See also* Global Eco-village Network
Ecuador 83, 90, 209, 297–299
edge / edge species 103, 145, 179, 188, 200, 297, 326–328
edible landscape 164, 195, 202, 306
Edible School Yards 164
Edmund Rice Centre (ERC) xvii, 246, 259
Education
    Accredited Permaculture Training (APT) xv, xvi, xvii, 40, 164, 173–175, 214, 224, 229, 239, 242, 262, 266–268, 334
    Cape Eleuthera Institute 318
    Certificate I 175, 224, 229, 234
    Certificate II 175
    Certificate III 174
    Certificate IV 174, 178, 185, 214, 224
    Course Orientation Workshop (COW) 174, 224
    Creative Adult Learning Facilitation (CALF) 174, 224
    Creative Facilitation 185
    Deep Ecology 181–182
    Diploma of Permaculture 26, 86, 93, 195, 213, 224, 241, 262, 267–268
    Dynamic Groups 178, 185
    Dynamic Learning 178
    Eco-Social Matrix (ESM) 46, 337
    Eltham College 224, 367
    Environmental Design (course) 22–23, 26, 113
    Environmental Planning and Landscape Architecture (course) 308–309
    Environmental Technology Centre 288
    Griffith University 306
    Island School 317–318
    Ladakh Ecological Development Centre 296
    Murdoch University 78, 126, 240, 288
    Outside Classroom 240
    Permaculture College Australia Inc 40
    Permaculture Design Certificate/Course (PDC). *See* Permaculture Design Certificate/Course (PDC)
    Registered Training Organisations (RTOs) 267–268, 335
    Rural Training Centres 301
    Schumacher College 308–311
    Seoul National University 318–319
    Tasmanian College of Advanced Education 22, 113, 339
    Time for an Oil Change (course) 164, 175
    Train the Trainer (course) 164, 174
    UNESCO Master of Sustainability program 320
    University of Melbourne 308
    Victorian Essential Learning Standards (VELS) 219
    Victoria University (VU) 38
    Vocational Education and Training (VET) xvii, 173–174, 224, 262, 335
Edwards, Susie 44
elder(s) vii, xv–xvi, xviii, 47, 66, 73, 183, 194, 205, 342
El-Hage Scialabba, Nadia 258, 337
Eltham College 224, 267
emotion / emotional xiv, 51, 72, 93, 97, 105, 153, 167, 193, 213, 280, 329–330
energy descent 28, 175, 334
Energy Descent Action Plan 164
environmental design 18, 284, 292, 337
Environmental Design (course) 22–23, 26, 113
Environmental Planning and Landscape Architecture (course) 308–309
Environmental Technology Centre 288

Epicentre 114–115, 119
ethical investment 81–82, 115, 118
ethical life 151
ethics ix, xvii, 4, 8, 53, 136, 138, 151, 176, 184, 206, 216, 218–219, 225–226, 230, 234, 254, 260, 308, 328, 330
Ethics of permaculture ix, xii, xiv, 4, 8, 104, 133–134, 136, 145, 151, 158, 160–161, 172, 182, 185, 188, 208, 229, 251, 253, 264, 286, 308, 314, 322, 326
    Earth care 4, 34, 104, 199, 201, 244, 261
    Fair share 225
    People care 34, 118, 151, 261
Ethiopia 146, 153
Ethos Foundation 306
Europe(an) xv, 18, 36, 42, 47, 55, 67, 82, 88, 135, 142, 168, 171, 196–197, 238, 320
European Union (EU) 320

F

Fair share. *See* Ethics of permaculture
Fair Trade 228, 249, 255
Fanton, Jude and Michel 194, 296, 337
feng shui 199–200
Field Guide to the Wild Flowers of Northern Europe (book) 42, 337
Finch, Joy 120, 201, 333
Findhorn 202, 310
Finnane, Jill xvii, 130, 246–261, 337, 340
first generation alienated 21, 23
Firth, Julie 214
Flannery, Tim 85, 337
Flow of Life 282
Food and Agriculture Organisation (FAO) 258–259
food forest 10–11, 84, 99, 196, 213, 245, 288
Food Forest, The xvi, 131–145, 211, 265
food scarcity / food security 2, 45, 50, 75, 110, 122–124, 151, 158, 244, 249, 256, 258, 282, 294, 316, 321, 337
footprint. *See* ecological footprint
Forest Farming (book) 78, 337
forest gardens 239, 255, 303
Forestry Commission 181

forest system(s) 101, 181, 184
Francis, Robyn xv, 40–51, 114, 119–120, 130, 173–174, 337
Fraser, Andrew 175
Freire, Paulo 248
Fryers Forest Ecovillage 269
Fukuoka, Manasobu 308, 337
Fuller, Buckminster xi, 32
future-oriented 315

G

Gaia Nursery 240
Gamble, Morag xviii, 130, 306–323, 338
Gami Seva Sevana Organic Farm (GSS) 255–256
Gardening Australia (ABC TV series) xviii, 18, 277, 284, 290–291, 338
geomancy 190, 193, 195, 197–201, 203
geopathic stress 192
Getting Started in Permaculture (book) 238, 340
Gillfedder, Gahan 55
glint / glint moment 9–10, 13, 15
Global Eco-village Network 52
Global Gardener, The (ABC TV) 118–119, 253, 274, 340
globalisation 11, 35, 320
Good Life, The (BBC TV series) 286, 342
Grameen bank 9
Grass Roots magazine 32, 180, 338
Gravestein, Hugh 208
Gravestein, Vries vii, xv, 14, 36, 66–75
Grayson, Russ xvi, 15, 110–130
Great Depression (1930s) 19–20, 41, 165, 263
Green Connections magazine 120, 201, 269, 335
green guerrilla 34
Green Harvest (company) 115
Green Home program 259
Green Life Soil Company 240
Greenpeace 45, 193
Green politics xiv
Green Team 316
Green Vouchers 240
greywater xvii, 94, 101, 236–238, 240–243, 292, 339

Griffith University 306
Growing Together (film) 253, 336
Grupp, Mischa 269
guru 24, 26–27, 260

H

Hamaker, J. D. 196, 338
Ha Noi, Vietnam 256–257
Harland, Maddy & Tim 239
Harrison, Lea 44, 55, 58, 180, 184
Hart, Robert J. 78, 239, 337
Havana, Cuba 33, 36, 50, 316
head, heart and hands 10, 51
Heat: How to Stop the Planet from Burning (book) 63, 340
Heinberg, Richard 28
heptachlor (chemical) 13, 80
herbicide 304
Hidden Connections, The (book) 310, 337
High, The (concept) 58–59
Hill, Professor Stuart vii, xii, xix, 8, 13, 324–333, 338–339, 341
Hills, Martha xvi, 12, 158–163
Himalaya 296, 311
hippy / hippies (era, people etc) 77, 87–88, 100, 159, 167–169, 171, 179, 192, 276, 278, 334
holism / wholism 14, 69–70
Holistic Life,The (book) 262, 266, 270, 339
Holistic Resource Management 279, 331, 341
Holland xv, 66, 135–136, 168–170
Holmgren, David vii, ix–xi, xiii, xv, xviii, xx, 18–29, 32, 34–35, 66, 113, 117, 127, 130, 138, 154, 158–159, 164, 175, 206, 208–209, 215, 262, 264–265, 269, 274, 278, 325–326, 329, 331, 333, 339, 340
Holmgren, Oliver 27, 211
Hopkins, Rob xiv, 175, 189, 339
horticulture 115, 119, 121, 127, 132, 134, 140, 150–151, 173, 228
Hortus Australia 173
human rights xvii, 21, 108, 259–260
human-scale 133, 176, 315
humility 7, 90, 93, 100, 276, 323

Hundred Yen store 64
hunting 23, 88, 148, 231, 261, 274

I

If You Love this Planet (film) 193, 336
Illich, Ivan 252, 339
India 31, 34, 43, 57, 153, 155, 253, 296, 303
indigenous 46, 47, 105, 140, 143, 145, 151, 155, 156, 173, 193, 197, 199, 200, 203, 205, 246, 274, 276, 288, 313, 314, 334
Indonesia 313, 320–322
Indonesian Development of Education for Permaculture (IDEP) 321
inevitable / inevitability (key concept) 6–7, 9, 10, 12, 13, 15
In Grave Danger of Falling Food (TV program). *See* Visionaries: In Grave Danger of Falling Food
inner landscape xii, xiv, xix, 3, 15, 184, 329, 331, 341
Inner Pod 125
Integrated Pest Management 72, 73
integrator indicators 331, 332
intentional communities xiii
intentional community/eco-community xv, 23, 41, 45, 118, 120, 215
intention(s) (key theme) 8
inter-cultural capital 332
Intermediate Technology Development Group 117
International Palm Society 83
International Permaculture Convergence. *See* Permaculture Convergence
International Society for Ecology and Culture (ISEC) 310, 341
interpersonal capital 332
invisible structures xvi, 9, 15, 89, 117–118, 125, 128
Islam 88
Island School 317–318
Italian Permaculture Academy 320
Italy 320
Izmir community 317

INDEX

J

Jansen, Tony xvi, xviii, 123, 253, 294–305
Japan 18, 59–60, 84, 100, 165, 303
Jarlanbah (property) xv, 46
Jeeves, Andrew 209, 340
Jerome, Burri 47, 193
Jewish 20–22, 207–208
Jewish Ecological Coalition 208
Jordan, Jill 65
Journals. *See* Magazines

K

Kandy forest gardens 255
Kastom Gaden Association (KGA) xviii, 111, 123–124, 294, 298, 300, 304
Kennedy, Declan & Margrit xii, 33, 339
Kenya 76, 84, 185–186
keyline xi, 33, 58, 66, 69, 329, 342
Kibutze 225
Kiewa Field Days 70
Kinsale, Ireland 175
Koppulla, Narsanna and Padma 303
Korean Permaculture Institute 318
Kumar, Satish 310

L

Ladakh Ecological Development Centre 296
Ladakh, India 296, 311–312, 322
Lane, Terry 32
Lang, Francis 115
Lawns into Lunch (book) 246, 258–259, 337
Lawton, Geoff xiv, xvi, 12, 83, 86–109, 171
Lawton, Nadia 88
Let's Have Healthy Children (book) 168, 337
Lillington, Ian vii, xviii, 8, 27, 158–159, 162, 173, 209, 262–270, 339
Lindegger, Max xv, xix, 52–65, 89, 115, 130
linear (one dimensional) responses xiv, 4, 11
listening (key concept) 14–15

Listening to the Land (book) 30, 36, 336
Local Agenda 21 (LA21) 319
Local Energy Trading System (LETS) 116–117, 265, 270, 334
local food viii, xiv, 2, 65, 117, 156, 261, 263, 275, 289, 312, 316, 320, 322
Long Bay Gaol 253–254
Lost Crops of the Incas (book) 83, 336
Lovelock, James 37, 339
Lovins, Amory 309, 317–318
Lynch, Damien 115, 118
Lyons, Kristen 322

M

Mack, Thomas 209
Macy, Joanna 49
Magazines
    Earth Garden 32, 77, 201, 337
    EarthSong journal viii, 337
    Grass Roots 32, 180, 338
    Green Connections 120, 201, 269, 335
    Organic Gardener and Farmer 54, 339
    Owner Builder 32, 341
    Permaculture International Journal (PIJ) xv, 23, 30, 33, 35–36, 40, 43, 48, 115, 120, 175, 195, 197, 201, 212, 251, 335, 341
    Permaculture (UK) 201, 341
    Permaculture Web 119–120
    Resurgence 312, 341
Mahzar, Farhad 256
mainstream ix, xi, xiv, xv, xvii–xviii, 3–4, 14–15, 58, 61, 129, 135, 160, 168–169, 172, 179, 185, 188, 201, 212, 214, 217, 219, 221, 226, 229, 254, 267–268, 291, 293, 303, 335
malaria 101–102, 107, 308
Manual for Teaching Permaculture Creatively (book) 186, 337
Maori 105–106
Mars, Ross xvi–xvii, 236–245, 339–340
Marxism 21–22, 24, 159
Maryborough Permaculture Association 33
Mason, Ian 119

McCarthy, Dennis 78
McCurdy, Robina 124, 126, 209, 211
McHarg, Ian 308, 340
McNeil, Barry 22
Media
   ABC Radio 32, 54, 249
   ABC Television xviii, 9, 18, 118, 206, 277, 284, 289–293, 342
   Gardening Australia (ABC TV series) xviii, 18, 277, 284, 290–291, 338
   Global Gardener (ABC TV) 118–119, 253, 274, 340
   Good Life, The (BBC TV series) 286, 342
   Growing Together (film) 253, 336
   If You Love this Planet (film) 193, 336
   Passive Solar Design of Buildings (video) 243, 340
   RAPS – Remote Area Power Supply (video) 243, 340
   Think Global: Eat Local – A Diet for a Sustainable Society (documentary) 306
   Today Tonight (TV program) 289
   Visionaries: In Grave Danger of Falling Food (TV program) 206, 296, 340
   West Australian, The (newspaper) 289
meditation/meditating 60, 63–64, 105, 269, 280
Melanesia xvi, 298–299, 302, 304–305
Melliodora (property) 18, 27, 208, 333
Method, Bradley 193
Metua, Anaheke 322
Michaels, Jeff 115
microclimate 70, 74
micro-credit 3
Miller, Ron 228, 340
Millington, Janet xvi, 164–177, 341
modern agriculture xvi, 258, 301
modernism 87
Mollison, Bill vii, ix–xi, xiii, xv, xvii–xx, 14, 18–20, 22–28, 32, 43, 47, 54–56, 58, 63, 66, 69–70, 76, 78, 86, 88–90, 108, 113, 115–118, 130, 136, 138, 151, 154, 160, 171–172, 197, 201, 206, 209, 213, 221, 226, 237, 239, 241, 243, 249, 251, 253, 256, 262–263, 268, 274, 276–277, 296, 325–326, 339–340
   carpet story 98
Monbiot, George 63, 340
Mondragon, Martha 209
Moore, Alanna xvi, xviii, 190–203, 340
Morrow, Rosemary xvi–xvii, 116, 127, 130, 146, 194, 251, 258, 260, 340–341
Moshav 225
Mountain District Permaculture group 275–276
Movement with No Name ix
Murdoch University 78, 126, 240, 288

N

Naess, Arne 310
Nambour 54–55, 89
National Association of Sustainable Agriculture Australia (NASAA) 140
National Federation of City Farms 263
National Permaculture Convergence. *See* Permaculture Convergence
Nayakrishi Andalon (farming movement) 256
negentropic relationships 332
Nepal 34, 253
Neuro-Linguistic Programming 185
New Age 104, 184
New Frontier 3, 15, 325
New Zealand 18, 55, 105, 185, 211
Ngare Ndare school 84
Ngare Ndare Solar Self Help Group 85
Nichols, Les 322
Nickels, Michael 83
Nimbin xv, 45, 49–50, 120, 192
No Dig Gardening and Leaves of Life (book) 249, 337
Noi, Ha 256
Non-Governmental Organisation (NGO) xviii, 110–111, 123, 255–257, 259, 294, 319, 334
Norberg-Hodge, Helena 296, 310, 312, 341
Northey Street City Farm 306, 311–312, 322–323

INDEX

Nott, Brad 119, 130
nuclear 6, 54, 63, 144, 165–166, 168, 193–194, 249, 282, 330
nuclear bomb 81. *See also* atomic bomb
Nugent, Jeff xv, 13, 76–85, 89, 341
Nuttall, Carolyn 126, 164, 175, 341

O

observation/observe xiv–xv, 4, 7–8, 12, 14–15, 27, 35–36, 43, 46, 55–56, 64, 69, 73, 95, 103, 105, 107, 136, 143, 152, 182, 196, 205, 208, 239, 243, 251–252, 258, 278, 280, 299, 301
Ochre, Glen 159
Odum, Howard and Eugene xi, 32, 341
olives 135, 143
One Straw Revolution (book) 64, 337
Open Garden Scheme (Australian) 289, 292
Organic Agriculture and Food Security (FAO report) 258, 337
Organic Association of WA 284
Organic Gardener and Farmer (journal) 54, 339
organic gardening 28, 122, 140, 206, 290
Organic Garden Project 256–258
Organisations
    Action for World Development (AWD) 246, 248–253
    Agriculture and Food Group 252
    Alternative Energy Development Board 243
    Appropriate Technology for Community and Environment (APACE) 110–111, 122–125, 298, 334
    Australian City Farms and Community Gardens Network 110, 115, 121, 126, 306, 312
    Australian Conservation Foundation 36, 259, 336
    British Society of Dowsers 192, 202
    Brunswick Valley Permaculture Group 180
    Campion Society 248
    Central Victorian Greenhouse Alliance 30, 37
    Centre for Alternative Technology 310
    Centre for Energy Research and Environmental Sustainability (CERES) 125, 264, 334
    Centro de Investigación de los Bosques Tropicales (CIBT) 297
    Commonground 158–159
    Common Ground 125
    Conservation Council of WA 284
    Down to Earth Association in Australia 77
    Down to Earth Association of WA 78
    Dowsers Society of NSW 193
    Dryland Institute 214
    Ecological Agriculture Projects 324, 337, 339
    Edmund Rice Centre (ERC) xvii, 246, 259
    Ethos Foundation 306
    Food and Agriculture Organisation (FAO) 258–259
    Forestry Commission 181
    Global Eco-village Network 52
    Greenpeace 45, 193
    Hortus Australia 173
    Indonesian Development of Education for Permaculture (IDEP) 321
    Intermediate Technology Development Group 117
    International Palm Society 83
    International Society for Ecology and Culture (ISEC) 310, 341
    Italian Permaculture Academy 320
    Jewish Ecological Coalition 208
    Kastom Gaden Association (KGA) xviii, 111, 123–124, 294, 298, 300, 304
    Local Energy Trading Scheme (LETS) 116–117, 265, 270, 334
    Maryborough Permaculture Association 33
    Mountain District Permaculture group 275–276
    National Association of Sustainable Agriculture Australia (NASAA) 140

National Federation of City Farms 263
Non-Governmental Organisation (NGO) xviii, 110–111, 123, 255–257, 259, 294, 319, 334
Open Garden Scheme (Australian) 289, 292
Organic Association of WA 284
Pacific Calling Partnership 246, 260, 337
Penn Wimochana Gnanadayam (PWG) 256
Permaculture and Environment Centre of WA (PECWA) 240
Permaculture Association 23, 33, 75, 114, 119–120
Permaculture Association of Western Australia (PAWA) 82, 89, 198, 239–240, 334
Permaculture Brisbane 312
Permaculture Institute 20, 212, 241, 297, 318
Permaculture Institute of WA (PIWA) 239
Permaculture International Limited (PIL) 120, 173, 335
Permaculture Nambour 88–90
Permaculture New Zealand 23
Permaculture Noosa 106, 164
Permaculture North (Sydney) 120, 164
Permaculture Oceania 121
Permaculture Research Institute (PRI) 86, 90
Permaculture Research Institute Jordan 90
Permaculture Research Institute USA 90
Permaculture Sydney 114–116, 119–120, 122, 125
Rainforest Information Centre 297
Randwick Community Centre 116, 118
Registered Training Organisations (RTOs) 267–268, 335
Satyodaya 255
Sea Shepherd 61
Stephanie Alexander Foundation 219
Sunshine Coast Energy Action Centre 164
Sustainable Gardening Australia 284
Sustainable Planning Unit 316
Sydney Food Connect 110
Sydney Food Fairness Alliance (SFFA) 127–128, 246, 260
Tea Group, The 249
TerraCircle Inc 110, 125–126, 342
Two Bays permaculture group 210
United Nations (UN) 33, 49, 306
Unnayan Bikalper Nitinird-haroni Gobeshona (UBINIG) 256
Western Alliance for Greenhouse Action 37
Willing Workers on Organic Farms (WWOOF) 28, 61, 84, 132, 187, 195, 204, 211, 213, 335
World Development Tea Co-op 249
Orr, David 231, 317
Outdoor Classrooms: A Handbook for School Gardens (book) 164, 341
Outside Classroom 240
Owner Builder magazine 32, 341

P

Pacific Calling Partnership 246, 260, 337
Packiam, Siva 256
Pakistan 57
Pappinbarra 114, 130
Papworth, John 312
Parajubaes (palm species) 83
Parks, Rosa 3, 10, 12, 325
Passive Solar Design of Buildings (video) 243, 340
passive solar / solar passive 23, 161, 237, 265, 269–270, 272, 307
pattern(s) xix, 4, 8, 12, 46–47, 97–99, 107, 149, 150, 159–160, 184, 188, 218, 277, 280–281, 315–316
Peak, Bill 55
peak oil xiv, xix, 18, 28, 55, 144, 176, 212, 244, 268, 293, 323, 334, 335

Penn Wimochana Gnanadayam (PWG) 256
People (also indexed by surname)
  Adele Davis 168, 337
  Alanna Moore xvi, xviii, 190–203, 340
  Alan Savory 279, 331, 341
  Amory Lovins 309, 317–318
  Anaheke Metua 322
  Andra Pradesh 303
  Andrew Denton 9, 342
  Andrew Fraser 175
  Andrew Jeeves 209, 340
  Arne Naess 310
  Badri Dahal 253
  Barry McNeil 22
  Bill Mollison vii, ix–xi, xiii, xv, xvii–xx, 14, 18–20, 22–28, 32, 43, 47, 54–56, 58, 63, 66, 69–70, 76, 78, 86, 88–90, 108, 113, 115–118, 130, 136, 138, 151, 154, 160, 171–172, 197, 201, 206, 209, 213, 221, 226, 237, 239, 241, 243, 249, 251, 253, 256, 262–263, 268, 274, 276–277, 296, 325–326, 339–340
  Bill Peak 55
  Bradley Method 193
  Brad Nott 119, 130
  Bronwyn Rice 116
  Buckminster Fuller xi, 32
  Burri Jerome 47, 193
  Caroline Smith viii–xix, 209, 326
  Carolyn Nuttall 126, 164, 175, 341
  Charles Rogers 299
  Chris Dixon 239
  Christopher Alexander 280, 336
  Christopher Day 310, 337
  Clive Blazey 62
  Colin Ball 114, 336
  Damien Lynch 115, 118
  Dave Coleman 211, 237, 240
  David Arnold 127
  David Holmgren vii, ix–xi, xiii, xv, xviii, xx, 18–29, 32, 34–35, 66, 113, 117, 127, 130, 138, 154, 158–159, 164, 175, 206, 208–209, 215, 262, 264–265, 269, 274, 278, 325–326, 329, 331, 333, 339, 340
  David Orr 231, 317
  Declan & Margrit Kennedy xii, 33, 339
  Denise Sawyer 115
  Dennis McCarthy 78
  Dr. Graham Phillips 121
  E.F. Schumacher 117, 308, 312, 341
  Esther Deans 43, 249, 337
  Evan Raymond xix, 311–313, 315, 322–323, 338
  Farhad Mahzar 256
  Fiona Campbell xvi, 15, 111–130
  Francis Lang 115
  Frank Zappa 230
  Fritjof Capra 308–310, 322, 336–337
  Gahan Gillfedder 55
  Geoff Lawton xiv, xvi, 12, 83, 86–109, 171
  Geoff Wallace xi, 69, 74, 342
  George Monbiot 63, 340
  George Sobol 266
  Glen Ochre 159
  Graham and Annemarie Brookman xvi, 132–145, 211, 265
  Graham Bell 27, 108, 239, 278, 336
  Gregory Bateson xi, 32, 336
  Guy Rischmueller 173, 267
  Helena Norberg-Hodge 296, 310, 312, 341
  Helen Caldicott 193, 336
  Hilary Whitehouse 225, 342
  Howard and Eugene Odum xi, 32, 341
  Hugh Gravestein 208
  Ian Lillington vii, xviii, 8, 27, 158–159, 162, 173, 209, 262–270, 339
  Ian Mason 119
  Ian McHarg 308, 340
  Isabell Shipard 104
  Ivan Illich 252, 339
  James Lovelock 37, 339
  Jane Scott vii, xviii, 36, 272–283
  Janet Millington xvi, 164–177, 341

Janine Benyus 310, 336
Jason Alexander 36
J. D. Hamaker 196, 338
Jeff Michaels 115
Jeff Nugent xv, 13, 76–85, 89, 341
Jill Finnane xvii, 130, 246–261, 337, 340
Jill Jordan 65
Jim Cairns 77–78
Joanna Macy 49
John & Nancy Todd 317–318
John Papworth 312
John Seed 181
John Walmsley 140
Josh Byrne xvi, xviii, 284–293
Joss Brooks 253, 336
Joy Finch 120, 201, 333
J. Quinney 326, 341
J. Russell Smith 78, 342
J. Sholto Douglas 78, 337
Jude and Michel Fanton 194, 296, 337
Judith Thurley 44
Julia Boniface 81, 341
Julie Firth 214
Kerry Dawborn xiv, 2–15, 158, 163, 325
Kristen Lyons 322
Lea Harrison 44, 55, 58, 180, 184
Les Nichols 322
Machaelle Small-Wright 202, 342
Maddy & Tim Harland 239
Manasobu Fukuoka 308, 337
Martha Hills xvi, 12, 158–163
Martha Mondragon 209
Max Lindegger xv, xix, 52–65, 89, 115, 130
Michael Nickels 83
Mischa Grupp 269
Morag Gamble xviii, 130, 306–323, 338
Morris Berman 309, 336
Muhammad Yunus (Professor) 3, 9–10, 12, 15, 325, 342
Nadia El-Hage Scialabba 258, 337
Nadia Lawton 88

Naomi Coleman xvii, 11, 174, 204–223
Narsanna and Padma Koppulla 303
Nigel Shepherd 125
Oliver Holmgren 27, 211
Pat Dare 82, 198
Patrick Whitefield 239
Paulo Freire 248
P.A. Yeomans xi, 33, 58, 69, 329, 342
Peter Cundall 277, 290
Phil Arundel 119
Phil Callahan (Professor) 196
Ranjith de Silva 255
Richard A. Slaughter 3, 342
Richard Heinberg 28
Robert J. Hart 78, 239, 337
Rob Hopkins xiv, 175, 189, 339
Robina McCurdy 124, 126, 209, 211
Robin Clayfield xvi, xviii, 58, 138, 174, 178–189, 210, 224, 337
Robyn Francis xv, 40–51, 114, 119–120, 130, 173–174, 337
Ron Miller 228, 340
Rosa Parks 3, 10, 12, 325
Rosemary Morrow xvi–xvii, 116, 127, 130, 146, 194, 251, 258, 260, 340–341
Ross Mars xvi–xvii, 236–245, 339–340
Russ Grayson xvi, 15, 110–130
Satish Kumar 310
Shann Turnbull 265
Siva Packiam 256
Sonya Wallace 175
Stuart Hill (Professor) vii, xii, xix, 8, 13, 324–333, 338–339, 341
Su Dennett 18, 27, 262
Susie Edwards 44
Ted Trainer 295
Terry Lane 32
Terry White xv, 23, 30–39, 278, 336
Thomas Berry xix, 336
Thomas Mack 209
Tim Flannery 85, 337
Tom Bentley 229, 336
Tony Jansen xvi, xviii, 123, 253, 294–305

Vandana Shiva 309–310, 312
Virginia Solomon vii, xvii, 173–174, 214, 224–235
Vries Gravestein vii, xv, 14, 36, 66–75
People care. *See* Ethics of permaculture
Peppermint Ridge Farm 276
Perelandra (garden) 202, 342
permablitz 37, 266, 317, 335
permaclutter 156
Permaculture: A Designers' Manual (book) 26, 47, 79, 85, 103, 172, 209, 226, 239, 251, 340
Permaculture and Environment Centre of WA (PECWA) 240
Permaculture Association 23, 33, 75, 114, 119–120, 239
Permaculture Association of Western Australia (PAWA) 82, 89, 198, 239–240, 334
Permaculture Brisbane 312
Permaculture College Australia Inc 40
Permaculture Convergence xi, 92–93, 97, 207, 335
    APC4 - Albury NSW, 1990 66
    APC5 - South Australia, 1995 126, 210–211
    APC8 - Melbourne, Victoria, 2005 xii, 188, 204, 214–215, 241, 243
    APC9 - Sydney, NSW, 2009 243
    Crystal Waters, 1993 197
    IPC1 - Pappinbarra, NSW, Australia, 1984 20, 24, 26, 114
    IPC3 - Scandinavia, 1993 209
    IPC6 - Perth & Bridgetown, WA, Australia, 1996 81–82, 171, 211–212, 238, 287–288
Permaculture Design Certificate/Course (PDC) x, xv–xviii, 20, 23, 26–27, 40, 43–44, 47, 49, 56, 66, 70–73, 78, 81, 89, 92, 97, 99–100, 110–111, 114–120, 122, 127–128, 133, 138–139, 143, 146, 150, 152–154, 158, 162, 171, 173–174, 178, 180, 181, 184–185, 188, 194–195, 204–205, 208–213, 211, 218, 220, 232, 237, 240–242, 252, 262, 264, 268, 287–288, 296–297, 311, 314, 318–320, 334

Permaculture for Peace 316
Permaculture Institute 20, 212, 241, 297, 318
Permaculture Institute of WA (PIWA) 239
Permaculture International Journal (PIJ) xv, 23, 30, 33, 35–36, 40, 43, 48, 115, 120, 175, 195, 197, 201, 212, 251, 335, 341
Permaculture International Limited (PIL) 120, 173, 335
Permaculture magazine (UK) 201, 341
Permaculture Nambour 88–90
Permaculture New Zealand 23
Permaculture Noosa 106, 164
Permaculture North (Sydney) 120, 164
Permaculture Oceania 121
Permaculture One (book) 18, 23, 27, 43, 64, 78, 81, 113–114, 127, 130, 136, 138, 158–159, 168, 171, 206, 264, 285, 308, 340
permaculture pioneers ix–xi, xv–xvi, xix, 4, 58, 152
Permaculture Plants: Agaves and Cacti (book) 76, 82, 341
Permaculture Plants: A Selection (book) 76, 341
Permaculture Plants: Palms and Ferns (book) 83, 341
Permaculture: Principles and Pathways Beyond Sustainability (book) 18, 20, 28, 175, 262, 339
Permaculture Research Institute (PRI) 86, 90
Permaculture Research Institute Jordan 90
Permaculture Research Institute USA 90
Permaculture Sydney 114–116, 119–120, 122, 125
Permaculture Two (book) 23, 36, 78, 114, 171, 285, 340
Permaculture Way, The (book) 108, 336
Permaculture Web 119–120
Permafest 72
permanent culture 32, 38
personal capital 332
Philippines 55, 61
Phillips, Dr. Graham 121
phosphorus 73–74

pioneer species 12
pistachio 141, 143, 211
Places. *See also* Countries
   Byron Bay 181, 195, 296
   Candlelight Farm 237–238, 241–242, 245
   Centre for Alternative Technology 310
   CERES (Centre for Energy Research and Environmental Sustainability) 125, 264
   Commonground 158–159
   Cooroy Butter Factory 171
   Crystal Waters Permaculture Village xv–xvi, xix, 56–57, 59, 89, 115, 119, 182–184, 188–189, 197, 210, 243, 306, 311–312, 322–323
   Djanbung Gardens 40, 49, 120
   Environmental Technology Centre 288
   Epicentre 114–115, 119
   Findhorn 202, 310
   Food Forest, The xvi, 131–145, 211, 265
   Fryers Forest Ecovillage 269
   Gaia Nursery 240
   Gami Seva Sevana Organic Farm (GSS) 255–256
   Himalaya 296, 311
   Jarlanbah (property) xv, 46
   Kandy forest gardens 255
   Long Bay Gaol 253–254
   Melliodora (property) 18, 27, 208, 333
   Nambour 54–55, 89
   Nimbin xv, 45, 49–50, 120, 192
   Northey Street City Farm 306, 311–312, 322–323
   Pappinbarra 114, 130
   Peppermint Ridge Farm 276
   Perelandra (garden) 202, 342
   Red Planet Plants 236, 245
   Southern Cross Permaculture Institute (SCPI) 204, 223
   Tagari farm 23, 86, 90
   Tasmania ix, xii, xx, 22–23, 43, 54, 70, 79, 89, 113–114, 126, 179, 201, 209, 213, 220
   Tyalgum 44, 90, 180, 243
   Warrawong (property) 140
   Western Australia xvi–xvii, 22, 76–82, 126, 211, 214, 236–243, 284, 287–290
   Willuna (property) 71–72
   Willunga 136, 269
Places of the Soul (book) 64, 337
Plants
   carobs 133, 135, 143
   olives 135, 143
   Parajubaes (palm species) 83
   pistachio 141, 143, 211
   Quito palm 83
   walnuts 143
plastic (bags/bottles ets) 45, 63, 101, 154, 244, 263
population 6, 83, 87, 101, 122, 140, 144–145, 155, 189, 259, 279, 299, 316
Power Down (movement) 96
Power Tower/Tower of Power 196, 198, 201
Pradesh, Andra 303
prejudice 15, 104–105, 151, 154, 233, 238
pride 12, 92, 94–95, 149, 217
principles ix–xi, xvi, 4, 8, 20, 27, 32, 34, 46, 69, 71, 100, 117, 121, 123, 133, 134, 136, 138, 145, 162, 182, 185, 194, 210, 214, 251–254, 256–257, 260, 264, 275, 277, 286, 296–298, 309–310, 314, 322, 326, 335
professions 22, 24, 26, 44, 51, 96, 107–108, 114, 118, 127, 138, 145, 193, 239, 241, 259, 313, 324
profit-maximising glasses 10
profound simplicity 5, 9, 12, 15
Project Branchout 30, 35, 37

Q

Quaker xvi, 146, 151, 153, 155
Quinney, J. 326, 341
Quito palm 83

R

Rainbow Power Company 45
Rainforest Information Centre 297
rainwater tanks 62, 236, 240–243, 245, 287, 292
Randwick Community Centre 116, 118
RAPS – Remote Area Power Supply (video) 243, 340
Raymond, Evan xix, 311–313, 315, 322–323, 338
reading the landscape 14, 24, 56
Red Planet Plants 236, 245
Re-enchantment of the World (book) 309, 336
Registered Training Organisations (RTOs) 267–268, 335
religious faith xvii, xviii, 20, 22, 32, 88, 136, 207, 260
relocalisation xvi, 310, 322
replicability 62, 122, 125, 129
resource consumption 41, 138, 200, 226–227, 230, 260–261, 287, 300, 304
resources ix, 11, 20, 37, 52, 107, 126, 141–142, 144–145, 150, 153, 155, 169, 172, 213, 216, 224, 227, 229, 233–234, 240, 253, 255, 257–258, 275–276, 298, 300, 304, 313, 318–319, 321, 324, 330, 332
responsibility 4, 24, 48, 51, 59, 94–95, 102, 121, 136, 139, 145, 168, 184, 245, 260, 275, 282, 293, 313, 332
Resurgence magazine 312, 341
Rice, Bronwyn 116
right livelihood 151
Right Livelihood award 26
Rischmueller, Guy 173, 267
rite of passage 95
Rogers, Charles 299
roof gardens 33, 36
Rosa Parks moment 14
Round Towers of Ireland 196
Rural Training Centres 301

S

Satyodaya 255
Savory, Alan 279, 331, 341
Sawyer, Denise 115
Schumacher College 308–311
Schumacher, E.F. 117, 308, 312, 341
Scott, Jane vii, xviii, 36, 272–283
Sea Shepherd 61
second generation alienated 21, 26
SEED International xix, 306, 322, 338
Seed, John 181
Seed Savers 124, 296
Seoul National University 318–319
sharing circles 183
Shepherd, Nigel 125
shifting cultivation 299
Shipard, Isabell 104
Shiva, Vandana 309–310, 312
Slaughter, Richard A. 3, 342
small-scale farming x, 2, 75, 140, 143, 149
Small-Wright, Machaelle 202, 342
Smith, Caroline viii–xix, 209, 326
Smith, J. Russell 78, 342
Sobol, George 266
social business glasses 10
social capital 117, 332
social ecology 324, 331, 338
social justice iii, ix–x, 2, 15, 21, 114, 205, 207, 260, 322, 326, 332
soil xv, 12, 14, 27, 50, 65–66, 69, 71, 73–75, 84–85, 97, 101, 116, 124, 135–136, 138, 141, 151–152, 154, 161, 170, 179, 184, 196, 200, 226, 238–239, 244, 250–252, 255–257, 263, 270, 278, 299, 301–302, 319, 321, 324, 329, 338, 342
solar hot water 63–64, 143, 161, 277
solar passive. *See* passive solar / solar passive
solar power 45, 85, 94, 133, 144, 187, 288, 330
Solomon Islands xviii, 110–111, 122–125, 253, 294, 298–299, 301, 303–305
Solomon, Virginia vii, xvii, 173–174, 214, 224–235
South Africa viii, 60, 124, 135, 150, 185

Southern Cross Permaculture Institute (SCPI) 204, 223
South Korea 315, 318–319, 322
Spain 320, 322
spiritual xi, xvi–xviii, 19, 46–47, 51, 56, 62–64, 138, 152, 167, 170, 172, 184, 188, 191–193, 197, 199–203, 310
Sri Lanka 62, 246, 255–256
starving 57–58, 101
State of the Environment (report) 30, 36, 336, 342
Stephanie Alexander Foundation 219
Stone Age Farming: Eco-Agriculture for the 21st Century (book) 201–202, 340
subtle energies 200
Sunshine Coast Energy Action Centre 164
superphosphate 73–74
sustainable culture xiii–xiv
sustainable development 49, 246, 295, 318, 337
Sustainable Gardening Australia 284
Sustainable Planning Unit 316
Swan Garden Centre 240
Sydney Food Connect 110
Sydney Food Fairness Alliance (SFFA) 127–128, 246, 260

T

Tagari farm 23, 86, 90
tangibles/intangibles 151, 154
Tanzania 101–102
Tao of Physics, The (book) 309, 337
Tasmania ix, xii, xx, 22–23, 43, 54, 70, 79, 89, 113–114, 126, 179, 201, 209, 213, 220
Tasmanian College of Advanced Education 22, 113, 339
Tea Group, The 249
temenos (wilderness patch) 199–200
TerraCircle Inc 110, 125–126, 342
Thailand 64, 153, 185, 192, 233, 289
Think Global: Eat Local – A Diet for a Sustainable Society (documentary) 306, 322, 338
Three Sisters (sacred site) 194
Thurley, Judith 44
Time for an Oil Change (course) 164, 175

Today Tonight (TV program) 289
Todd, John & Nancy 317–318
Totnes, United Kingdom 129, 175
Trade Winds (company) 249, 255
traditional protocols 105, 107–108
Trainer, Ted 295
Training. *See* Education
Train the Trainer (course) 164, 174
transformation x, xii–xiv, 3, 5, 9, 37, 84, 152, 155, 178, 184, 196, 213, 272, 275, 279–280, 291, 314–315, 317, 324–325, 327
Transitioners xiv
Transition Town Maleny 323
Transition Town (movement) xiv, 45, 96, 129, 145, 164, 175, 189, 270, 283, 323, 335, 342
Travels in Dreams (book) 108, 340
Tree Crops (book) 78, 342
Trees on the Treeless Plains (book) 35, 339
Tuntable Falls Co-op 192
Turkey 60, 317, 322
Turnbull, Shann 265
Turning Point, The (book) 309, 336
Two Bays permaculture group 210
Tyalgum 44, 90, 180, 243

U

UNESCO Master of Sustainability program 320
United Kingdom 27, 60, 117, 149, 164, 175, 192, 239–240, 262, 264, 266, 273
United Nations (UN) 33, 49, 306
United States of America (USA) 18, 44–45, 50, 55, 60, 72, 142, 159, 202, 309, 316, 322, 329
Universal Knowledge 200
University of Melbourne 308
Unnayan Bikalper Nitinird-haroni Gobeshona (UBINIG) 256
Urban Agriculture Report 306
urbanisation xvi
urban permaculture 110, 114, 118, 127, 286, 311–312, 318–319, 336
urban sprawl 150
urgency addiction (concept) 15, 125

V

values-oriented 315
Vanuatu 299
vernacular wisdom 4
Victorian Essential Learning Standards (VELS) 219
Victoria University (VU) 38
Vietnam 21, 41, 135, 145–146, 149, 152–153, 246, 256–258
Vietnam Friendship Village 257
Vietnam War 21, 77–78, 134, 167, 199
Visionaries: In Grave Danger of Falling Food (TV program) 206, 296, 340
Vocational Education and Training (VET) xvii, 173–174, 224, 262, 335

W

Wallace, Geoff xi, 69, 74, 342
Wallace plough. *See* Aeration plough
Wallace, Sonya 175
Walmsley, John 140
walnuts 143
war xv, 21, 67, 69, 75, 77, 102–103, 107, 152–153, 155, 159, 165–167, 193, 199, 248, 308, 315, 319. *See also* Vietnam War; *See also* World War 2 (WWII)
Warrawong (property) 140
wastewater 236, 288. *See also* greywater
Water for Every Farm (book) 33, 342
Weather Makers, The (book) 85, 337
Web of Life, The (book) 309, 337
weeds 11–12, 56, 97, 105, 132, 228, 245, 250–251
West Australian, The (newspaper) 289
Western Alliance for Greenhouse Action 37
Western Australia xvi–xvii, 22, 76–82, 126, 211, 214, 236–243, 284, 287–290
Western Treatment Plant 234
When You Grow Up (book) 246, 340
Whitefield, Patrick 239
Whitehouse, Hilary 225, 342
White, Terry xv, 23, 30–39, 278, 336
wholism / holism 14, 69–70
wicked problems 4
wild-harvest 87
Willing Workers on Organic Farms (WWOOF) 28, 61, 84, 132, 187, 195, 204, 211, 213, 335
Willuna (property) 71–72
Willunga 136, 269
World Development Tea Co-op 249
World War 2 (WWII) xv, 19, 66–67, 87, 134, 165, 251, 263

Y

Year of Design 319
Yeomans, P.A. xi, 33, 58, 69, 329, 342
You Can Have Your Permaculture and Eat It Too (book) 178, 187, 337
young men (growing up) 95, 134, 230
Yunus, Professor Muhammad 3, 9–10, 12, 15, 325, 342

Z

Zappa, Frank 230
Zones (permaculture design) 99, 142, 222, 303
    Zone 0 222
    Zone 1 71, 302, 304
    Zone 5 200

Other media from Melliodora Publishing

Permaculture: Principles and Pathways Beyond Sustainability

This book uses permaculture principles as a framework for an empowering but challenging vision of creative adaptation to a world of energy descent. It is relevant to every aspect of how we reorganise our lives, communities and landscapes to creatively adapt to ecological realities which shape human destiny.

If the 'Permaculture Principles' that David Holmgren discusses in this extremely important book were applied to all that we do, we would be well on the road to sustainability, and beyond. - Professor Stuart B. Hill

Permaculture Ethics and Design Principles Teaching Kit

The *Permaculture Ethics & Design Principles Teaching Kit* (second edition) is designed as an aid to teaching ethics and principles on Permaculture Design Courses, but will also be useful in introducing permaculture to a wide range of audiences from school children to design professionals. This kit will be useful for anyone involved in permaculture and environmental education wanting a condensed way to communicate and remember the essence of the powerful concepts explained in *Permaculture: Principles & Pathways Beyond Sustainability*. The teaching kit consists of a Teachers' DVD, 15 fridge magnets, 52 cards and an explanatory booklet printed on recycled card stock, packed as a boxed set.

Permaculture Ethics and Design Principles DVD

In this presentation, David Holmgren explains permaculture ethics and design principles as thinking tools for creatively responding to the energy descent future on a limited planet.

Future Scenarios

In *Future Scenarios*, permaculture co-originator and leading sustainability innovator David Holmgren outlines four scenarios that bring to life the likely cultural, political, agricultural, and economic implications of peak oil and climate change, and the generations-long era of "energy descent" that faces us.

David Holmgren: Collected Writings 1978-2006 eBook (2nd edition)

This collection of magazine articles, conference papers, public lectures, book reviews and presentations provides source material on the origins, influences, applications and evolution of the Permaculture concept from its younger co-originator. Together they provide an historical record of the work of one of Australia's most influential environmental thinkers.

Flywire House

In 1983 permaculture co-originator David Holmgren responded to the tragic Ash Wednesday fires with a project called the *Flywire House*. While there has been a great deal of research and publications on this subject in the years since 1983, the *Flywire House* is still compatible with latest understandings and provides a unique case study approach. A new foreword reviews the material in the context of the 2009 fires.

Trees on the Treeless Plains: Revegetation Manual for the Volcanic Landscapes of Central Victoria eBook

This design manual is a result of years of research and observation into the role and potential of trees and shrubs on farms. It addresses the transformation of broader farm landscapes through the application of permaculture principles to revegetation. The manual includes revegetation strategies and design solutions relevant to increasing and diversifying farm productivity while stabilising the landscape. It also address the public land on roadsides, stream sides and reserves.

Permaculture Ethics and Design Principles fridge magnets

Show how permaculture is more than gardening! Bring permaculture ethics and principles into everyday life with this set of fifteeen magnets each with a simple graphic representing a core permaculture concept. Turn the magnet over to read the principle and proverb. Great memory tool for permaculture teaching, daily living and spreading the permaculture message to friends and family including children.

Melliodora (Hepburn Permaculture Gardens): A Case Study in Cool Climate Permaculture 1985-2005

The book was written inside the property it is describing - it's a kind of autobiography. It was written over a five year period, and with property now "established", the text has been revised with the benefits of hindsight. It is ideal for anyone seriously interested in sustainable living - both at a practical level and with a good dose of Holmgren holistic thinking. - Ian Lillington

The eBook contains the original 1995 A3 format book *Melliodora: Ten Years of Sustainable Living* in digital form as well as new and updated material about this leading cool climate permaculture demonstration site.

Relocalisation: How Peak Oil Can Lead to Permaculture DVD

In this presentation David Holmgren explains permaculture as a design system to relocalise our economies and communities in the face of the twin threats of Peak Oil and Climate Change.

The DVD includes introductions by Christine Carroll of Permaculture Noosa and Dr Anne Miller of the University of the Sunshine Coast Queensland as well as an extended Q & A session with the audience of over 120 in August 2006.